Fish Piracy

COMBATING ILLEGAL, UNREPORTED AND UNREGULATED FISHING

OECD

ORGANISATION FOR ECONOMIC CO-OPERATION AND DEVELOPMENT

ORGANISATION FOR ECONOMIC CO-OPERATION AND DEVELOPMENT

Pursuant to Article 1 of the Convention signed in Paris on 14th December 1960, and which came into force on 30th September 1961, the Organisation for Economic Co-operation and Development (OECD) shall promote policies designed:

- to achieve the highest sustainable economic growth and employment and a rising standard of living in member countries, while maintaining financial stability, and thus to contribute to the development of the world economy;

- to contribute to sound economic expansion in member as well as non-member countries in the process of economic development; and

- to contribute to the expansion of world trade on a multilateral, non-discriminatory basis in accordance with international obligations.

The original member countries of the OECD are Austria, Belgium, Canada, Denmark, France, Germany, Greece, Iceland, Ireland, Italy, Luxembourg, the Netherlands, Norway, Portugal, Spain, Sweden, Switzerland, Turkey, the United Kingdom and the United States. The following countries became members subsequently through accession at the dates indicated hereafter: Japan (28th April 1964), Finland (28th January 1969), Australia (7th June 1971), New Zealand (29th May 1973), Mexico (18th May 1994), the Czech Republic (21st December 1995), Hungary (7th May 1996), Poland (22nd November 1996), Korea (12th December 1996) and the Slovak Republic (14th December 2000). The Commission of the European Communities takes part in the work of the OECD (Article 13 of the OECD Convention).

FOREWORD

In its 2003-2005 programme of work, the OECD's Committee for Fisheries decided to undertake research examining illegal, unreported and unregulated (IUU) fishing issues. As part of the project, the Committee hosted a Workshop which took place in Paris on 19-20 April 2004. The Workshop was attended by 120 participants and included representatives from OECD Member and non-member countries, as well as international governmental and non-governmental organisations. The Workshop was made possible by funding from a number of OECD Member countries.

The Committee for Fisheries will continue the analytical work on illegal, unreported and unregulated fisheries issues and plans to publish its findings in 2005.

The views and opinions expressed in these *Proceedings* are those of the individual authors and do not necessarily represent the views of the OECD Committee for Fisheries or the OECD Member countries. It is published on the responsibility of the Secretary-General of the OECD.

Acknowledgements

The OECD and the Committee for Fisheries express appreciation to all participants for contributing to the success of the Workshop on Illegal, Unreported and Unregulated Fishing Activities. This publication has been made possible through voluntary funding by a number of OECD Member countries. It has been prepared and edited by Kathleen Gray, Fiona Legg and Emily Andrews-Chouicha.

TABLE OF CONTENTS

INTRODUCTION

The issue of Illegal, Unreported and Unregulated (IUU) fishing has moved to the forefront of the international fisheries policy agenda in recent years. Governments around the world have recognised the negative effects of IUU fishing activities on resource sustainability, biodiversity and economic and social sustainability. This situation led the OECD Fisheries Committee to address this problem in its 2003-05 work programme, focusing on the environmental, economic and social issues surrounding IUU fishing, both in terms of the incentives for engaging in IUU operations as well as their environmental, economic and social impacts. The Workshop hosted by the OECD on 19-20 April 2004 was a step forward in bringing together information, analysis and debate on this topic, and proposing new approaches to combating it. Around 120 experts from OECD and non-OECD countries, regional fisheries management organisations (RFMOs), international governmental organisations, non-governmental organisations and academia attended the Workshop.

The Fisheries Committee, as with other committees in the OECD, focuses on analytical work, mainly from an economic point of view, in support of high-profile policy issues and challenges for Member countries, and the broader international policy domain. Consistent with sustainable development objectives, the Committee seeks to pursue sustainable management of fisheries resources, to ensure sustainable livelihoods and other benefits, while minimizing the extent of possible distortions to domestic and global markets for goods and services.

The Fisheries Committee's activities are mainly based on data compilation and dissemination, informed dialogue, raising the awareness of common challenges and opportunities, and building support for different policy approaches. This includes *i)* the identification of characteristics of new innovations and best practices and policy mixes that have the potential to move the international community forward; *ii)* improving the conditions for implementation and overcoming impediments to change; and *iii)* adopting approaches to meet the particular needs and context of individual national experiences.

To develop policy capacity the Committee seeks empirical evidence, information and analysis of both the problems and implications of different fisheries policy perspectives and tools, as well as establishing policy coherence across numerous disciplines, tools and institutions.

IUU Fishing and the OECD Fisheries Committee

It is in the context of ensuring the sustainability of resources that IUU fishing has arisen as a priority global issue. IUU fishing not only undermines the sustainability of fisheries management regimes both domestically and internationally, affecting broader oceans' biodiversity, but also has undesirable economic and social implications.

The issue acts across all kinds of institutions and organisations with numerous players, both public and private. This is why, in 2002, the Fisheries Committee decided – as part of its 2003-05 work programme – to undertake work on the environmental, economic and social issues surrounding IUU fishing, focusing not only on the *drivers* of IUU behaviour but also on its environmental, economic and social *impacts*. The goal is to compare these drivers to the current range of actions, to identify synergies and gaps in these efforts, and to confirm or recommend the actions and approaches that will help combat IUU fishing. The work is intended to be a broad view, looking at the range of incentives and factors governing this behaviour, and affecting both national exclusive economic zones (EEZs) and the high seas.

Requirements of this work

In working on the IUU issue it became clear that empirical evidence on the nature and extent of IUU activities was scarce and that there is a need for systematic and consolidated information that will move the agenda beyond the fragmented and anecdotal. There is a role for data and information based on direct experience as well as estimation.

Second, more transparency is needed to understand the range of direct and indirect drivers that lead to IUU behaviour, and place them in an analytical framework that will help build a framework for action. This framework will focus on expected benefits from IUU fishing compared to legal fishing in light of expected costs (including the risks and costs of being caught).

Third, an inventory of approaches and tools already in place, or being put into place, to address IUU fishing has been compiled, along with an assessment of their focus. This includes such aspects as international and national legal frameworks and monitoring, surveillance and enforcement. Other tools considered are:

- Activities and frameworks of organisations such as RFMOs.
- The range of national measures in OECD member countries some of which are in place within the framework of the Food and Agriculture Organisation's (FAO) International Plan of Action (IPOA) on IUU fishing, as either coastal, port or flag states.
- Economic measures (including the role of investment rules – on re-flagging for instance – trade rules, financial transfers and the like).
- Non-economic and social mechanisms to discourage engagement in IUU fishing.
- Capacity building and other foreign aid.

Fourth, the focus with respect to the most important implications to identify an effective and feasible integrated response has been widened. A multifaceted issue needs a multifaceted response, especially as some take longer, or are more difficult to realise than others, and some are direct while others are more indirect. No single perspective, institution, approach or tool will have a corner on the issue; ultimately, combating IUU fishing needs the strengths and tools of a number of players to be mobilised effectively and coherently.

The role of the Workshop on IUU Fishing Activities

The initiative to host the Workshop was a first step in bringing together available information, analysis and debate on this topic, drawing on a diversity of perspectives and experience on the issue, however, with a particular emphasis on the economics of the activity. These *Proceedings* provide visibility for ongoing deliberations in the Committee's work, pending the publication of a final report in 2005. This work will benefit and provide food for thought for other institutions including international organisations, RFMOs and non-governmental organisations.

KEY OBSERVATIONS AND FINDINGS

The Workshop was organised around four sessions addressing: the state of play of IUU fishing; data and information needs; economic and social drivers; and possible future actions. The Workshop Chairs[1] compiled the following list of Observations and Findings that provide a brief overview of the main outcomes of the Workshop.

The State of Play on IUU Fishing

- IUU fishing is a worldwide problem, affecting both domestic waters and the high seas, and all types of fishing vessels, regardless of their size or gear.

- IUU fishing is harmful to fish stocks and undermines the efficiency of measures adopted nationally and internationally to secure fish stocks for the future.

- IUU fishing activities also have adverse effects on the marine ecosystem, notably on the populations of seabirds, marine mammals, sea turtles and bio-diversity as a whole (discards, etc.).

- IUU fishing distorts competition and jeopardizes the economic survival of those who fish in accordance with the law and in compliance with relevant conservation and management measures.

- There are important social costs associated with IUU fishing as it affects the livelihoods of fishing communities, particularly in developing countries, and because many of the crew on IUU fishing vessels are from poor and underdeveloped parts of the world and often working under poor social and safety conditions.

- The impact of IUU fishing for some species (primarily tuna and tuna-like species) is global, whereas that for other species (*e.g.*, Patagonian toothfish and Orange roughy) is specific to those areas where such species occur. This means that global and local solutions are required, as well as solutions tailored to specific species.

- There is concern that excess capacity in fisheries in OECD countries can lead to a spillover of capacity into IUU fishing activities.

- IUU fishing is a dynamic and multi-faceted problem and no single strategy is sufficient to eliminate or reduce IUU fishing — a concerted and multi-pronged approach is required nationally, regionally and internationally, and by type of fishery. The full range of players should be involved in helping bring forward solutions to the IUU problem.

- Many developed and developing states have not been fully responsible in complying with their responsibilities as flag states, port states, coastal states, states of vessel owners and trading nations.

---

[1] The Workshop Chairs were Mr. Ignacio Escobar, Mr. Jean-François Pulvenis de Seligny, Mr. Nobuyuki Yagi, Ms Jane Willing and Ms Lori Ridgeway.

- The FAO International Plan of Action to combat IUU fishing contains tools to tackle the IUU issue. The question is to find ways to better implement such tools.

Information and Data Needs

- In spite of recent improvements in information collection, there remains a lack of systematic and comprehensive information on the extent of IUU fishing operations and impacts. This is compounded by the varying level in quality, accessibility, reliability and usefulness of the available data.

- There are a number of international instruments addressing the collection of fisheries information and statistics. However, these need to be integrated and further, there remains a need for improvement in national statistics on trade in fish and fish products, especially in relation to IUU fishing.

- There is a diversity of actors involved in gathering, processing and disseminating information on IUU fishing activities — governments, intergovernmental organisations, RFMOs, regional fisheries bodies (RFBs), NGOs and industry.

- Trade-tracking and the resulting accumulation of information by market countries are an enormous task but it is very important for the creation of effective measures to combat IUU fishing.

- There is a need to broaden the scope of the information gathered so it covers activities and situations "upstream" and "downstream" of the IUU fishing operations themselves. This will help to better define the nature and scope of IUU fishing and to improve knowledge of the economic and social forces which drive IUU fishing in order to help target future actions.

Economic and Social Drivers

- Under current conditions, IUU fishing activities can be extremely profitable due, amongst other factors, to lower cost structures than for compliant fishing activities. Strategies to combat IUU fishing need to include measures that reduce the relative benefits and raise the costs of IUU fishing.

- The demonstration effect achieved by government and RFMO efforts in fighting IUU fishing activities is significant. This will provide positive signals to legal fishers and send the message to IUU fishers that their products will be excluded from the international market and that their activities will not be tolerated.

- Inefficient domestic fisheries management may work as a driver for IUU fishing activities; the more economically efficient management is, the higher the fisher income will be, thus lessening the incentive to engage in IUU activities.

- The size of penalties and the risk of being apprehended is not generally a sufficient deterrent to IUU fishing activities. This is complicated by the ease of re-flagging vessels and the difficulties in tracking company structures and identifying beneficial owners of IUU vessels. The lack of harmonisation of penalties across countries is also a concern.

- IUU fishing inflicts damage on a law abiding fishing industry aiming at sustainable exploitation.

- IUU fishing activities also make it harder for countries to strike a balance between food security and protection of the marine environment.

Possible Actions

- There is a wide range of possible measures that can be undertaken to address the problem of IUU fishing. These will need to cover legal, institutional, economic and social dimensions and will require the involvement of multiple players in the national, regional and international fisheries sectors.

- Determining the cost-effectiveness of alternate approaches to addressing IUU fishing problems should be undertaken to help identify priorities amongst the possible options so that the best results can be obtained from the limited resources that are available to national governments and international organisations.

Flag State actions

- Links between flags of convenience and tax havens have been established and a more concerted approach towards both could be undertaken.

- There is a need to improve transparency on the procedures and conditions for re-flagging and de-flagging.

- More countries could usefully investigate the possibilities for applying extra-territorial rules for their nationals.

- The penalties for IUU fishing offences should be significantly increased and harmonised between jurisdictions.

Port State actions

- The development of minimum guidelines for port state controls and actions against IUU fishers, particularly with respect to the use of prior notice and inspection requirements (including health and safety conditions), should be encouraged. The harmonisation of these controls and actions should be a priority.

- There is a need to ensure a broader use of port state control measures including inspections, preventing access to services and goods of IUU vessels.

- There needs to be an agreement to make it illegal to tranship, land and trade in IUU fish.

- There is also a need to improve the monitoring of the provision of at-sea services and transhipment of fish and fish products.

Coastal State actions and international trade responses

- It is necessary to augment monitoring, control and surveillance capacities and improve fisheries management across the board, but in particular in developing countries.

- Improving and extending the use of catch and trade documentation schemes could help provide additional information on IUU fishing activities.

- Fair, transparent and non-discriminatory countermeasures should be adopted, consistent with international law, against countries that do not comply with the conservation and management measures adopted by RFMOs, or fail to effectively control the vessels flying their flag, in order to ensure they comply with the conservation and management measures adopted by RFMOs.

- Countries should identify the area of catch, name of fishing vessels and their past history (of name and flag) in order to collect information necessary for better fisheries management and elimination of IUU fishing.

RFMO actions

- Strengthening the mandate and role of RFMOs and RFBs, in particular their possibilities for tracking IUU fishing, is an important requirement.

- There is a need to improve information sharing and co-operation among RFMOs, particularly in terms of linking and integrating their data on IUU fishing activities.

- More RFMOs should consider publishing lists of companies and vessels engaged in high seas IUU fishing activities and lists of vessels that are authorized to fish. The use of positive and negative lists of IUU fishing vessels and companies is strongly encouraged in this regard.

- The creation of a global record/register of authorised fishing vessels that are technically capable of engaging in high seas fishing should be considered.

International co-ordination

- Resources matter: more technical and financial resources are needed for capacity building, in particular in the developing states, for monitoring, control and surveillance, and in all activities to combat IUU activities.

- The international community should move to ratify relevant international treaties on labour and working conditions in the maritime sector in order to strengthen international hard and soft laws to protect fishing crews in general.

- Improved monitoring of foreign direct investments (out-going and in-coming) in the fishing sector will assist in tracking potential IUU fishing operations.

- Work should be undertaken nationally and multilaterally to lift the veil of corporate secrecy surrounding the companies undertaking IUU fishing activities and related services. Partnerships between public authorities and businesses offer important scope in the fight against IUU fishing. In this regard, the OECD Guidelines for Multinationals offer some possibilities that could be followed-up by national regulatory authorities.

- A major effort is required, in particular by regional fisheries management organisations and market countries, to collect and disseminate relevant information.

- The efforts already underway to improve information at all levels and mechanisms to share information need to be supported and strengthened.

NGO and private sector actions

- Whenever possible, governments should consider bilateral consultation with businesses engaged in IUU activities to determine if alternative means of getting IUU vessels out of the business can be found.

- There should be continued efforts to communicate the IUU problem, for example through promotional/educational campaigns with the market, including intermediate buyers, processors, distributors and consumers. Such activities will help raise awareness of the problem and improve the knowledge of the social, economic and environmental consequences of IUU activities.

- Industry and NGOs should be encouraged to continue to self-organise their response to IUU fishing and information collection.

PART I

OVERVIEW OF THE STATE OF PLAY ON ILLEGAL, UNREPORTED AND UNREGULATED FISHING

The first session of the workshop provided participants with an overview of the state of play and the political, economic and environmental problems that we face. It set the stage for more detailed discussion of the social and economic aspects of IUU fishing and new and alternate ways to combat it.

CHAPTER 1

REGULATING IUU FISHING OR COMBATING IUU OPERATIONS?

Olav Schram Stokke and Davor Vidas, Fridtjof Nansen Institute, Norway

Introduction

Why is this study needed? The past decade has produced a large number of measures aimed at combating the phenomenon now commonly referred to as 'illegal, unregulated and unreported' (IUU) fishing. Most of these measures are contained in legal instruments falling within the sphere of the law of the sea, including fisheries management and conservation. Among the global instruments, major milestones following on the 1982 UN Law of the Sea Convention were the 1993 FAO Compliance Agreement, the 1995 Fish Stocks Agreement,[1] as well as the 2001 International Plan of Action against IUU Fishing. Regional fisheries bodies also adopted a great many specific measures. Various national measures have been adopted as well.

-However, there has been no significant reduction in the IUU fishing activity against which those numerous measures are targeted. Indeed, in some regions it is even on the rise. Where sharp decreases of IUU fishing have been documented, this seems to be in areas where fish stocks have been exposed to over-fishing, so that incentives for (IUU) fishing have ceased to exist.

What is the reason for the weak correspondence between the measures adopted and their impact? Should we start by studying the measures? Or should we return to 'square one' and ask: Do we have the right 'diagnosis' of the problem?

The next section of this study re-examines the diagnosis, asking: Is our current understanding of the problem comprehensive enough? Does it focus on all the segments we need to address in order to deal with it effectively? This discussion is followed by three sections that review various existing measures to combat IUU fishing and examine the extent to which they respond to the diagnosis. Might it be that the main thrust of present measures has focused on curing the symptoms rather than addressing the causes? In each of those sections, we seek to identify potentials for improvement. How can the effect of current measures be enhanced, and which areas merit more attention? Our ambition

---

[1]   Agreement for the Implementation of the Provisions of the United Nations Convention on the Law of the Sea of 10 December 1982 relating to the Conservation and Management of Straddling Fish Stocks and Highly Migratory Fish Stocks.

here is not to enter into detailed proposals for new measures, but rather to pinpoint those areas where we see potential for improvements, and identify some of the actors who could be engaged.

The problem: do we have the right diagnosis?

What is our current understanding of the problem? While no mandatory definition of the problem is available, a commonly accepted one is found in the 2001 FAO International Plan of Action to Prevent, Deter and Eliminate Illegal, Unreported and Unregulated Fishing (IPOA-IUU). The 'nature and scope' of the problem is defined as being illegal, unreported and unregulated fishing. Here, 'illegal' fishing refers to 'activities conducted by vessels operating in contravention to national laws or international measures'. 'Unregulated' refers to 'fishing activities conducted by vessels that, while not in formal conflict with laws and regulations, are nevertheless inconsistent with conservation measures or broader state responsibilities to this effect'. This diagnosis therefore describes 'fishing activity' and 'vessel operations' – which are either illegal, unregulated or unreported (or all at the same time) – as being the constituent elements of the problem. Accordingly, the recommended measures to 'prevent, deter and eliminate' this problem primarily concern vessels and their (IUU) fishing activity.

The operation of vessels involved in IUU fishing is indeed an important *manifestation* of the problem, and has visible impacts on the status of fish stocks. In this study, however, we wish to offer several hypotheses about the diagnosis of the problem. First, fishing vessel activities engaged in IUU fishing are not the origin of the problem. Second, that IUU fishing has proven resilient to regulatory efforts is not only because of jurisdictional obstacles in regulating the activities of fishing vessels at sea. Third, vessel operations and their fishing activity are not the ultimate purpose of IUU operators' engagement.

If those hypotheses prove correct – as will be argued in this section – they would suggest that the main effort so far has involved treating symptoms rather than causes; dealing with manifestations of the problem rather than the purposes of those who create it. Moreover, this has often been done by relying on means that are relatively costly, such as enforcement at sea; or on concepts that have proven controversial, such as attempting to define what constitutes a 'genuine link' between the vessel and the flag state.

The scope of the problem is, we maintain, far broader than indicated by the commonly accepted diagnosis of the problem as 'IUU fishing'. Accordingly, the prevailing focus of the currently available measures needs to be re-examined. While one should indeed combat IUU fishing, it is not necessarily the case that this can be done exclusively and directly in the area where such activity occurs – its main drivers, just as its facilitators, are to be found elsewhere.

Fishing *per se* constitutes only one segment of the overall problem. In Figure 1 below, the sphere of IUU fishing is indicated by dotted lines. As can be seen, this is clearly only a part of a larger whole. It seems more correct to understand the problem as an inter-related *chain* of various links – of which 'at sea'[2] operations are only a part. What we need to do is to expose the problem by defining and analysing various links in the chain of an 'IUU operation' – a more accurate term than 'IUU fishing'.

---

[2]   'At sea' we understand here in sense used in the Law of the Sea, thus from vessel registration to the landing of catch in a port.

As illustrated in Figure 1.1, an IUU operation *for the purpose of international trade* can be understood as a chain composed of several main links:[3]

1. Purchase of a fishing vessel and its transfer from the real (beneficiary) to the declared (registered) owner.

2. Vessel registration in a national registry, so that vessel acquires a flag state.

3. Vessel involved in IUU fishing at sea (including refuelling at sea, and transhipment of catch at sea).

4. IUU catch landed at a port.

5. Catch/product imported, then often reprocessed and re-exported, as a rule through an intermediary state.

6. Catch/product imported by final importing state.

7. Fish product reaching retailers, distributors and end-consumers.

Figure 1.1 The IUU Operation

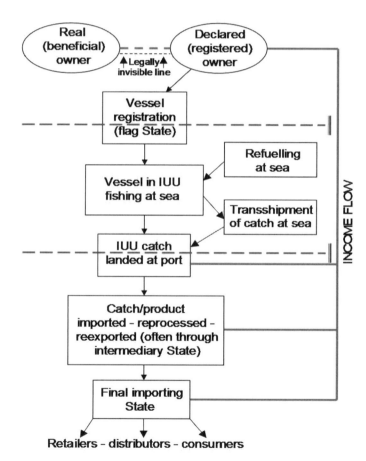

IUU fishing can be conducted either for the market of the port state or for international trade. Our study focuses on international trade only, which generally applies to lucrative IUU fishing for high-value fish species.

21

Source: D. Vidas, speech at the University of Berkeley, California, 21-22 February 2003.

Those links cluster in *three* segments of an IUU operation, each of which can be targeted by measures designed to combat IUU operations:

- First, *fishing vessel activity*, from vessel registration to landing of fish at a port. This is the international segment 'at sea', and corresponds largely to what is understood as 'IUU fishing'. However, this is in many ways a manifestation of the problem.

- Second, the *logistical* aspect of an IUU operation addresses the organisation of supplies and services, and is largely played out in a transnational sphere.[4] This is where the main strength of any IUU operation is created: its flexibility.

- The third segment is *catch/product* in international trade and market. This is where income-flows occur and net incomes are generated; this is the main purpose and the driving force for IUU operations.

Those three segments, then, constitute our diagnosis of the problem. Its manifestation is fishing vessel operations; its resilience and flexibility are enhanced by the transnational mode of its logistical activities; and its ultimate purpose is to generate net income. Measures that primarily address 'at sea' activities, as do most of the measures elaborated so far, are hampered by the considerable flexibility available to IUU operators – all the way from vessel registration to the landing of the catch at a port. Such measures have only a limited potential to impact on the main purpose of any IUU operation: the generation of net income.

Measures to address an IUU operation effectively will need to deal with all three segments of the phenomenon. In addition, they must exploit potentials to cut across those three segments. This is in line with the perspective enshrined in the general objectives of IPOA-IUU. There, a 'comprehensive and integrated approach' is formulated, according to which 'States should embrace measures building on the primary responsibility of the flag State and using all available jurisdictions in accordance with international law, including port State measures, coastal State measures, market-related measures and measures to ensure that nationals do not support or engage in IUU fishing' (para. 9.3 of IPOA-IUU). This comprehensive and integrated approach, while perhaps not yet elaborated in all aspects, corresponds to our understanding of the problem as being one of IUU *operations* rather than IUU *fishing* only.

According to the Introduction to IPOA-IUU, '[e]xisting international instruments addressing IUU fishing have not been effective due to a lack of political will, priority, capacity and resources to ratify or accede to and implement them.' There is no reason to dispute this view. Rather, the issue is whether we today have measures suited to deal with the complexity of an IUU operation. And what is the best way to proceed: More measures? Better integration among existing ones? Or a shift of emphasis among such measures?

---

[4]   'Transnationality' is marked by *direct* involvement of individuals and/or companies from one state in the jurisdictional sphere of another state or states, and is thus different from the 'international' sphere, where subjects of international law, such as states, interact. This transnational element provides many options for flexibility of an IUU operation, by utilising the comparative advantages, and loopholes, of varying legal systems.

In the following sections, we will explore measures as responding to the three main segments of the IUU problem: the vessels at sea; the transnational logistics, and the catch in trade. We will not enter into descriptive details of the measures devised so far, as the intention here is to examine whether various categories of measures are responsive to the diagnosis of an IUU operation. Further, we want to pinpoint the main reasons for their (in) effectiveness, and explore ways and conditions for overcoming existing limitations. An additional aim here is to indicate institutions and stakeholders that may have a potential to contribute to such enhanced effectiveness.

Measures targeting IUU vessels: the Law of the Sea domain

The sphere covered by the Law of the Sea governs an IUU operation from vessel registration to landing in a port. Here, we will focus on three main stages:

- vessel registration, through which IUU operators acquire a flag state (vessel nationality);

- jurisdiction, control and enforcement regarding fishing vessel operation at sea – the balance of flag state and coastal state competences; and between the flag state jurisdiction, on the one hand, and measures of regional fisheries organisations, on the other;

- landing in port and port state jurisdiction regarding fisheries.

In the following section, we take a closer look at each of those three stages of 'at sea' IUU operations, inquiring as to the reach of measures addressing these stages.

Vessel registration and acquiring of nationality of a flag state

Vessel registration can be described by various legal definitions; essentially, based as a rule on registration, a state grants its nationality to a ship. Every state has the right to sail vessels under its own flag. This is a fundamental right under the Law of the Sea, and in itself is not disputable. So far, states have not been able to reach any widely accepted agreement on whether this basic right can be made conditional by internationally agreed requirements that specify the nature and content of the link between a vessel and a state.[5] Consequently, conditions for registration are today determined by states largely at their own discretion.[6] When a vessel acquires the nationality of a certain state, that state becomes its flag state and thereby assumes primary responsibility and jurisdiction over the vessel. This is, in very simplified terms, how vessel registration, nationality, and flag state principle operate – as seen from the perspective of states.

There is another perspective to the same issue: that of the operator. This can be a physical person, though as a rule it is a juridical person, *e.g.* a company. Numerous companies have the opportunity to register business activity in more than one state. This is a core feature of international business and trade, and is in itself not controversial. However, a company may well have a perspective on vessel registration that differs considerably from that of a state. If the company is an IUU operator, vessel registration will be understood as a formal step by which that operator equips a vessel at its disposal with a *suitable* flag. Whether a flag is a suitable one will depend on circumstances, which in the case of fishing are more fluid than those related to the use of 'flags of convenience' in world shipping.

---

[5] The contents and fate of the (stillborn) 1986 UN Convention on Conditions for Registration of Ships is good proof to that effect.

[6] See Art. 91 of the UN Law of the Sea Convention. For a discussion, see Vukas and Vidas (2001).

23

When the two perspectives are combined, the result is that many companies – whether IUU operators or not – may choose from among many national arenas where to conduct their businesses. Setting up a one-ship company in one country and registering a vessel there, in order to obtain nominal nationality and a flag on a vessel, is essentially an initial phase of a business operation which at that stage cannot easily be considered to be illegal, unregulated or unreported. Even if the 'company' may consist of a post-box address only, and this may remain its main connection to the 'host' country, in many countries this does not contravene national law. Likewise, having a vessel registered in a registry without any real attachment to the country, other than formal registration and payment of fees, is in many countries not contrary to national law. It is therefore not illegal, not unreported, and – albeit somewhat unregulated – it is not prohibited.

From here, an IUU operation will start its voyage. What can international law, or for that matter the law of the sea, do to assist in combating IUU operations at the stage of vessel registration and, subsequently, the licensing of a vessel to fish? Instead of re-opening the eternal discussion about 'genuine link' and 'flags of convenience', let us start by identifying the elements that an IUU operator needs at this stage. First, he needs to find a suitable flag state. Second, he needs to have at his disposal a suitable fishing vessel that can be entered in that country's register and thereafter licensed. Those are the two firm elements. The rest (like setting up a company) may be an abstraction only, or generally too difficult to trace (*e.g.,* the hiring of crew). We will therefore focus on those two firm elements: a state and a vessel.

Is international law, or international co-operation, entirely impotent here? Or is there still some potential for further action in the sphere of vessel registration and licensing?[7] Can international co-operation help to make some *states* less suitable for the purposes of IUU operators? Similarly, is it possible to make *vessels* less suitable for the purposes of IUU operators?

States less suitable for IUU operators. While there may be numerous companies, the number of states in the world is limited, and many states are simply not suitable for IUU operators. Those that are, fall into two categories. One group consists of states not members of a certain regional fisheries management organisation; among those, only states that do not exercise their flag state responsibility will qualify as suitable for IUU operators. The other group is usually quite limited, but also a significant feature in IUU operations: states members of regional fisheries management organisations that lack either the will or the capability to exercise their flag state responsibility.

Common to all states suitable for IUU operators is, therefore, the absence of flag state responsibility. Applying the commonly accepted label of 'flags of convenience' for those states is neither correct nor productive.[8] A recent FAO study noted that the flags used in IUU fishing are actually 'flags of non-compliance'; soon afterwards, that term was adopted by the Commission for the Conservation of Antarctic Marine Living Resources (CCAMLR).[9] While possibly attractive in the

---

[7] Here we will not enter into discussion of economic measures (such as subsidies) or national legislative measures (such as vessel registration denial by some countries), but will remain on the level of international co-operation and international law. Issues of subsidies and denial are discussed later in this chapter.

[8] Essentially, the term as such is also misleading, due to its relative nature. The notion of 'convenience' is accurate only from the perspective of IUU operators; for all others, these are essentially 'flags of inconvenience'.

[9] See: Port State Control of Fishing Vessels, FAO Fisheries Circular No. 987 (Rome: UN Food and Agriculture Organisation, 2003). See also CCAMLR, Resolution 19/XXI: 'Flags of Non-Compliance', adopted in November 2002; text in: Commission for the Conservation of Antarctic

context of duty to co-operate, this reasoning is nevertheless open to one (formal) objection: not all states are obliged to comply with the conservation measures of RFMOs – only those that are members of the RFMO in question, or parties to the UN Fish Stocks Agreement. Other states, if they so wish, may remain in non-compliance as long as that does not conflict with duties they have accepted or are bound to under general international law. However, there is one minimal requirement that remains valid for all flag states: All states are to be responsible for exercising some degree of control over vessels flying their flag. That is their *flag state responsibility*. Those who flag vessels without exerting any form of control over their activities, fail to exercise their basic responsibility as states in relation to vessels having their nationality. The flags of such states deserve to be labelled *flags of no responsibility*.

Some states may accept the label 'convenient' but hardly any state will accept being branded irresponsible. In international co-operation, 'naming and shaming' can be a powerful measure.[10] This can be done through a range of steps – from direct correspondence to the flag state by secretariats, through diplomatic demarches, etc. The more states (and with higher prominence in the particular context) join in exerting such pressure, the greater will be the sense of exposure, and thus embarrassment for the state in question. Greater transparency of this action will result in increased embarrassment. The use of an appropriate label may further add to the convincing strength – and a label related to the lack of 'flag state responsibility' would be firmly based on the development of international law over the past decade.

Any such label will be essentially relative, being linked to the context of a particular fishery only. However, it may easily become perceived as absolute. This is a dilemma that regional organisations, such as CCAMLR, have had to face when discussing proposals for the listing of flags. Enhanced co-ordination between RFMOs should be able to assist in making this label less relative.

Vessels less suitable for IUU operators. A vessel will be seen as less suitable for an IUU operator if registering it in various national registers is difficult, or if it can be expected that the vessel will be denied a license to fish. For this, a vessel needs a 'history', a bad record of involvement in IUU fishing. Herein lies a potential for international co-operation: it can become a vehicle for establishing a record of IUU fishing for some vessels. Recently, CCAMLR parties agreed to prohibit issuing a license to fish to vessels appearing in the newly established CCAMLR–IUU Vessel List, both for fishing in the Convention Area and in any waters under the fisheries jurisdiction of the parties.[11] While the CCAMLR Secretariat compiles this list, the Commission approves it; however, the list is available only on password-protected pages of the CCAMLR website.[12]

Echoing the FAO Compliance Agreement, the IPOA–IUU contains clear limitations. While it holds that flag states should avoid flagging vessels with a history of non-compliance, the IPOA–IUU allows exceptions where ownership of the vessel has subsequently changed, or if the flag state determines that flagging the vessel would not result in IUU fishing.[13]

Marine Living Resources, Schedule of Conservation Measures in Force, 2002/03 (Hobart: CCAMLR, November 2002), pp. 125–126.

[10]   See also section on shaming below.

[11]   CCAMLR Conservation Measure 10-06 (2002).

[12]   Para. 15 of CCAMLR Conservation Measure 10-06 (2002).

[13]   Para 36 of IPOA-IUU.

Ultimately, where is the problem with all the measures that can be used through international co-operation in this area? While they do exert some effect, gradually narrowing down the scope of movement for IUU operators, they share one pervasive feature of international co-operation: they are slowed down by cumbersome procedures. Many RFMOs meet only once a year, and while their secretariats may operate year-round, decision-making occurs at an annual pace – and in organisations where consensus is the rule, it may take several years before a decision is agreed upon by all.

It will take far less time for an IUU operator to change a flag on a vessel, or to otherwise adjust to the emerging situation. Today, vessels can be re-flagged by a few clicks on a PC connected to the Internet. There are several specialised websites that offer full services, from Q & A to assisting in prompt company setting and vessel flagging, probably the best-known of these being (www.flagsofconvenience.com).

While international co-operation is slow and operates through firm principles of international law, business – such as setting up an IUU operation – is swift and operates not according to these principles but in the loopholes between them. This may be contrary to moral norms, but today – a decade after the adoption of the FAO Compliance Agreement and the UN Fish Stocks Agreement – IUU operators can still easily obtain flags and fish licenses for their vessels from several states. From there, the IUU operation can set sail.

Jurisdiction, control and enforcement at sea

At sea, the Law of the Sea operates through a balance of sovereignty, sovereign rights and jurisdiction between the coastal state and the flag state. On the one hand, the rights of the coastal state decrease as the zones are more remote from its coasts or baselines; and in respect of fisheries management, individual coastal state rights cease at the outer limit of that state's EEZ. On the other hand, the rights of the flag state in respect of fisheries are valid to their full extent on the high seas, where the freedom of fishing governs; correspondingly, the rights of the flag state over the vessel flying its flag decrease in the direction of any coast other than its own. In between this balance are RFMOs, which can adopt conservation and management measures on the high seas (as well as in coastal zones) within their area of application. Enforcement capability, however, rests with states.

From the legal perspective, the coastal state is entitled to exert control and enforcement over fisheries activities in its various *coastal zones*. In this connection, it has often been said that the only truly effective means against IUU fishing is a patrol boat at sea.[14] While the coastal state can indeed arrest a foreign fishing vessel involved in IUU fishing in its EEZ, there are still legal limitations: the flag state can require the prompt release of a vessel from detention upon the posting of a 'reasonable bond'.[15]

From a practical perspective, in areas where this is possible, a patrol boat at sea can indeed be an effective means of control and enforcement. However, in many coastal waters, especially in EEZs and even in the territorial seas of many developing countries, this is difficult due to the combination of poor capacity, high costs and extensive fishing grounds. Difficulties are also encountered in areas of

---

[14]   In reality, this is comparable with the view that the only effective way to fight crime is a police constable patrolling the street. Neither the causes nor most of the consequences can be dealt with in this way; moreover, it is very costly.

[15]   Arts. 292 and 73(2) of the UN Law of the Sea Convention. Several prompt release cases have been decided upon in recent years by the International Tribunal for the Law of the Sea, all originating in IUU fishing for Patagonian toothfish in EEZs around sub-Antarctic islands under French and Australian sovereignty.

disputed sovereignty, or in remote areas such as the coastal zones around the various sub-Antarctic islands.

For an IUU operator, the abstract legal construction of coastal state jurisdiction in coastal zones matters only to the extent that effective physical control at sea can be expected. Where this expectation is higher, IUU fishing will depend on a simple risk assessment: probable net income from fish likely to be caught in a season *vs*. the value of a vessel likely to be sacrificed in the case of arrest.[16] Where the likelihood of arrest is negligible and fish resources well identified, an IUU operation will emerge from the risk assessment as a safe and good investment.

In this area, it is not realistic to contemplate any more significant conceptual legal developments in the foreseeable future, other than perhaps more rigorous ITLOS interpretation of what should be understood as a 'reasonable bond'.[17] In respect of international co-operation, one available avenue is more intensive co-operation between the coastal state and the flag state – for instance, in cases where observation has enabled identification of a vessel, but without other control or enforcement interventions taking place.

On the *high seas*, the situation is different, both from the legal and, as a rule, from the practical perspective as well. Unfortunately, both work in favour of an IUU operator. Here, what applies is one of the basic legal principles of international law of the sea: *freedom of fishing*, which all states enjoy. Today, this is a freedom subject to conservation and management of marine living resources. RFMOs are a mechanism increasingly used to specify conservation and management measures. However, those measures are legally binding only on members of an RFMO; all other states remain 'third parties'. Here one other basic principle of international law comes into play: *pacta tertiis*, the principle that international treaties do not oblige third states without their consent.[18]

On the high seas, thus, not only practical impediments but also basic legal principles work in favour of IUU operators. Fishing here is free for all, and although there has been an increase in conservation measures by RFMOs, these are not binding on third states and, accordingly, on the vessels under their jurisdiction.

In this area, post-UNCLOS law of the sea has seen some important developments, prompted primarily by innovative regional solutions. These needed global sanction, which was acquired through the 1993 FAO Compliance Agreement and, especially, the 1995 UN Fish Stocks Agreement, now both legally in force. The development here can be summed up as going in two directions: extending the effect of measures adopted by RFMOs to third parties; and extending the reach of the 'patrol boat' from zones under national jurisdiction to the high seas. For international law, those were significant, almost revolutionary developments. As to their practical impact, however, in many areas this has remained moderate, with few prospects for improvement.

As to the first of these developments, Article 8(3) of the UN Fish Stocks Agreement specifies how a flag state fishing on the high seas, where conservation measures adopted by RFMOs apply, is to

---

[16]   Also for this reason, many IUU operators use fishing fleets in which vessels have different roles (fuel supply, storage etc). One of these roles may, sometimes, be that of the vessel to be sacrificed in order that other, more valuable, vessels can escape. This was likely the role of 'Lena', apprehended in the same action together with 'Volga', both under Russian flag; the rest of that fleet, comprising more advanced vessels flying flags of third parties, escaped with the fish that had been caught.

[17]   This trend can be observed in ITLOS, especially after the 'Volga' case in December 2002.

[18]   Art. 34 of the Vienna Convention on the Law of Treaties.

give effect to its otherwise general duty to co-operate: by becoming a member to the RFMO or by agreeing to apply the measures in question. Moreover, Article 8(4) provides that only those flag states who act accordingly shall have access to the fishery resources to which the measures by the RFMO apply. Many RFMOs have followed up with more specific requirements. However, among the parties to the UN Fish Stocks Agreement, there are only a small number of flag states truly addressed by those provisions. And, perhaps of even graver concern, many problems of IUU fishing are caused by states that are parties to various RFMOs, but that fail to implement their conservation measures or to exercise their flag state responsibility.[19] In such cases, as has been demonstrated, the resort to persuasion by other members of that RFMO may require years of systematic follow-up – with the burden of proof regularly resting on those seeking to prove the offence.

As to the second major legal breakthrough, Article 21 of the UN Fish Stocks Agreement authorises states parties to the Agreement that are members of a RFMO to board and inspect fishing vessels flying the flag of any other state party to the Agreement, regardless of whether this state is a member of the RFMO in question. This means moving a 'patrol boat' to the high seas, though it is limited to inspections. While certainly a useful solution in the specific regional context from which it originates,[20] and in areas of geographic and geopolitical proximity (*e.g.*, the Barents Sea), or potentially in a semi-enclosed/enclosed sea not divided into EEZs (such as the Mediterranean Sea), in many other cases this innovation is of little practical value.[21] In the Southern Ocean, for instance, this would mean patrolling high seas fishing areas like the Ob and Lena Banks, several thousand kilometres away from the nearest harbours – only to carry out inspections, not arrests (and only in respect of vessels flying the flag of a party to the UN Fish Stocks Agreement). Moreover, inspections in the Southern Ocean are done almost exclusively in maritime zones under (disputed or not) sovereignty, and those cover only a small fraction of the entire toothfish fishing area.

This is not to say that RFMOs have no role to play in high seas control: on the contrary, information collection, its transparency,[22] and collective pressure on the flag state are all important mechanisms. This system, however, may function only in respect of those states that do exercise their flag state responsibility, or those who may decide to exercise it when faced with increased international pressure.

In addition, for those areas where internationally agreed management and conservation measures apply, RFMOs do have a role to play by introducing and implementing catch certification and trade documentation schemes. Their operation begins at sea, and it is often at this stage that the fraud regarding documentation originates.[23]

---

[19]   Let alone being unwilling or unable to control the activities of their *nationals* pursued under jurisdictions of other states.

[20]   The provision is in many respects modelled after the Bering Sea Doughnut Hole Convention.

[21]   However, that provision may be an additional impediment for some states to ratify the UN Fish Stocks Agreement. As to regions such as the Mediterranean, where this type of compliance mechanism can be conceived of, there is as yet little evidence that it would be relevant in practice.

[22]   It is, however, transparency which is often difficult to achieve, with information about fisheries often being comprised by commercial privacy of data. A further obstacle is reliability of information, and thus an additional reason for caution when transparency is required. See the next two sections, and the Conclusions of this chapter.

[23]   Catch certification and documentation are discussed further below.

Port state jurisdiction and control regarding fisheries

The final point where an IUU operator meets the Law of the Sea is while landing a catch in a port. Port state control in respect of fisheries is a relatively new development. After some initial regional experiments, it first emerged on the global level in the 1993 FAO Compliance Agreement. Under that Agreement, however, the power of the port state is quite limited: if it has reasonable grounds for believing that a vessel has been involved in IUU fishing, all the port state can do is to promptly notify the flag state about this.[24] The 1995 UN Fish Stocks Agreement goes further: it is 'the right and the duty' of the port state to take non-discriminatory measures against IUU fishing.[25] The Agreement entitles (and instructs) the port state to, *inter alia*, inspect documents, fishing gear and catch on board the fishing vessel. If it is established that the catch originates in IUU fishing, the port state may, pursuant to its laws, prohibit landings and transhipment. Its power stops short of detaining the vessel, however.[26]

At present, fighting IUU operations in ports would seem another weak point of the Law of the Sea. True, waiting for the catch to arrive in port is far cheaper than chasing the fishing vessel on the sea. Nevertheless, in the world there are many port states, and many more ports, and it is difficult to know in which of those an IUU catch will be landed. The history of landings of IUU catches of Patagonian toothfish can serve as an illustration. When this IUU fishing started on a larger scale in the early to mid-1990s, the initial ports used for landing were in South America. Then, as IUU fishing moved to the Indian Ocean sector, initially Southern African ports were used, first in Namibia and Mozambique and, then increasingly, Mauritius. Although Mauritius is still cited today, this is largely 'outdated' – the major landings have now moved to ports in Asia.

We may compare the effectiveness of unilaterally implemented port state control measures with the effectiveness of traffic police waiting at the very end of a highway, hoping to apprehend here all those who have gone too fast on the entire highway. Just as there are many exits from a highway, there is always 'some other port' (and port facilities may be under private control). Second, just as one can slow down before passing a speed control, IUU operators can adjust the usage of the flag on the vessel, or even adjust the vessel itself, before appearing in port. The landing of an IUU catch can be done by 'some other flag', due to re-flagging, or by 'some other vessel', due to the prevalence of transhipment at sea.

Despite such practical limitations, port state measures seem to be an area with potential for development, perhaps more than any other Law of the Sea mechanism. There are probably three areas in which – based on the development of RFMO practice, indications from IPOA–IUU, and the on-going processes in the FAO – we can expect further elaboration of port state measures as a mechanism against IUU fishing.[27]

First, any meaningful port state control must be based on co-ordinated efforts, resulting in compatible measures. Recently, this understanding has led to the process towards developing such

---

[24]   Art. V(2) of the FAO Compliance Agreement.

[25]   The exact wording is given in Art. 23(1) of the UN Fish Stocks Agreement.

[26]   Some states, like the United States under the Lacey Act, do have stronger national measures; many other states deny access under some circumstances. However, those measures largely lack co-ordination.

[27]   The resulting measures will need to be fair, transparent and non-discriminatory, as stated in IPOA–IUU.

measures at the FAO, first through an Expert Consultation in November 2002, while a Technical Consultation is scheduled for the second half of 2004.

Second, broadening the extent of port state measures is a discernible trend in state practice, in RFMO measures and in consecutive global instruments. The direction here is towards not merely sitting and waiting for a vessel to arrive in port, but also undertaking port state measures before that. Through state practice some requirements have developed in this respect, now formulated in IPOA–IUU: reasonable advance notice before entry into port, providing a copy of the authorisation to fish, and specifying details of the fishing trip and quantities of fish on board.[28] If this would lead to 'clear evidence' that the vessel has been involved in IUU fishing, landing or transhipment can be denied. Since re-directing of the vessel may add to the financial burden for the IUU operator, this approach is worth considering for wider global sanction.

Third, strengthening of the content of port state measures, as well as further specification of these, is also a trend evident from recent practice and reflected in IPOA–IUU. Reversal of the burden of proof, placing it on the vessel to establish that the catch was taken in a manner consistent with conservation measures, is already enshrined in IPOA–IUU (para. 63). Attention can also be drawn to the degree to which RFMOs need to provide proof of a vessel being involved in IUU fishing: actual 'sighting' of a non-member vessel in an area of conservation measures is gradually becoming replaced by a non-member vessel being 'identified' as engaged in fishing activities.[29]

Finally, there is the economic aspect. Due to greater cost-efficiency, the advantage of port state measures over enforcement at sea is especially attractive for developing countries. On the other hand, implementation of port state measures requires adequate training in fishery inspection: this is an area where international assistance projects should be stimulated.[30] This could also be an additional mechanism to persuade some states to forgo the benefits from transhipment activities related to IUU fishing.[31]

What general conclusions can be drawn about the reach of the Law of the Sea measures that are applicable 'at sea' – from vessel registration, to the landing of catch in port? First, the Law of the Sea as an effective tool for combating IUU fishing is clearly limited by general legal principles otherwise necessary for upholding legal security. These principles, however, provide IUU operators with ample room for manoeuvre. While international law by its nature needs to be stable, IUU operators, by the nature of their business, need to be efficient, flexible and creative. Second, the development of legal measures, whether through regional or through global international co-operation, is a slow process; and when it brings results, these tend to come in small portions. Furthermore, today's IUU operators have access to modern information technology, enabling them to react and adjust to changes quickly. Third, enforcement at sea is a costly operation; even for states with good enforcement machinery at their disposal, the financial cost can exceed the value of the fish resources to be protected. Moreover,

---

[28]   Para. 55 of IPOA–IUU, stressing also due regard to confidentiality of data. For an overview of state and RFMO practice, see: 'Implementation of the International Plan of Action to Prevent, Deter and Eliminate Illegal, Unreported and Unregulated Fishing', *FAO Technical Guidelines for Responsible Fisheries*, No. 9 (Rome: FAO, 2002), pp. 41–45.

[29]   See *ibid*, comments at p. 46.

[30]   The FAO Fish Code Programme is one vehicle for such assistance; see (www.fao.org/fi/projects/fishcode/aboutfishcode.html), especially the project 'Support for the Implementation of the International Plan of Action to Prevent, Deter and Eliminate Illegal, Unreported and Unregulated Fishing (IUU Fishing)'.

[31]   On the latter aspect, see also comments in *ibid*, p. 45.

states may operate on the basis of various policy considerations, not only economic ones. For an IUU operator, the cost-benefit analysis is simpler, and a risk assessment rather straightforward; moreover, the relevant areas are vast, measured in millions of square kilometres, without any legal possibility of direct enforcement. All this combines to give clear advantages to IUU operators.

Nonetheless, the measures developed so far to combat IUU fishing have been predominantly in the Law of the Sea sphere of regulation. After some advances on the harmonisation of port state control measures likely in the near future, the arsenal of the Law of the Sea will largely be exhausted for some time. However, the real impact of those measures so far has not been in direct enforcement, but in their indirect effects. With more information available about IUU operations and with increased pressure from states, often through RFMOs, some flag states have improved the exercise of their flag state responsibilities. With some waters being more effectively patrolled, IUU operators have found it necessary either to change their fishing grounds or become involved in higher-risk operations.[32] With greater international attention focused on IUU fishing, some loose grips – such as a 'reasonable bond' under the Law of the Sea Convention – are now becoming firmer through judiciary practice. With fewer ports fully open to IUU operators, for such operators there is less flexibility and often higher costs involved in circumventing new regulations, either by fraud or by changing port. All the same, these are rather modest outcomes in view of the sizeable investments in time, resources and political attention directed to the problem of IUU fishing throughout the whole of the past decade.

There is thus an obvious need to target an IUU operation at links where there is less opportunity for avoidance of regulation, where the implementation of measures is less costly, and where the measures can more directly target the basic profit-earning purpose of an IUU operation (not only its visible manifestation), and its flexible transnational character.

Measures targeting IUU logistical activities

Such a complex operation as an IUU activity involves the organisation of capital, manpower, supplies and services. Accordingly, this section will discuss governmental and private initiatives to create *frictions* by reducing the availability, or enhancing the cost, of various resources needed for the smooth operation of IUU activities. Such resources include access to national waters, equipment and bunkering, and financial, legal, insurance, freight and processing services. Three sets of tools are addressed here: specific and hard measures that seek to restrict access to desired input factors; softer means that target the reputation of companies associated with IUU operations; and more general efforts aimed at reducing the overcapacity in world fisheries, which is believed to be a root cause of many IUU operations.

Denial

The strategy underlying the first set of measures discussed here is denial: IUU operators, or those who co-operate with and support them, can be denied access to inputs or outlets that are controlled by actors prepared to use access as leverage. Government blacklists of vessels with a history of IUU fishing is an instructive example. Such lists can serve as a basis for refusing access to national resources, ports or services. More generally, three questions arise when classifying denial measures and considering expansion of existing measures. First, who is the denier: governmental or private actors? Second, what is being denied: port access, landing rights, fishing rights, particular services, or any combination of these? Third, who is targeted for denial: the flag state, the beneficiary vessel

---

[32] However, increased patrolling in some areas, including around some sub-Antarctic islands, is often a result of political considerations, not necessarily primary prompted by the needs of marine living resources management and conservation, and can thus change if the motivation changes.

owner, only the vessel, or only the cargo believed to stem from IUU fishing? Or is denial extended to 'IUU complicits', such as those who provide transhipment services, bunkering, insurance etc.?

To illustrate, the CCAMLR IUU Vessel List is an instance of multilaterally co-ordinated denial that makes use of member states' authority to licence individual vessels for harvesting in the CCAMLR area and in national waters. For its part, the Norwegian blacklist system[33] implemented in order to close the Barents Sea Loophole was a unilateral initiative that extended beyond licensing to cover port access too. The result was to reduce the second-hand value of vessels with a history of contravention of rules created by the Norwegian-Russian Fisheries Commission, especially on the European Community market. Corporate-level denial has also occurred in this region and has on some occasions even targeted companies or vessels that had provided inputs to IUU activities. For instance, during the peak years of the Loophole fishery, a series of private *boycott* actions were introduced, aimed at strangling Norwegian supplies of provisions, fuels, and services to Loophole vessels, as well as punishing domestic companies that failed to adhere to such boycotts (Stokke 2001). The Russian Fisheries Committee put similar pressure even on the ports of the most active high-seas fishing state by encouraging the Murmansk-based trawler industry to discontinue landings of cod in Iceland.

As discussed further in the section, Measures targeting IUU, denial can also be exercised indirectly by making landings and transhipment conditional on documents substantiating that the fish has been caught legally. While both blacklists and the 'white list' approach of documentation schemes can be circumvented by means such as document fraud, re-registering of vessels under new names, and laundering an illegal catch by mixing it with legal harvest, even such circumvention can be costly and will generally add friction to IUU operations.

Some reservations have been expressed with regard to the denial strategy, especially when applied by governments operating unilaterally. On one occasion, Iceland filed a complaint to the surveillance authority under the European Economic Area Agreement over Norway's refusal to render repair services to an Icelandic vessel that had been engaged in Loophole fishery.[34] More generally, the due process concerns articulated for instance by the United States with regard to blacklists[35] highlight the importance of transparency regarding criteria for being placed on such lists, the accuracy and verifiability of information on which such placement occurs, and opportunities for the targets of denial to be permitted to present their case.

Although the relationship is not unequivocal, such means to ensure due process can be hampered by the prevalent confidentiality that surrounds information about IUU operations compiled within governmental management regimes.[36] Lists of IUU vessels compiled within one co-operative framework are in some instances, such as CCAMLR, not available to other management regimes or to

---

[33] Norway, *St.prp.* 73 (1998-99), Sec. 2.2; legislation providing for blacklisting was introduced in 1994 but not used in practice until 'around 1997'; *ibid.*

[34] The Authority indicated the occurrence of such a violation, but no further action was taken because 'the underlying conflict concerned a dispute between Norway and Iceland over Icelandic fishing rights in the Barents Sea' 'Freedom to Provide Services', *EFTA Surveillance Authority: Annual Report 1998* (http://www.efta.int/structure/SURV/efta-srv.cfm). Art. 5 of Protocol 9 to the EEA Agreement provides for access to ports and associated facilities but exemption is made for landings of fish from stocks, the management of which is subject to severe disagreement among the parties.

[35] See Draft for Public Review and Comment of the National Plan of Action of the United States of America to Prevent, Deter and Eliminate Illegal, Unreported and Unregulated Fishing, 2003, Sec. 7.3.

[36] The ambiguity arises from the argument that could be made that due process is best served if information about IUU operations is only acted upon in the context in which it was compiled.

the wider public. From one perspective, such confidentiality may be seen as constraining the effectiveness of the blacklist approach. Improved dissemination of the information contained in the list would enhance the ability of governments to act on it also in other geographic areas. On the other hand, awareness that information will be broadly exposed may significantly obstruct the provision of information to the regime secretariat.

If access to government-compiled IUU information becomes more broadly available, this would facilitate the mobilisation of private actors, including insurance and financial service providers or freighters, that might see it in their interest to refrain from doing business with IUU operators or even support the development of lists by volunteering information about the identity of IUU actors and the extent of their operations. One group of actors with such incentives are legitimate fishers, for instance the list of allegedly *rogue* vessels published by the Coalition of Legal Toothfish Operators (COLTO) in the Southern Ocean.[37] The same is also true for companies with strong brand names that are concerned with corporate environmental responsibility and their reputation.[38]

The effectiveness of the denial strategy is obviously enhanced if the number of deniers, or more accurately their share of the object desired by IUU operators, is high. As illustrated by the Northwest Atlantic Fisheries Organization (NAFO) and CCAMLR, regional fisheries management regimes are natural vehicles for co-ordinated denial: the challenge is often to persuade non-party providers to join a boycott. For some government-level measures, such as refusal of resource or port access, this can be done by *ad hoc* diplomatic means. In the 1990s, for instance, Norway ensured that annual fisheries agreements drawn up with states neighbouring the Barents Sea included provisions to prohibit landing of fish taken in international waters without a quota under the regional fisheries regime (Stokke 2001). As discussed in the previous section, broader options include memoranda of understanding among coastal states in conjunction with procedures for harmonised or even co-operative maintenance of lists of vessels or companies with a history of IUU engagement.

Turning to the objects being denied to IUU actors, any expansion from government-controlled objects, like port and resource access, into privately provided supplies, such as refuelling, freight and financial services, is constrained by the frequently fragmented structure of supply for such inputs. It has been argued that some important equipment, like means for satellite navigation and certain safety equipment, is sufficiently concentrated in supply to enable restrictions on access that might make acquisition more costly for IUU operators. Similarly, one study indicates that the number of reefers likely to be engaged in the transport of Japan-bound sashimi-grade tuna, a key IUU product, is not overwhelming (Gianni and Simpson 2004). Nevertheless, since most of the input factors needed by IUU operations have many potential providers based in many jurisdictions, the transparency of supply is low and collective action difficult. This is one of the reasons why the recent resolution by the International Coalition of Fisheries Associations – that governments, importers, freighters, traders and distributors should refrain from dealing with IUU catches (Wynhoven 2004:16) – cannot be expected to have much impact. Indeed, some of the input factors mentioned are probably of less importance to IUU operations than to legitimate fishers. Many vessels registered under flags of convenience are not fully insured or not insured at all; and equipment designed to improve environmental and worker safety is frequently sparse. Beyond this, many of the IUU operations in the tuna and toothfish sectors are parts of vertically integrated structures that, although sometimes loosely connected, ensure access to both supplies and outlets.

---

[37]   See (www.colto.org).

[38]   See also the discussion of shaming below.

Denial measures may even extend to the manpower of IUU operations. Since wages make up a high proportion of the running costs of IUU operations, crews tend to be recruited in low-income countries where lack of alternative employment opportunities will continue to ensure stable supply of low-cost labour. Fishing masters and especially captains, however, are in many instances residents of wealthier countries, some of which are prepared to introduce measures to reduce the leeway for their nationals to take part in IUU fishing operations. Thus, in 2002 Spain introduced legislation that constrains the involvement of Spanish citizens in fishing operations of vessels flying flags of convenience.[39] While such measures are difficult to enforce, they may have some effect and over time strengthen the social norm among respected fishers that IUU involvement is unacceptable. That said, unemployment too is frequently perceived as unacceptable and will place limits on the effectiveness of this strategy.

All three dimensions of the denial strategy - the agent, the object, and the definition of the target - may be relevant to the compatibility of various denial measures with trade rules. If governmental denial of landing rights is applied at the level of flag states, for instance by targeting certain flags of convenience, this may contravene international trade rules. This is because such measures in effect would discriminate against vessels that have operated in consistence with RMFO regulations but fly a 'wrong' flag. If this happens, it could be seen as a violation of the national treatment and most-favoured nation principles of the World Trade Organisation (WTO).[40] That said, no complaint has been filed under WTO on the import bans implemented under the International Commission for the Conservation of Atlantic Tunas (ICCAT) on states whose vessels have been determined as harvesting bluefin tuna or swordfish in a manner not consistent with that regime (Chaytor et al. 2003). For its part, denial of access to national fish resources based on blacklists of individual vessels with a history of IUU harvesting is unlikely to be challenged under trade rules, since access to EEZ resources is usually not among the entitlements flowing from international trade regimes. Resource access does not fall within the category of a 'good' or a 'service' as understood under the WTO. As demonstrated in the Barents Sea Loophole case, measures that also prohibit port calls may become contested. An intermediate option could be to deny access to all vessels owned or operated by a blacklisted IUU company. This would probably be compatible with international trade rules, provided that national and foreign firms are treated identically, but is likely to be intractable in practice due to complex and rapidly shifting ownership situations. Nor would this option add much to effectiveness due to the prevalence among IUU operations of the one-vessel company structure. Exclusively private-level denial initiatives, like those implemented in the Barents Sea Loophole case, are not constrained by international trade rules since only states are bound by such rules.

Shaming

The naming and shaming of participants in IUU operations by actors who do not themselves control any input factors desired by IUU operators is a strategy that targets the reputation of named companies. Indirectly, it may also support denial measures, to the extent that public or private suppliers act on the information provided. The typical agents of shaming are business or environmental NGOs that provide vessel- or company-specific information about IUU operations. Sometimes, shaming can be extended to those who supply IUU operations with goods and services. Activities such as these have been undertaken in other environmental areas as well, starting in the 1970s but becoming more prominent in the 1990s (Haufler 2003). Underlying this 'corporate

---

[39]   'Royal Decree 1134/2002 of 31 October 2002, on the application of penalties to Spanish nationals employed on flag-of-convenience vessels', along with other national measures on the part of Spain and other OECD members, is summarised in OECD (2004).

[40]   WTO agreements are downloadable at (www.wto.org).

accountability' movement is the belief that information that indicates lack of environmental or social responsibility may harm the companies involved by reducing their net incomes – either directly by influencing input access and outlets, or indirectly through loss of reputation or subsequent government regulation. A frequent problem with such initiatives, however, is that incriminating information, especially when it involves claims about illegal activities, can be very difficult to substantiate.

In the IUU context, the International Southern Oceans Longline Fisheries Information Clearing House (ISOFISH) initiative is notable. Established in 1997 by an Australian NGO and funded by legal toothfish operators and Australian authorities (Agnew 2000:369), this initiative aimed at compiling and disseminating information about the harvesting operations and corporate ownership of IUU fishing vessels in the region. More recently, COLTO has become the major vehicle for the shaming of unregulated harvesting in the Southern Ocean. In general, activities such as these can be argued to follow up on the encouragement articulated in the IPOA-IUU of efforts to 'promote industry knowledge and understanding… and… co-operative participation in, MSC activities to prevent, deter, and eliminate IUU fishing' (para. 24.6).

Normally, IUU operators are not particularly vulnerable to this kind of social pressure, but it is nevertheless of interest to pinpoint factors likely to shape its potential. For instance, it is widely believed that a number of Norwegian vessel owners disengaged from IUU operations in Antarctic waters largely as a consequence of ISOFISH publications having named them, drawn public attention to their activities and rendered such engagement socially unacceptable in the domestic vessel-owner community. A second factor is concern with brand name and reliance on environmentally conscious markets, and this could become relevant for IUU fishing operators. Pacific Andes, for instance, a large transnational claimed to be central in the Kerguelen Plateau fishery for toothfish, is reportedly planning to expand its market presence in Europe and Japan, where environmental awareness and political attention to the IUU fishing problem is higher than in its present stronghold, China.[41] This company has rejected any allegations of involvement in IUU operations.

A third factor is the prominence of the shamer. There is much to suggest that lists based on information compiled by an international organisation would be the most credible, since such shaming would usually require that a number of governments have decided to back the criticism. Being named and shamed by an individual government would also be severe. Although private advocacy groups are generally seen as less accountable and more confrontational than are governments and international bodies, there is considerable diversity among them with regard to public stature. It would be of interest to explore the possibilities for mobilising NGO heavyweights with extensive attention to fisheries matters but no economic stakes in the activity, such as Greenpeace or the World Wide Fund for Nature (WWF), in specific naming and shaming efforts. Legal issues would be relevant here, including the vulnerability of list makers to being sued by companies that reject charges of IUU involvement. In the United States, where resort to court action is a frequent aspect of environmental controversies, many states have passed legislation to ensure that the freedom of speech and the right to petition government policies is not unduly constrained by so-called 'strategic litigation against public participation' (SLAPs).[42] Where individuals or advocacy groups have been able to demonstrate that public statements brought to court for alleged defamation is a part of, or in support of, petitioning activity, charges have usually been dismissed even in cases where statements are found to be partially false, deceptive, or unethical (Potter 2001). Major NGOs with ample legal resources of their own are rarely

---

[41] *The Standard* (Hong Kong newspaper), 12 January 2004, available at (www.thestandard.com.hk).

[42] See generally (www.clasp.net).

targeted by SLAPs; and there are many examples where they have upheld shaming campaigns despite law suits by major companies.[43]

There are also more indirect causal pathways between private shaming and the resilience of transnational IUU operations. For example, company-level information compiled by private organisations such as ISOFISH and TRAFFIC (Lack and Sant 2001) has influenced the approach of international management bodies. By encouraging the examination of trade statistics, it has thus assisted in the development of CCAMLR's 'blacklist' system.

Efforts to reduce overcapacity

Overcapacity aggravates the problem of IUU operations in at least three ways. It reduces the opportunity and profitability of legal operations; the periodic idleness associated with it provides incentives for individual vessel owners to pursue IUU options; and overcapacity drives down the price of vessels, especially second-hand vessels but presumably new ones as well, thereby reducing the overall costs of illegitimate (as well as legitimate) harvesting operations. Efforts to reduce capacity and curb investments in vessels destined for IUU fishing are of several kinds but they have two features in common: counterforces are strong, and progress is likely to be limited, slow, or both.

One type of possible measure involves reduction or redirection of government subsidies. Figures on the amount of subsidies provided to the fisheries sector vary widely, a reflection partly of scattered knowledge and partly of different definitions or operationalisations (Milazzo 1998). Recent estimates suggest a level somewhere between 7 and 14 billion USD each year (Ruckes 2000). The effect of subsidies on capacity is particularly relevant in cases where management policies are unsatisfactory (Hannesson 2001:17–19; Cox 2003), including in many high-seas areas and developing-country zones where IUU harvesting is pervasive.[44] Demands for stronger disciplines on fisheries subsidies have been strong in recent years; the 2001 Doha Ministerial Declaration, which provides the mandate for the new 'Millennium Round' of multilateral trade negotiations, aims to 'clarify and improve WTO disciplines on fisheries subsidies'.[45] The 1994 Subsidies and Countervailing Measures (SCM) Agreement under the WTO umbrella provides detailed and legally binding rules concerning subsidies, supported by an elaborate compliance system that includes compulsory and binding procedures for dispute settlement and authorisation of countervailing trade sanctions. To date, however, no fisheries subsidy has been challenged under WTO rules, an important reason being that only a limited subset of direct or indirect financial transfers to the fisheries industry is clearly disciplined under present rules.[46]

Conceptual vagueness contributes to a general lack of information regarding the extent, nature and objective of subsidies. All proposals for enhanced checks on subsidies emphasise transparency and the need for improved information and notification measures (Grynberg 2003:503). Several

---

[43]   For instances involving Greenpeace and Friends of the Earth respectively, see (http://archive.greenpeace.org/pressreleases/arctic/1997aug18.html) visited 29 February 2004 and (www.foe.co.uk/resource/press_releases/19990419174235.html), visited 29 February 2004.

[44]   Access conditions are generally believed to be the most important factor explaining cross-state variation in excess capacity (Cunningham and Gréboval 2001).

[45]   Doha Declaration, Art. 28; available at (www.wto.org/english/tratop_e/dda_e/dda_e.htm); fisheries subsidies are also addressed in Art. 31.

[46]   A study commissioned by the Asia-Pacific Economic Co-operation, which includes several of the world's foremost fisheries subsidy nations including Japan and South Korea, concluded that only 10 out of an inventory of 162 instances of fisheries subsidies in this region stood a high chance of being successfully challenged under the SCM Agreement (PricewaterhouseCoopers 2000).

international organisations, including the OECD and the FAO, have work programmes on the matter. Efforts to reduce fisheries subsidies are complicated by the fact that governments may have a whole range of worthy reasons for providing them, including employment in shipbuilding, harvesting or processing sectors, food security, or protection of settlements in sparsely inhabited or economically disadvantaged coastal regions.[47]

Related to the subsidies issue is a second possible measure: the development of governmental buyback schemes aimed at reducing harvesting capacity. The overall efficiency and environmental impact of buyback schemes have been questioned, even when they require scrapping of the vessels withdrawn from national fisheries.[48] Further reservations are appropriate with regard to arrangements that are parochial in their approach by allowing the vessels involved to be exported. The recent change in EU regulations of government subsidies, which imply that Community-financed buyback schemes can no longer permit disposal of vessels by sales to third countries,[49] reflects the growing appreciation of the global nature of the overcapacity problem and its role in threatening sustainable management. It also reflects the fact that fisheries subsidies have been a priority issue among European environmental organisations throughout the past decade.

A third measure in this category is regulation of foreign direct investments, notably with regard to flag-of-convenience countries. Many, if not most, IUU operations are believed to have beneficiary owners who are residents in OECD countries, and Wynhoven (2004) discusses how, among others, the OECD 1961 Code on Liberalisation may impact on efforts to curb IUU operations. The overall effect of that investment instrument, implemented by member states subject to OECD peer review procedures, may even be to constrain such efforts, since the guiding principle of the Code is non-discriminatory removal of restrictions on capital flows. Thus, the introduction of new restrictions targeting vessel investments in flag-of-convenience states would run counter to the spirit of this agreement, although reservations are permitted under certain conditions. According to Wynhoven (2004:10-12), only Japan maintains a reservation permitting it to restrict outward fisheries investments by its nationals, applying to enterprises engaged in fishing regulated by Japan or international treaties to which it is a party. More generally, in today's increasingly liberalised world economy, the tendency is for fewer rather than more restraints on global and regional capital flows.

On balance, the causal chain that may connect these various means to reduce the capacity of world fisheries to higher costs of IUU vessel purchase is a long one, and there is considerable opposition to the strengthening of international rules. Although reductions of fisheries subsidies will be a positive contribution, this is likely to be a slow process rather than an easily-mobilised policy measure. Subsidy reform and other capacity initiatives are relevant and important within a long-term strategy to combat IUU fishing, but they cannot be expected to yield rapid results.

This section has dealt with measures designed to make IUU operations more difficult and more expensive by seeking to constrain their access to various inputs and outlets. The effectiveness of these measures – whether denial, shaming, or various efforts to complicate new investments in IUU fishing capacity – will depend critically upon the flow and management of information about IUU activities. The same is true for measures to reduce the incomes flowing from such activities, addressed in the next section.

---

[47]   See e.g. WT/CTE/W/175, 24 October 2000, available at www.docsonline.wto.org/gen_search.asp

[48]   See Porter (2002: 16-22); see also the discussion in Cox (2003).

[49]   EU, Council Regulation amending Regulation (EC) No 2792/1999 laying down the detailed rules and arrangements regarding Community structural assistance in the fisheries sector, COM(2002) 187 final.

Measures targeting IUU catch

Co-ordinated trade measures against non-members of international conservation regimes have been used since the early 1990s as inducements to join existing regimes, and later also as a compliance mechanism. One problem with early versions of this instrument is that they operated on a flag basis and thus did not permit differentiation between vessels that fish legitimately and those engaged in IUU fishing. In this regard, blacklisting of individual vessels was an important step forward.

This section addresses three categories of measures that seek to reduce incomes from IUU operations by targeting the products they bring to the markets. The first two categories are governmental, permit-based restrictions on imports and exports of certain commodities. Documentation schemes under regional fisheries regimes have been mentioned already; additionally we will discuss the possible use of a broader instrument, the Convention on International Trade in Endangered Species of Wild Fauna and Flora (CITES). The third set of measures discussed here concerns eco-labelling. This can be privately organised, and seeks to mobilise environmental awareness among retailers and consumers for purposes of enhancing the sustainability of harvesting operations.

Catch documentation schemes

Several fisheries regimes have developed schemes for documentation of catches, to promote better management and conservation of particular species. This represents a further step forward in differentiating between legal and IUU catches; these schemes target neither the flag state nor the vessel – only the cargo. Such schemes are especially relevant for IUU fishing carried out for international trade, as is the case with high-value tuna species and toothfish stocks.

ICCAT introduced trade documentation for bluefin tuna in the early 1990s. This has been followed by several other 'trade documentation' schemes developed on that model, especially within the tuna trade: those by CCSBT, IOTC, and by ICCAT for bigeye tuna and swordfish. The 'catch certification' system, as developed by CCAMLR since 2000, differs from these. In trade documentation systems, documents are issued at the point of *landing* and only for products that enter international trade; by contrast, in a catch certification system, the documents are issued at the point of *harvesting*, and are related to *all* fish to be landed or transhipped.[50] The CCAMLR catch documentation scheme (CDS)[51] covers toothfish catches taken in the Convention area as well as on the high seas outside that area. Participation in the CDS is open to CCAMLR parties and non-parties alike; to date, several non-parties with significant roles in various stages of toothfish catch movement between vessel and market have joined the CDS: China, Seychelles, Singapore and, partly, Mauritius. Most of the toothfish market is currently covered by countries participating in the CDS, including the United States, the European Union and Japan; other sections, however, are not (especially Canada). It has been estimated that countries involved in the CDS constitute about 90% of the market for international trade of toothfish; and that it is being applied to an area that is home to 90% of the global population.[52]

The purpose of the CDS is to place obstacles in the way of trade in IUU catches in several ways. First, toothfish caught in the Southern Ocean without a 'paper' should become more difficult to export

---

[50]   See discussion in Miller, Sabourenkov and Slicer (forthcoming 2004).

[51]   CDS is currently based on CCAMLR Conservation Measure 10-05 (2003), 'Catch Documentation Scheme for *Dissistichus* spp.' On CDS see especially Agnew (2000).

[52]   Miller, Sabourenkov and Slicer (forthcoming 2004).

and import, and therefore less attractive to the market – which would mean diminished net income to IUU operators. Soon after the CDS was introduced, it was estimated that the price of toothfish not accompanied by a valid catch document was as much as 25–40% lower;[53] and even higher differences have been cited.[54]

Second, the CDS operates in tandem with other CCAMLR measures, and with national legislation in some countries. Port state measures are especially relevant. On the basis of CDS information, landing and transhipment in ports can be denied. The burden of proof is placed on the operator, who must establish that the toothfish has been caught legitimately outside the Convention area or within the CCAMLR area in accordance with applicable conservation measures.[55] Such denial targets both exports and imports, and is strengthened by national legislation in major market countries, such as the United States.

Third, an important purpose of the system is to supply parties and the CCAMLR secretariat with data on toothfish trade and to assist in verification of such data. With the obligatory Vessel Monitoring System (VMS) for parties fishing in the CCAMLR area,[56] against the backdrop of license requirements authorising fishing in the Convention area, the flag state can determine the catch location and certify the catch *before* it is landed or transhipped. The introduction of electronic, web-based CDS, currently as a pilot project, aims at almost real-time data and at further facilitating cross-checking and verification capabilities.

While the CDS targets a weak spot of an IUU operation, some loopholes remain. After CCAMLR introduced the CDS, an increasing amount of toothfish has been reported as caught in FAO Statistical Areas 51 and others, in the Southern Ocean just beyond the area of application of CCAMLR conservation measures. Current scientific knowledge suggests, however, that it is unlikely that such amounts of toothfish can in fact be found in those areas. Difficulties related to VMS verification and the fact that VMS data are not sent directly to the CCAMLR secretariat, but only *via* the flag state (and coastal state, for fishing licensed within its EEZ), have facilitated this situation. Some CCAMLR parties have advocated the adoption of a centralised reporting system, modelled after NAFO or North East Atlantic Fisheries Commission (NEAFC), which would enable direct (parallel) sending of satellite data to the CCAMLR secretariat, but no consensus has been reached. Several CCAMLR parties are, however, now participating in a voluntary centralised system as a 'pilot project'.

Import restrictions such as documentation schemes, co-ordinated under regional management regimes and pertaining to fish caught in violation of regional conservation measures, could be challenged under WTO rules, especially by non-parties to the relevant management regime, as implying discrimination against 'like products' (Chaytor *et al.* 2003). In designing the CCAMLR documentation scheme, the parties were highly attentive to this possibility and drew upon the dispute settlement reports on the tuna/dolphin cases and the more recent shrimp/turtle case (Agnew 2000:369-

---

[53]   Para. 2.3 of the 'Report of the Standing Committee on Observation and Inspection (SCOI)' (Hobart: CCAMLR, 2000).

[54]   Miller, Sabourenkov and Slicer (forthcoming 2004) indicate prices at 8.40 USD/kg for fish with catch document against 3 USD/kg for fish not accompanied with the document.

[55]   CCAMLR Conservation Measure 10-03 (2002), 'Port Inspections of Vessels Carrying Toothfish'; in accordance with that conservation measure, advance notice is required, as well as a declaration of not being engaged or supporting IUU fishing, and access to the port can be denied. On trends in port state measures, see relevant section above.

[56]   See CCAMLR Conservation Measure 10-04 (2002), 'Automated Satellite-Linked Vessel Monitoring System (VMS)'.

70). Like ICCAT before it, the CCAMLR Secretariat has also presented and discussed its documentation scheme with the WTO Committee on Environment and Development, with a view to minimising tensions.[57] The conservation measure that established the documentation scheme placed it explicitly in the range of policies that may be justified under the WTO environmental exceptions. Moreover, the non-effectiveness of less trade-restrictive measures was emphasised, as was the placement of the scheme in an inclusive and transparent multilateral process that would render usage for protectionist purposes difficult. Failure to exhaust measures that would impinge less on international trade, notably under multilateral environmental regimes, has been severely criticised in WTO dispute settlement reports. Moreover, to avoid charges of discrimination, the CCAMLR scheme is implemented on domestic as well as foreign vessels; it is open for participation by non-parties to CCAMLR; and it extends also beyond the CCAMLR area.

Use of a broader instrument: species-oriented trade restrictions

The objective of the Convention on International Trade in Endangered Species of Wild Fauna and Flora (CITES) is to remove or reduce the pressure exerted by profitable international trade on the survival of threatened species. This goal is pursued through a set of appendices containing lists of species that are subject to varying degrees of restrictions on export, import, and introduction from the sea, involving national permits, quotas, or a combination of the two.[58] The prominence of CITES in discussion of IUU measures is due to the attempt by Australia, encouraged by domestic advocacy organisations, to muster support for Annex II listing of species of toothfish and the opposition that was mounted against this initiative. Such listing would imply that export or re-export of toothfish would require a national permit that, according to CITES provisions, can be granted on two conditions only. First, a nominated scientific authority must confirm that trade will not be detrimental to the survival of the species; and second, a nominated management authority must confirm that the toothfish has been acquired lawfully (Art. IV). Correspondingly, imports of toothfish by a CITES party would require presentation of an export or re-export permit. For catch 'introduced from the sea', *i.e.* 'taken in the marine environment not under the jurisdiction of any State', the requirements are somewhat softer, as no lawfulness assessment is necessary.[59] Landings of catch taken in national waters for domestic consumption do not fall within the scope of the convention.

The term 'species' in the CITES Convention is defined as 'any species, subspecies, *or geographically separate population thereof*',[60] thus permitting the listing of individual stocks. The main rationale for proposing listing of stocks subject to IUU fishing would be threefold. First, as CITES has a membership of 162, applying its provisions would constrain more flag states, port states, export states, and import states than does any relevant regional fisheries regime. Although CCAMLR has successfully expanded participation in its measures to combat IUU operations, for instance by the accession of new parties and the participation in its catch documentation scheme of non-parties important in the toothfish trade, the availability of flags and ports that do not require any catch documents remains a limitation of the system.[61] Second, the geographic scope of CITES includes high-seas harvesting areas that fall outside the ambit of regional regimes; and third, CITES has a more

---

[57] WT/CTE/W/148, 30 June 2000, The Commission for the Conservation of Antarctic Marine Living Resources, Communication from the CCAMLR Secretariat.

[58] The Convention, appendices, and resolutions are available at (www.cites.org).

[59] CITES Convention, Arts. I (definition) and IV (substantive requirements).

[60] CITES Convention, Art. I (a), italics added.

[61] But, according to Miller and Sabourenkov (2004), the overall coverage of the CCAMLR catch documentation scheme is more than 90% of the world trade in toothfish.

forceful compliance system than those of most fisheries regimes. The Conference of the Parties of CITES has on several occasions recommended effective suspension of trade in one or more listed species with states that had failed to implement its obligations under the Convention.[62]

That said, CITES listing of fish species has been highly controversial, both within CITES and in other international organisations. One set of objections focuses on the *appropriateness* of CITES as an instrument for management of commercially exploited marine species. The suitability of the listing criteria for fisheries management has been questioned, especially the guidelines on how to apply the population decline criterion. Two FAO expert consultations have been held on the matter (FAO 2001), and CITES is presently reviewing its criteria and guidelines in response to, *inter alia,* FAO input.[63] Concern has also been expressed about the CITES process of scientific evaluation, including the role played by non-governmental organisations. The forging of stronger links to the scientific bodies of existing regional fisheries regimes has been advocated.[64] Finally, the decision-making procedures of CITES, especially the infrequency of meetings and the high procedural threshold for de-listing species, have been criticised as inadequate for adaptive fisheries management.

A second set of objections concern certain *indirect effects* of CITES listing. In particular, many fishing nations perceive CITES as an excessively blunt management tool that would be likely to elevate trade barriers not only for products that originate in IUU operations but for those extracted from well-managed stocks as well. There are several reasons for this concern. First, the difficulties associated with differentiating products in trade according to the stocks from which they originate suggest considerable implementation problems for any stock-specific listing (FAO 2000:48). Second, the Convention provides that, if necessary to ensure the effective control of trade of a threatened species, other species that 'a non-expert, with reasonable effort, is unlikely to be able to distinguish' from the listed species shall also be listed.[65] Many fishing states worry about the possible impacts of this expansive 'look-alike' provision if any future CITES listing should involve a stock of a commercially important species such as cod or other major whitefish. Another type of indirect effect of CITES listing was prominent in the heated CCAMLR debate on Australia's proposal for listing of toothfish. Many delegations expressed deep concern that CITES listing of species falling under the competence of CCAMLR would undermine the legitimacy of this regional regime in the world community.[66]

For stocks that are threatened by extensive IUU fishing, proposals for listing under CITES are likely to be forwarded also in the future. In some cases, such listing would enhance the possibility to monitor and regulate trade in products that originate from threatened stocks. On the other hand, political impediments to such listing, based on a perception among many fishing nations that CITES is not an appropriate instrument for fisheries management, will not be easily overcome without substantive or procedural changes in the CITES regime itself.

---

62   *Yearbook of International Co-operation on Environment and Development 2003/2004*, p. 209; on the procedure, see CITES Convention, Arts. XI and XIII.

63   CITES Decision 12.7 provides for the drafting of a Memorandum of Understanding between CITES and FAO; see (www.cites.org).

64   See for instance CCAMLR (2002), item 10.

65   CITES Convention, Art II 2 (b); citation is from Annex 2b to Resolution Conf. 9.24 (available at www.cites.org) which clarified the interpretation of Art. II.

66   See CCAMLR (2002), item 10. For a broader discussion of this aspect of resource management in the Antarctic, see Stokke and Vidas (eds. 1996).

Eco-labelling

Eco-labelling schemes are a third set of market-oriented measures that could be relevant in combating IUU operations. Unlike the permit-based schemes discussed above, eco-labelling is 'a voluntary multiple-criteria-based third-party programme that... authorises the use of environmental labels on products indicating overall environmental preferability... based on life cycle considerations'.[67] Its history in fisheries is relatively brief. The most prominent example is the US government-backed 'dolphin safe' tuna label issued in conjunction with the decision of the major US tuna processing companies that they would buy fish only from harvesters who adhered to by-catch provisions based on the US Marine Mammals Protection Act (Carr and Scheiber 2002). This particular initiative is widely seen as highly effective – but the conditions were also unusually favourable (Teisl *et al.* 2002).

Multi-criteria, global, third-party certification schemes are even more recent, starting with the initiative taken by WWF and Unilever in 1996 to establish the Marine Stewardship Council (MSC) (Schmidt 1998). To date, only rather small fisheries have been certified under this scheme, but this is now changing, especially with the ongoing Alaska pollack process.[68] If such schemes manage to establish themselves in major seafood markets, they can provide a competitive edge for legal fishers. Under MSC, certification is conducted by means of criteria based on three key principles. First, the harvesting pressure must be consistent with the precautionary approach; second, ecosystem impacts must be considered; and third, effective management structures must be in place. As shown in the South Georgia toothfish longline fishery application, the effective management principle indicates that measures to deal with IUU fishing can be an important criterion for awarding a certificate.[69]

A general limitation of eco-labelling initiatives is their geographic scope: they feed on 'green consumerism', and that is a phenomenon largely restricted to certain parts of the world. Thus, the MSC is firmly established only in some Northern European markets, especially the UK; and its area of expansion is, predictably, Australasia and North America (*cf.* British Columbia salmon and Alaska pollack). MSC officials are much less optimistic about Japan, for instance (May *et al.* 2003:28). Even within environment-conscious markets, the effectiveness of eco-labelling programmes may be jeopardised by the presence of several green labels. This fact can be exploited by industry whenever existing labelling schemes are seen as detrimental to their interests. For instance, the National Fisheries Institute – which, despite its name, is the primary trade association of the US commercial fishing industry – has set up the Responsible Fisheries Society charged with developing an alternative programme to MSC (Carr and Scheiber 2002). From the perspective of combating IUU, such a proliferation of labels need not be problematic, provided that other labels too include among their certification criteria that firms and management authorities take adequate measures against IUU operations and have structures adequate for implementing such criteria.

---

[67] WT/CTE/GEN/1, 19 November 2002, Progress in Environmental Management Systems (EMS) Standardization. Statement by the International Organization for Standardization (ISO); the definition is contained in ISO 14024:1999, 'Environmental labels and declarations – Type I environmental labelling – Principles and procedures.' The life-cycle approach implies an assessment of environmental impacts not only from the use and disposal of a product but also from its production – sometimes referred to as 'cradle-to-grave' analysis.

[68] Information about past and ongoing certification processes is available at (www.msc.org/).

[69] Annual catches in the 2000-2002 period were around 5,000 tons. While IUU fishing is an issue also in the South Georgia area, it is much less pervasive than on the Kerguelen Plateau of the Indian Ocean. Agnew *et al.* (2002:4) estimate the IUU share of the 2000/2001 South Georgia IUU catch at only 5% and on its way down.

Another specific challenge to management-oriented labelling lies in the diversity, complexity and length of the chains of custody associated with most seafood products (May *et al.* 2003:15). This is amplified in the IUU context by the unlawful activities frequently associated with it, such as 'laundering' of illegally obtained fish, and bribing customs officials. Accordingly, under MSC a chain-of-custody certification distinct from the fishery certification is designed to ensure that products carrying the MSC logo actually originate in a certified fishery. Particular attention is directed at the processing stage, and production plants must document satisfactory control systems for keeping MSC produce apart from other inputs (Scott 2003: 89-91). Main components are MSC-endorsed chain-of-custody certificates issued by suppliers, physical or temporal separation of certified and non-certified products, product labelling, output identification, and adequate record keeping.

A third challenge is the potentially trade-distortive effect of eco-labelling schemes (Vitalis 2001). The Doha Declaration also mentioned environmental labelling as one of the areas where WTO rules might be in need of clarification.[70] However, while most eco-labelling schemes are non-state and voluntary, WTO rules have focused on mandatory governmental labelling schemes. The Technical Barriers to Trade (TBT) Agreement explicitly acknowledges that unrestricted trade may sometimes collide with other legitimate objectives such as national security or protection of human health and the environment.[71] If this happens, measures like labelling regulations or standards may be introduced. To ensure that such rules are non-discriminatory and not unnecessarily restrictive, however, the Agreement obliges governments to ensure a high level of harmonisation and transparency of such regulations and standards. Accordingly, even governmental labelling schemes are explicitly permitted, provided they include reasonable operational safeguards against protectionist abuse. Harmonisation and transparency provisions under the WTO are softer for regulations and standards upheld by local government or non-governmental bodies, such as MSC. Notification rules are not as strict and the role of member states is indirect. Governments are required only to 'take such reasonable measures as may be available' to ensure that harmonisation and transparency rules are accepted and complied with by those other bodies, and to refrain from measures that 'require or encourage' violation of those rules.[72]

Eco-labelling programmes are in line with a few other measures to improve environmental sustainability in the fisheries sector, including shaming of IUU activities, the active involvement of private organisations and even individual consumers. As such, this measure may enhance societal awareness about the problem of IUU and support more extensive public efforts to combat it. Eco-labelling in fisheries is still a fairly new phenomenon and one that has yet to take off. The recent MSC certification processes involving larger fisheries may change that situation: it is encouraging to note that IUU activities receive considerable attention when certification criteria are operationalised.

Conclusions

Focusing on three segments of an IUU operation – vessel at sea, transnational logistics, and catch in trade – this paper has examined the varieties and limitations of measures designed to combat this problem. An underlying theme is that if they are to succeed, efforts aimed at dealing with such a complex, transnational, and evasive phenomenon must apply the broadest range of tools. When seen alone, each of the measures in question has severe limitations and cannot be expected to deliver the

---

[70] WT/MIN(01)/DEC/1, 20 November 2001, Ministerial Declaration. Adopted 14 November 2001; see Art. 32.

[71] Agreement on Technical Barriers to Trade (TBT), Arts. 2.2, 2.10, 5.4 and 5.7, available at (www.wto.org). Labelling is also addressed in the Agreement on Sanitary and Phytosanitary (SPS) measures but only in the context of food safety.

[72] TBT Agreement, Arts. 3, 7, and 8.

goods. When the various measures are seen in conjunction and given time to mature, the accumulated costs they impose on IUU operations and their complicits can become substantial and thus make such activities less lucrative and limit their scope.

The following conclusions seem warranted. First, the range of global and regional instruments developed within the sphere of the *Law of the Sea* to address IUU fishing is quite impressive, especially given the short time that has passed since this issue gained prominence on the political agenda. Nevertheless, it is clear that measures that primarily target the vessel at the stage of registration and at sea attack the chain of an IUU operation at its most robust links. Activities conducted here enjoy a high degree of insulation from those who may seek to constrain them. This is due to general legal principles, especially the primacy of flag state jurisdiction and the rule that treaties do not create obligations for third states without their consent – as well as the physical remoteness of much IUU harvesting.

It is necessary to target IUU operations at links where there are fewer possibilities of avoiding regulation and where enforcement can be made in more cost-efficient ways. After all, the basic purpose of an IUU operation is not fishing *per se*, or for avoidance of legal measures: it is a profit-making venture that seeks to maximise net income. Further development of port state measures would seem to be a promising avenue, especially with regard to regional harmonisation and pre-entry documentation procedures that reverse the burden of proof by obliging vessels to show that a catch has been taken legally.

Second, measures targeting the *logistical* activities of IUU operations have the potential to involve a large number of states and non-governmental actors. There is, however, a need to improve the generation and management of relevant information. The denial strategy, frequently in the form of 'blacklists' of vessels with a history of IUU fishing and subsequently denied licensing or even port or supply access, relies upon information that must be both extensive and reliable – two requirements that are sometimes difficult to combine. Due process concerns and the need to comply with international trade rules dictate transparency and harmonisation of the procedures that guide various denial measures, and regional fisheries management regimes can be important vehicles in achieving this.

Mobilising non-governmental organisations, including other harvesters and environmental advocacy groups, to generate and disseminate information about IUU activities has been important also for exposing corporate irresponsibility on the part of individual firms and vessel-owners. When the amount and quality of information permits, this shaming strategy can be extended to those who provide necessary inputs to IUU operations. Both flexible company structures and rapidly shifting ownership situations place limits on the effectiveness of such measures. However, the number of IUU vessels engaged over extended periods of time in a given fishery is usually not very high. There is much to suggest, therefore, that time will work in favour of strategies involving denial and shaming.

Third, long-term efforts aimed at reducing or checking the growth of fishing capacity face strong counterforces, including the resilience of governmental subsidies in some countries and liberalisation of capital flows. That said, some progress has been made in recent years, and the issue remains high on the political agenda.

Fourth, measures targeting the final segment of IUU operations, the *commodities* brought to market, are promising also because they are less dependent upon costly monitoring and physical surveillance activities. Still, catch documentation schemes work best in practice when other components of the monitoring and enforcement system, especially port state co-ordination and VMS coverage, are well advanced. The design of recent schemes involves minimal tension with international trade rules. The use of CITES in the combat of IUU operations could expand the coverage of permit-based documentation schemes based in fisheries regimes, but it remains politically contested by many fishing states. Eco-labelling schemes in the fisheries sector are still at a rather early stage and it is too early to pass judgement on the role they may come to play in combating IUU fishing. It is promising, however, that procedures for certification under the Marine Stewardship Council include assessment of the level of IUU fishing and the adequacy of measures taken to combat it.

Finally, in all the segments we reviewed in this paper, the impact of *information* about IUU operations is a crucial factor. Regarding the vessel at sea, where the size of the marine area is huge but the number of flag states involved is actually relatively small, international pressure, when based on accurate information, can support the exercise of flag state responsibilities. Regarding the logistics of an IUU operation, its resilience is enhanced by the 'grey zones' of transnationality and becomes considerably diminished when exposed by means of accurate information. And regarding the flows of IUU catch in international trade, if current catch documentation schemes are backed up by timely and accurate information, fraud can be significantly reduced. Technology limitations do play a role here, but these are not the main concern. The strength of information as a tool for combating IUU operations is enhanced if it can be made transparent. Among the impediments should be mentioned the fact that commercial data are involved, and some stakeholders will be less willing to provide information knowing that it can become public. Moreover, other stakeholders may provide information that, at times, is not sufficiently substantiated. Improving the quality and management of information about IUU operations is a key task, and one that involves governments, international institutions, as well as non-governmental organisations.

REFERENCES

Agnew, D.J., G.P. Kirkwood and J. Pierce (2002), "An Analysis of the Extent of IUU Fishing in Subarea 48.3" (A report for the UK Government in respect of South Georgia and the South Sandwich Islands).

Agnew, David J. and Colin T. Barnes (2004), "Economic Aspects and Drivers of IUU Fishing: Building a Framework". OECD [AGR/FI/IUU(2004)2].

Agnew, David J. (2000), "The Illegal and Unregulated Fishery for Toothfish in the Southern Ocean, and the CCAMLR Catch Documentation Scheme", *Marine Policy,* Vol. 24, 361–74.

Carr, Christopher J. and Harry N. Scheiber (2002), "Dealing with a Resource Crisis: Regulatory Regimes for Managing the World's Marine Fisheries", *University of California International and Area Studies Digital Collection,* Vol. 1, No. 3, http://repositories.cdlib.org/uciaspubs/editedvolumes/1/3/

CCAMLR (2002), Report of the Twenty-First Meeting of the Commission. Hobart: Commission for the Conservation of Antarctic Marine Living Resources.

Chaytor, Beatrice, Alice Palmer and Jacob Werksman (2003), "Interactions with the World Trade Organization: The Cartagena Protocol on Biosafety and the International Commission for the Conservation of Atlantic Tunas". Berlin: *Ecologic,* http://www.ecologic.de/projekte/interaction/results.htm.

Cox, Anthony (2003), "Environmental Aspects of Fisheries Subsidies", *OECD Technical Expert Meeting on Environmentally Harmful Subsidies,* , [SG/SD(2003)12] Paris: OECD,

Cunningham, Steve and Dominique Gréboval (2001), "Managing Fishing Capacity: A Review of Policy and Technical Issues", *FAO Fisheries Technical Paper,* No. 409.

FAO (2000), "An Appraisal of the Suitability of the CITES Criteria for Listing Commercially-Exploited Aquatic Species", *FAO Fisheries Circular* 954.

FAO (2001), "Report of the Second Technical Consultation on the Suitability of the CITES Criteria for Listing Commercially-Exploited Aquatic Species", *FAO Fisheries Reports* 667.

Gianni, Matthew and Walt Simpson (2004), "Flags of Convenience, Trans-shipment, Re-supply and At-sea Infrastructure in Relation to IUU Fishing" (Paper in progress, 16 April 2004, submitted by WWF), Paris: OECD, AGR/FI/IUU(2004)22.

Grynberg, Roman (2003), "WTO Fisheries Subsidies Negotiations: Implications for Fisheries Access Arrangements and Sustainable Management", *Marine Policy,* Vol. 27, 499–511.

Hannesson, Rögnvaldur (2001), "Effects of Liberalizing Trade in Fish, Fishing Services and Investment in Fishing Vessels", *OECD Papers,* Vol. 1, No. 1.

Haufler, Virginia (2003), "Corporations and Conflict Prevention: Normative Change in the International Community", Lysaker: The Fridtjof Nansen Institute: Report for the project Oil Companies in the New Petroleum Provinces, http://www.ecologic.de/projekte/interaction/results.htm.

Lack, M. and G. Sant (2001), Patagonian Toothfish: Are Conservation and Trade Measures Working?", *TRAFFIC Bulletin,* Vol. 19, No. 1, 1–18.

May, Brendan, Duncan Leadbitter, Mike Sutton and Michael Weber (2003), "The Marine Stewardship Council (MSC)", in B. Phillips, T. Ward and C. Chaffee (eds.), *Eco-labelling in Fisheries: What is it All About?,* Oxford: Blackwell Science, 14–33.

Milazzo, Matteo (1998), "Subsidies in World Fisheries: A Reexamination", *World Bank Technical Paper: Fisheries Series,* No. 406, Washington, DC: The World Bank.

Miller, Denzil, Eugene Sabourenkov and Natasha Slicer (2004), "Unregulated Fishing – the Toothfish Experience", in M. Richardson and D. Vidas (eds.), *The Antarctic Treaty System for the 21st Century* (London: UK Foreign and Commonwealth Office, forthcoming).

OECD (2004), "National Measures Against IUU Fishing Activities" (Work in progress, 12 April 2004), Paris: OECD [AGR/FI/IUU(2004)6/PROV].

Porter, Gareth (2002), "Fisheries Subsidies and Overfishing: Towards a Structured Discussion", Geneva: United Nations Environment Programme, www.unep.ch.

Potter, Lori (2001), "Strategic Lawsuits Against Public Participation and Petition Clause Immunity", *Environmental Law Reporter News and Analysis,* Vol. 31, 10852–56.

PricewaterhouseCoopers (2000), "Study into the Nature and Extent of Subsidies in the Fisheries Sector of APEC Members Economies" (Prepared for Fisheries Working Group, Asia Pacific Economic Co-operation [APEC]), www.apecsec.org.sg.

Ruckes, E. (2000), "International Trade in Fishery Products and the New Global Trading Environment", *Multilateral Trade Negotiations on Agriculture: A Resource Manual,* Rome: FAO, www.fao.org/DOCREP/003/X7351E/X7351E00.HTM.

Schmidt, Carl-Christian (1998), "Marine Stewardship Council: The Role of Eco-labelling in Helping to Achieve Sustainable Fisheries", Conclusions and Papers Presented at the International Conference: Green Goods V – Eco-labelling for a Sustainable Future, Paris: OECD Doc.ENV/EPOC/PPC(99)4/FINAL, 123–24.

Scott, Peter (2003), "MSC Chain-of-custody-Certification", in B. Phillips, T. Ward and C. Chaffee (eds.), *Eco-labelling in Fisheries: What is it All About?,* Oxford: Blackwell Science, 86–93.

Stokke, Olav Schram (2001), "Managing Fisheries in the Barents Sea Loophole: Interplay with the UN Fish Stocks Agreement", *Ocean Development and International Law,* Vol. 32, 241–62.

Stokke, Olav Schram and Davor Vidas (eds.) (1996), *Governing the Antarctic: The Effectiveness and Legitimacy of the Antarctic Treaty System.* Cambridge University Press.

Teisl, Mario F., Brian Roe and Robert L. Hicks (2002), "Can Eco-labels Tune a Market? Evidence from Dolphin-Safe Labeling", *Journal of Environmental Economics and Management,* Vol. 43, 339–59.

Vitalis, Vangelis (2001), "Eco-Labelling and WTO Rules: What Needs to Be Done?", Paris: OECD, www.oecd.org.

Vukas, Budislav and Davor Vidas (2001), "Flags of Convenience and High Seas Fishing: The Emergence of a Legal Framework", in O. S. Stokke (ed.), *Governing High Seas Fisheries: The Interplay of Global and Regional Regimes,* Oxford: Oxford University Press, 23–52.

Wynhoven, Ursula A. (2004), "OECD Instruments and IUU Fishing", Paris: OECD [AGR/FI/IUU(2004)1].

CHAPTER 2

GLOBAL REVIEW OF ILLEGAL, UNREPORTED AND UNREGULATED FISHING ISSUES: WHAT'S THE PROBLEM?

David A. Balton, Deputy Assistant Secretary for Oceans and Fisheries,
U.S. Department of State

IUU fishing, and the related issue of fishing by vessels flying flags of convenience, is not a single phenomenon. As noted by the Food and Agriculture Organization of the United Nations, IUU fishing "occurs in virtually all capture fisheries, whether they are conducted within areas under national jurisdiction or on the high seas." Examples include re-flagging of fishing vessels to evade controls, fishing in areas of national jurisdiction without authorisation by the coastal state and failure to report (or misreporting) catches. But the list of activities encompassed by the term "IUU fishing" is really much broader.

Just as IUU fishing is a multifaceted phenomenon, the problems caused by IUU fishing are many and diverse. Among the obvious adverse consequences are:

1) Diminished effectiveness of fisheries management.

2) Lost economic opportunities for legitimate fishers.

3) Reduction in food security.

Those who conduct IUU fishing are also unlikely to observe rules designed to protect the marine environment from the harmful effects of some fishing activity, including, for example, restrictions on the harvest of juvenile fish, gear restrictions established to minimise waste and by-catch of non-target species, and prohibitions on fishing in known spawning areas. To avoid detection, IUU fishers often violate certain basic safety requirements, such as keeping navigation lights lit at night, which puts other users of the oceans at risk. Operators of IUU vessels also tend to deny to crew members fundamental rights concerning the terms and conditions of their labour, including those concerning wages, safety standards and other living and working conditions.

In addition to its detrimental economic, social, environmental and safety consequences, the very unfairness of IUU fishing raises serious concerns. By definition, IUU fishing is either an expressly illegal activity or, at a minimum, an activity undertaken with little regard for applicable standards. IUU fishers gain an unjust advantage over legitimate fishers. In this sense, IUU fishers are "free riders" who benefit unfairly from the sacrifices made by others for the sake of proper fisheries conservation and management. This situation undermines the morale of legitimate fishers and, perhaps more importantly, encourages them to disregard the rules as well. Thus, IUU fishing tends to promote additional IUU fishing, creating a downward cycle.

Given the diversity of the phenomenon we call IUU fishing and the multiple problems it causes, we must take a multi-tiered approach to combating it. The FAO International Plan of Action on IUU Fishing sets forth such an approach. The IPOA is conceived of as a "toolbox" – a set of tools for use in dealing with IUU fishing in its various manifestations. Obviously, not all tools in the toolbox are appropriate for use in all situations. Still, it is now incumbent on all FAO Members to fulfil their commitments under the IPOA, both in their general capacity as states as well as in their more particular capacities as flag states, port states, coastal states, market states and as members of regional fishery management organisations.

Other international institutions, including the OECD, also clearly have a role to play in the fight against IUU fishing. Through workshops such as this and follow-up activities, the OECD can shed further light on the economic drivers of IUU fishing, help refine the tools currently being used in response to IUU fishing and contribute to the development of new tools as well.

CHAPTER 3

IUU FISHING AND STATE CONTROL OVER NATIONALS[1]

David A. Balton, Deputy Assistant Secretary for Oceans and Fisheries,
U.S. Department of State

"What giants?" said Sancho Panza.

"Those thou seest there," answered his master, "with the long arms, and some have them nearly two leagues long."

"Look, your worship," said Sancho, "what we see there are not giants but windmills, and what seem to be their arms are the sails that turned by the wind make the millstone go."

"It is easy to see," replied Don Quixote, "that thou art not used to this business of adventures; those are giants; and if thou art afraid, away with thee out of this and betake thyself to prayer while I engage them in fierce and unequal combat."[2]

Introduction

The negotiation of the FAO International Plan of Action on Illegal, Unreported and Unregulated Fishing in many ways brought to mind the adventures of Don Quixote. Like Cervantes' hero, some of us involved in that negotiation saw ourselves as engaged "in fierce and unequal combat" against the bad actors of world fisheries, as we tried to restore a system of ethical rules to guide human activity in this field. Perhaps other saw us as tilting at windmills.

The IPOA takes an approach to the problem of IUU fishing that would have made Don Quixote proud, one that is universal in scope and resolute in temperament. All FAO Members have undertaken meaningful commitments under the IPOA, both in their general capacity as states as well as in their more particular capacities as flag states, port states, coastal states, market states and as members of regional fishery management organisations.

---

[1]   This paper was submitted to the IUU Workshop as a background paper.

[2]   Miguel de Cervantes, *Don Quixote* (1605), John Ormsby, trans.

One aspect of the IPOA that has not received much attention – state control over nationals – merits closer study. One reason why IUU fishing has been such a persistent problem is that many states have not been successful in controlling the fishing activities by their nationals that take place in the waters of other states or aboard vessels registered in other states. Admittedly, it may be difficult for many states to control, or even to be aware of, such activities. States may also have difficulty in preventing their nationals from re-flagging fishing vessels in other states with the intent to engage in IUU fishing.

The IPOA nevertheless calls on all states to take measures or co-operate to ensure that their nationals do not support or engage in IUU fishing. This paper will consider a number of measures that states have taken in this regard and will also take another look at the "re-flagging problem" that, unfortunately, remains with us to this day. It will also suggest some additional steps for addressing IUU fishing.

Existing measures

Under international law, a state is free to enact laws prohibiting its nationals from engaging in IUU fishing, even if the activity in question would take place aboard a foreign vessel or in waters under the jurisdiction of another state.[3] Some states have already done so.

For example, Japan requires its nationals to obtain the permission of the Japanese government before working aboard non-Japanese fishing vessels operating in the Atlantic bluefin tuna and southern bluefin tuna fishing areas. The goal of this measure is to prevent Japanese nationals from becoming involved in IUU fishing aboard foreign vessels. Japan also intends to deny permission to any Japanese national to work aboard a foreign fishing vessel in any other fishery, if the vessel's flag state is not a member of the regional fishery management organisation (RFMO) regulating that fishery.[4] New Zealand and Australia have also enacted legislation restricting the activities of their respective nationals aboard foreign vessels registered in states meeting certain criteria.

In the United States of America, the Lacey Act makes it unlawful for any person subject to U.S. jurisdiction to "import, export, transport, sell, receive, acquire, possess or purchase any fish ... taken, possessed or sold in violation of any ... foreign ... law, treaty or regulation." Hence, a U.S. national may be prosecuted for engaging in certain forms of IUU fishing aboard foreign vessels or in waters under the jurisdiction of another state.[5]

---

[3]   The principle that a state may apply its law to its nationals wherever the may be found is generally accepted. See, *e.g.*, *Bartolus on the Conflict of Laws* 51 (Beale, trans. 1914); *Restatement (Third) of the Foreign Relations Law of the United States*, §402(2) (1987). For further discussion, see "Tools to Address IUU Fishing: The Current Legal Situation," by William Edeson, one of a series of papers prepared as background documents for the Expert Consultation on Illegal, Unreported and Unregulated Fishing organised by the Government of Australia in co-operation with FAO, Sydney, Australia, 15-19 May 2000.

[4]   See "The Importance of Taking Co-operative Action Against Specific Fishing Vessels that are Diminishing Effectiveness of Tuna Conservation and Management Measures," by Masayuki Komatsu, one of a series of papers prepared as background documents for the Expert Consultation on Illegal, Unreported and Unregulated Fishing Organised by the Government of Australia in co-operation with FAO, Sydney, Australia, 15-19 May 2000.

[5]   See United States Code, Title 16, Chapter 53. For further discussion of how the Lacey Act might be adapted for other situations involving IUU fishing, see "National Legislative Options to Combat IUU Fishing," by Blaise Kuemlangan, one of a series of papers prepared as background documents for the

A return to the "re-flagging problem"

The European Union now also appears to be moving to control IUU fishing by nationals of its member states in a way that is bringing renewed attention to the "re-flagging problem." In May 2002, the European Commission issued a "Community Action Plan for the Eradication of Illegal, Unreported and Unregulated Fishing." In considering measures to control nationals of EU member states, this paper presents the following objective:

> *to discourage Community member state nationals from flagging their fishing vessels under the jurisdiction of a state which is failing to fulfil its flag-state responsibilities and from committing infringements.*

The articulation of this goal represents a positive development in the attitude of the European Commission toward the problem of vessel re-flagging. We must recall that the international community recognised the gravity of this problem more than ten years ago. Agenda 21, adopted by the United Nations Conference on Environment and Development in Rio de Janeiro, called upon states to:

> *take effective action, consistent with international law, to deter re-flagging of vessels by their nationals as a means of avoiding compliance with applicable conservation and management rules for fishing activities on the high seas.*[6]

Following the Earth Summit in Rio, the FAO served as the forum for the development of a new treaty to address the re-flagging problem, which ultimately became the 1993 FAO Compliance Agreement. An original draft of this treaty would have required Parties to prohibit their nationals who owned fishing vessels from re-flagging those vessels to other nations for the purpose of avoiding compliance with conservation and management measures adopted by RFMO. The original draft would also have required Parties to take practical steps to enforce this prohibition.

The European Union opposed this fundamental approach at that time. The EC delegation argued that fishing vessel owners frequently re-flag their vessels for perfectly legitimate reasons, and that re-flagging also often occurs legitimately when fishing vessels are sold to owners in other countries. At the time a fishing vessel is about to be re-flagged, a government cannot know whether the vessel owner is re-flagging the vessel with the intent to avoid compliance with conservation and management measures. Certainly, a fishing vessel owner on the verge of re-flagging a vessel is unlikely to announce such intent. Many governments are not even aware of when vessels subject to their jurisdiction are in the process of being re-flagged, making the regulation of re-flagging quite difficult for them.

These concerns forced the negotiation of the FAO Compliance Agreement on to a different track. The Agreement, as adopted by FAO, imposes no obligations on Parties to take any action to deter their nationals from re-flagging fishing vessels to notorious flag-of-convenience states. Instead, the Agreement focuses solely on the responsibility of flag states to control the fishing activities of their vessels.

Expert Consultation on Illegal, Unreported and Unregulated Fishing organised by the Government of Australia in co-operation with FAO, Sydney, Australia, 15-19 May 2000.

[6] Report of the United Nations Conference on Environment and Development, UN Doc A/CONF. 151/26, 1992, Agenda 21, ch. 17, para. 17.52.

The elaboration of specific flag-state responsibilities in the FAO Compliance Agreement (and in a number of other international instruments, particularly the 1995 UN Fish Stocks Agreement) has contributed significantly in the fight against IUU fishing. The international community now has a well-recognised set of standards by which to measure the actions of flag states in exercising control over their fishing vessels.

Unfortunately, the elaboration of these standards is not enough. The FAO Compliance Agreement is not yet in force. The UN Fish Stocks Agreement, though it entered into force in 2001, has only 32 parties,[7] none of which could be considered notorious flag-of-convenience states. Meanwhile, there are still quite a few such states who offer their flag to fishing vessels without any real ability, or even intention, to control the fishing activities of those vessels.

As evidenced by the IPOA on IUU Fishing, the international community has come to realise that reliance on flag-state responsibility alone will not solve the problem of IUU fishing being committed by re-flagged vessels. The "flagging out" states (that is, the states whose nationals are seeking to re-flag their vessels) should take steps to control such re-flagging. We cannot depend exclusively on the actions of the "flagging in" states (that is, the new flag state).

Of course, the concerns relating to the ability of the "flagging out" states to regulate re-flagging remain, but there are ways to address them. Governments face similar circumstances in trying to regulate or prohibit any activity of their nationals, where one necessary element is the intent of the person undertaking the activity. In such situations, governments can adopt laws or regulations prohibiting persons from undertaking the activity in question, then penalise those who subsequently undertake the activity if evidence exists that such persons had the requisite intent. Accordingly, if a government has evidence that a re-flagged fishing vessel owned or operated by one of its nationals is committing IUU fishing, the government would have at least a *prima facie* case that the vessel owner or operator re-flagged the vessel for that purpose.

On the strength of such evidence, the government could prosecute the owner and operator, assuming the government could obtain jurisdiction over such individuals. The government might also be able to take certain actions against the vessel directly (*e.g.*, by prohibiting the vessel from ever being re-registered in the original flag state or by prohibiting it from landing or transhipping fish in its ports). In particularly egregious cases, it might even be possible for a government to take action against other vessels owned by the same owners that have not yet been re-flagged (*e.g.*, by revoking fishing permits applicable to them).

RFMOs can also play a role in this effort, particularly by identifying flag states whose vessels are undermining the effectiveness of their conservation and management measures.[8] States can then take measures to deter their nationals from re-flagging fishing vessels, or from initially registering new vessels, in the identified states. Such measures could include controls on deletion of vessels from national registers, controls on the export of fishing vessels,[9] publicity campaigns to make vessel

[7]   The European Union and its member states, despite many statements of intention to become party to the UN Fish Stocks Agreement, have still not done so.

[8]   The International Commission for the Conservation of Atlantic Tunas, for example, has been identifying flag states in this way for several years. *Cf.*, article IV(3) of the Convention for the Conservation of Anadromous Stocks in the North Pacific Ocean ("Each Party shall take appropriate measures aimed at preventing vessels registered under its laws and regulations from transferring their registration for the purpose of avoiding compliance with the provisions of this Convention").

[9]   Japan, for example, has since 1999 denied all requests to export large-scale tuna longline vessels. In addition, Japan has worked through industry channels to develop understandings that certain former

owners aware of those states that have been so identified, and a prohibition on allowing vessels that are or have been registered in such states ever to be re-registered in the initial flag state.

Accordingly, it is to be hoped that the European Community and all other members of the international community vigorously pursue efforts to control the re-flagging of fishing vessels by nationals for the purpose of engaging in IUU fishing.

New initiatives

However, states must do more to control the activities of their nationals than merely regulate the re-flagging of fishing vessels. Owners and operators of fishing vessels sometimes register their vessels in responsible foreign states, but use those vessels to commit IUU fishing anyway. The flag state, of course, has responsibility to take action against such IUU fishing, as do any other coastal states, port states or market states if the IUU fishing involves them.

But the state of nationality of the owner or operator of the vessel can also act. For example, the state of nationality can make it a violation of its law for its nationals to engage in fishing activities that violate the fishery conservation and management laws of any other state or that undermine the effectiveness of conservation and management measures adopted by a RFMO. Such a law could be drafted as follows:

A person subject to the jurisdiction of [state] who:

a) on his or her own account, or as partner, agent or employee of another person, lands, imports, exports, transports, sells, receives, acquires or purchases; or

b) causes or permits a person acting on his behalf, or uses a fishing vessel, to land, import, export, transport, sell, receive, acquire or purchase, any fish taken, possessed, transported or sold contrary to the law of another state or in a manner that undermines the effectiveness of conservation and management measures adopted by a Regional Fisheries Management Organisation shall be guilty of an offence and shall be liable to pay a fine not exceeding [insert monetary value].

Sanctions against nationals who have engaged in such IUU fishing could include, for example, monetary fines, confiscation of fishing vessels and fishing gear and denial of future fishing licenses.[10]

As detailed in paragraphs 73 and 74 of the IPOA, each state should ensure that its nationals (as well as other individuals under their jurisdiction) are aware of the detrimental effects of IUU fishing and should find ways to discourage such individuals from doing business with those engaged in IUU operations.

To complement the actions of states in controlling their nationals, we must also see greater efforts to press flag states to fulfil their responsibilities. As one step in this process, the United States has provided funding to FAO to host an event designed to remind governments that maintain open vessel registers of the measures that need to be taken to help control IUU fishing and to urge them to take those measures.

Japanese vessels owned in Chinese Taipei should be scrapped, and that others constructed in Chinese Taipei should either be registered and regulated there or scrapped.

[10] Spanish legislation, for example, provides for the suspension of a captain's licence for up to five years for committing certain offences aboard flag-of-convenience vessels.

FAO also hosted a meeting of experts earlier this month to consider further action that port states might take to combat IUU fishing. A number of ideas surfaced at this meeting that are worth pursuing, particularly the possibility of developing regional port state memoranda of understanding (MOUs) in the field of fisheries, drawing on the experience we have gained through the regional port state MOUs that are in force in the fields of vessel safety and pollution.

RFMOs must also continue to adopt strong measures to control IUU fishing. The United States was pleased that ICCAT, at its most recent meeting in Bilbao, adopted decisions to enhance its use of a vessel "blacklist" and also to develop a complementary vessel "white list." Since the ICCAT blacklist will now be used to take action against individual vessels (and not only flag states), we believe that ICCAT acted properly in making the process for listing and de-listing vessels more rigorous, so as to provide greater due process and certainty. CCAMLR also took steps at its most recent meeting to control IUU fishing further, including through the creation of a pilot programme for electronic control of its toothfish Catch Documentation Scheme, a commitment not to allow vessels with bad records to re-register in the territories of CCAMLR members and movement toward a centralised Vessel Monitoring System.

Finally, we also must recognise than any effective action to combat IUU fishing cannot take place in isolation from other related initiatives underway in the field of international fisheries. In particular, efforts to reduce fishing capacity in oversubscribed fisheries and efforts to eliminate subsidies that contribute to overcapacity and overfishing must be key parts of our overall strategy. Governments must use available public funds to reduce overcapacity, not to exacerbate it. Governments have no justification, for example, in providing assistance toward the construction of new fishing vessels that are likely to seek to enter fisheries that are already fully subscribed.

Conclusion

Don Quixote de la Mancha represented the bold idealism of the human spirit untarnished by realism. To succeed in the struggle against IUU fishing, we must tap the well of this bold idealism, but channel our efforts in realistic ways. In a very real sense, the world has shrunk in the years since Cervantes wrote his masterpiece. People can move from place to place with an ease that Cervantes probably never even imagined. People who own or operate fishing vessels can also move their vessels from ocean to ocean – and from registry to registry – with remarkable ease today. In such a world, governments must use all the tools at their disposal to ensure that all people subject to their jurisdiction use fishing vessels responsibly.

CHAPTER 4

DEALING WITH THE "BAD ACTORS" OF OCEAN FISHERIES[1]

David A. Balton, Deputy Assistant Secretary for Oceans and Fisheries,
U.S. Department of State

Introduction

The great British poet, William Wordsworth, once wrote in praise of "a few strong instincts, and a few plain rules." The international community has begun to develop a few strong instincts in the face of declining ocean fisheries. Politicians, fisheries managers, environmental organisations – and responsible industry leaders – now instinctively call for a stronger conservation ethic to govern marine fishing activities. Their instincts also tell them to act upon sound scientific advice, rather than merely to pay lip service to science. They also know, instinctively, that to achieve sustainable fisheries, we must support the "good actors" of ocean fisheries: those flag states and vessel owners who play by agreed rules.

To support the good actors of ocean fisheries, the international community has also begun to develop a few plain rules to deal effectively with the "bad actors." Today, I hope to describe briefly who those bad actors are, how their actions jeopardize sustainable fisheries, and how the international community has, in fits and starts, been creating a few plain rules for dealing with them.

The bad actors

Just who are these bad actors? They take several forms and their actions are also diverse, making a simple definition elusive. But as a U.S. Supreme Court Justice once said about pornography: although it's difficult to define, I know it when I see it. Similarly, those of us engaged in the effort to achieve sustainable fisheries through international co-operation know the bad actors when we see them, even if their activities are not easy to describe concisely.

Just a few weeks ago, the United Nations Commission on Sustainable Development adopted some language to describe some of the bad actors of ocean fishing:

---

[1]   This paper was submitted to the IUU Workshop as a background paper.

57

... States which do not fulfil their responsibilities under international law as flag states with respect to their fishing vessels, and in particular those which do not exercise effectively their jurisdiction and control over their vessels which may operate in a manner that contravenes or undermines relevant rules of international law and international conservation and management measures.

As we say in the United States, this is quite a mouthful. To help further the discussion, I will try to give some concrete examples of the bad actors in action.

The classic bad actor is a fishing vessel owner who re-flags his vessel for the purpose of avoiding internationally agreed fishery regulations. When fishing vessels are re-flagged for this purpose, we say that they have obtained "flags of convenience," because the states who allow such vessels to fly their flags offer a convenient way for the vessels to avoid being bound by the agreed rules. These "flag of convenience states" are often unwilling or unable to control the fishing activities of the re-flagged vessels; indeed, such lack of control is precisely what makes these states so attractive and convenient to irresponsible vessel owners. The vessels typically have no real connection to such a flag state. The master, crew and real financial control all derive from elsewhere.[2]

In such situations, the governments of flag of convenience states are bad actors, too. Without them, this type of re-flagging could not occur.

Not all vessels operating under flags of convenience are re-flagged vessels. Some vessels are registered in flag of convenience states from the time they are built. When such vessels, and their re-flagged cousins, fish for stocks that are under the regulation of a regional fishery management organisation, they produce the phenomenon of "non-member" fishing.

Why are owners of these non-member vessels such bad actors? As you may know, a family of regional fisheries organisations and arrangements now exists around the world. Some, such as the Northwest Atlantic Fisheries Organisation and the International Commission for the Conservation of Atlantic Tunas, are formal bodies; others are less formal arrangements. But formal or informal, these organisations are the best means – really the only means – available to the international community to regulate fishing for shared marine stocks.

Unfortunately, given the present depleted status of such stocks, fishing opportunities are – or should be – limited. It thus follows that the regional fisheries organisations have had to become more and more parsimonious in the quotas they adopt and more and more restrictive in the other fishery rules they set.

These smaller quotas and tighter restrictions, in turn, require significant sacrifice on the part of the member states of regional fishery organisations. Every year the member states work hard at the meetings of these organisations to adopt agreed fishing rules. The negotiations are often arduous, and only succeed – if they succeed – through the application of considerable political will. At the end of these meetings, the member states then have the unenviable task of enforcing upon their unhappy fishing industries the smaller allocations and more onerous regulations just adopted.

Responsible vessel owners accept the smaller allocations and tighter regulations in the hope that today's conservation efforts will yield greater fishing opportunities tomorrow. Other owners, however,

---

[2]   Of course, not all vessel owners re-flag their vessels in order to avoid fishing restrictions. Many times fishing vessels are re-flagged for completely legitimate reasons, including to gain legal access to regulated fisheries.

re-flag their vessels (or initially flag their new vessels) in states that are not members of the organisation in question precisely to avoid these restrictions. These vessels then proceed to fish for the very same stocks in the very same region, unbound by the agreed rules. These non-member vessels are essentially free riders – enjoying the benefits of conservation efforts and scientific research undertaken by member states without bearing any of the costs. Not only is this grossly unfair – it also greatly compromises the integrity of the agreed rules and undermines the willingness of the remaining "good actors" to comply with them.

And when the good actors – those fishing vessel owners who do not change flags – start to violate the agreed rules, they become bad actors too.

I would include as a final category of bad actors those vessels that fish illegally within waters under the fishery jurisdiction of coastal states. The advent of Exclusive Economic Zones (EEZs) several decades ago placed vast areas of the planet's surface under the fisheries jurisdiction of the world's coastal states. For many of these states, however, their regulatory control over their EEZs remains nominal – they have little ability to police fishing activities occurring more than a short distance from shore. In the face of dwindling stocks, the temptation to fish illegally in these areas often becomes too great to resist. The phenomenon of such illegal fishing is certainly growing; the only question is: by how much?

From these examples, perhaps we can distil a working definition of the bad actors of ocean fisheries: fishing vessel owners who do not observe agreed fishing rules (or EEZ fishing rules) and the flag states that fail to take action against them.

International law framework

Although the bad actors have undoubtedly been around for some time, their activities have only begun to draw serious political attention in the last decade or so, when a number of the world's key fish stocks began to collapse from overfishing. Until this decade, however, few international law tools existed to deal with the bad actors. The 1982 United Nations Convention on the Law of the Sea calls upon states to prevent overfishing within their EEZs, to ensure that their vessels only fished in other state's EEZs with permission, and to co-operate with other states in the conservation of high seas fisheries. The general obligations constitute a vital regulatory framework, but have not proved specific or comprehensive enough to achieve sustainable fisheries overall.

The 1982 Law of the Sea Convention also reaffirmed the well-established principle of *exclusive flag state jurisdiction* over vessels on the high seas. Under the Convention, generally speaking, only the flag state may exercise fisheries jurisdiction over vessels operating on the high seas. In recent years, this principle has become something of a safe haven for the bad actors. The flag states that are unable or unwilling to regulate their fishing vessels on the high seas often hide behind the principle of exclusive flag state jurisdiction to deny any *other* state the ability to take action against such vessels when they undermine agreed fishery rules. What results is an unfair dual system – smaller quotas and stricter fishing regulations for the good actors and a regulatory vacuum for the bad actors.

Virtually all members of the international community continue to endorse the principle of exclusive flag state jurisdiction as reaffirmed in the Law of the Sea Convention. However, as I hope to demonstrate, the international community has now articulated a related principle: the exclusive jurisdiction over high seas fishing vessels enjoyed by flag states necessarily implies a corresponding duty. Flag states must ensure that their fishing vessels on the high seas do not undermine agreed fishery rules. Failure of flag states to fulfil this duty will have consequences, including, in some cases, some loss of exclusive authority over those vessels.

1993 FAO Compliance Agreement

The first treaty of global application that sought to address this problem of bad actors is the 1993 FAO Compliance Agreement, whose formal name is the Agreement to Promote Compliance with International Conservation and Management Measures by Fishing Vessels on the High Seas. The Compliance Agreement is an integral part of the Code of Conduct for Responsible Fisheries and is the only part of the Code that is legally binding.

The FAO Compliance Agreement in fact began specifically as an effort to combat the practice of the re-flagging of fishing vessels to avoid agreed fishing rules. As the negotiations on the Compliance Agreement proceeded, the scope of its provisions became broader. Instead of dealing solely with the re-flagging phenomenon, the Compliance Agreement elaborates a set of specific duties for *all* flag states to ensure that their vessels do not undermine conservation rules.

Under the Compliance Agreement, a flag state may only permit its fishing vessel to operate on the high seas pursuant to specific authorisation. A flag state may not grant such authorisation unless it is able to control the fishing operations of the vessel. If a vessel undermines fishery rules established by a regional fishery organisation, the flag state must take action against the vessel including, in many cases, rescinding the vessel's authorisation to fish on the high seas – even if the flag state is not a member of the regional fishery organisation.

In elaborating these duties, the Compliance Agreement does not explicitly alter the principle of exclusive flag state jurisdiction. Indeed, one might say that the Compliance Agreement is *premised* on the principle of exclusive flag state jurisdiction. Implicitly, however, the Compliance Agreement is sending another message to the bad actors: if flag states do not bring their high seas fishing vessels under control, the international community will be forced to find other ways to deal with the problem.

1995 UN Fish Stocks Agreement

The 1995 UN Fish Stocks Agreement basically incorporates these provisions of the Compliance Agreement in Article 18, concerning "Duties of the Flag State," and in Article 19, concerning "Compliance and Enforcement by the Flag State." One explanation for this overlap between the two treaties is that the negotiations on both of them took place at roughly the same time (although the Fish Stocks Agreement took considerably longer to conclude) and were conducted by many of the same individuals.

The Fish Stocks Agreement nevertheless takes matters a step farther than the Compliance Agreement in dealing with bad actors.

Rather than review the entirety of the Fish Stocks Agreement, with which the participants in this workshop are already familiar, I would like to highlight a few key provisions that are already proving helpful in dealing with the bad actors of ocean fisheries.

Articles 8(3) and 8(4) of the Fish Stocks Agreement seek to promote the integrity of regional fisheries organisations and the measures they adopt. To this end, they set forth "a few plain rules" that are particularly pertinent to the phenomenon of "non-member fishing." The first rule is that all states whose vessels fish for marine stocks regulated by regional fishery organisations should either join those organisations or, at a minimum, apply the fishing restrictions adopted by those organisations to their flag vessels. The second rule follows from the first: regional fishery organisations should be open to all states with a real interest in the fisheries concerned. The final rule also builds on the others: only member states of regional fishery organisations (or other states that apply the fishing restrictions adopted by those organisations) shall have access to the regulated fishery resources.

When President Clinton transmitted the Fish Stocks Agreement to the U.S. Senate, he stated that these rules, "if properly implemented, would greatly reduce the problems of 'non-member' fishing that have undermined the effectiveness of regional fishery organisations." I believe this assessment remains true today. If all flag states took these few plain rules to heart, non-member fishing would, almost by definition, largely disappear.

To bolster these few plain rules, the Fish Stocks Agreement also includes Article 17, concerning "Non-Members and Non-Participants." This Article provides quite simply that states which do not join regional fishery organisations, and which do not apply the fishing restrictions adopted by those organisations to their flag vessels, are not discharged from their obligation to co-operate with other states. In particular, they shall not authorise their vessels to fish for the regulated stocks.

Article 17 further requires the member states of the relevant organisation to take affirmative measures to deter non-member fishing, providing such measures are consistent with the Fish Stocks Agreement and international law in general. As I will discuss below, this notion of joint action to deter non-member fishing is already taking root in a number of regional fishery organisations.

But, as professors used to ask in seminars on arms control, what if deterrence fails? For such situations, the Fish Stocks Agreement contains Articles 21 and 22. These articles are a set of carefully negotiated provisions that permit, under certain circumstances, states other than flag states to board and inspect fishing vessels on the high seas and, where they find evidence that the vessels have engaged in serious violations of agreed fishing restrictions, to take limited enforcement action to prevent further violations.

A number of governments that have not yet ratified the Fish Stocks Agreement have expressed concerns that these provisions stray too far from the principle of exclusive flag state jurisdiction. The more I have considered these provisions, however, the more I have come to see how they mostly codify existing international practice.

First, a number of regional fishery organisations and arrangements, including NAFO, the North Pacific Anadromous Fish Commission and the Central Bering Sea Pollock Convention, had set up joint boarding and inspection regimes even before the Fish Stocks Agreement was negotiated. Second, the Fish Stocks Agreement retains the very crux of exclusive flag state jurisdiction: no other state may take action against a fishing vessel on the high seas without the consent of the flag state. However, like the NAFO, NPAFC and Central Bering Sea Conventions that preceded it, the Fish Stocks Agreement gives flag states a mechanism to provide such consent in advance – by becoming party to the Fish Stocks Agreement.

Finally, and perhaps most importantly, the Fish Stocks Agreement expressly recognises the authority of the flag state to require any other state that may be taking enforcement action against one of its vessels to turn over that vessel to the flag state – provided that the flag state is ready, willing and able to take effective enforcement action against the vessel itself.

In short, the Fish Stocks Agreement secures the rights and prerogatives of responsible flag states, while giving other responsible states certain limited authority to deal with bad actors who have not been deterred from their bad actions.

At least two other provisions of the Fish Stocks Agreement that are designed to address illegal fishing in EEZs merit attention. In cases where there is evidence of such fishing, Article 20(6) requires the flag state to co-operate with the coastal state in taking enforcement action. Moreover, Article 25, which provides for co-operation with developing states, calls specifically upon Parties to render assistance to developing coastal states to help them achieve greater enforcement capacity within their EEZs.

Finally, Article 23 of the Fish Stocks Agreement calls upon port states to exercise their prerogatives in ways that can address the problems caused by the bad actors. Along these lines, some RFMOs have already adopted schemes, discussed below, to prevent the landing of fish caught by non-member vessels in ways that undermine agreed fishing rules.

Examples of regional fishery organisation actions

Today, neither the FAO Compliance Agreement nor the Fish Stocks Agreement is yet in force. But the principles and approaches contained in those treaties are already having effect, and a number of the regional fishery bodies are beginning to take decisive action against the bad actors involved in their fisheries.

To date, two approaches have been adopted to deal with the problem of non-member fishing. One approach uses trade as a lever. This approach was developed by the International Commission for the Conservation of Atlantic Tunas (ICCAT) in response to growing evidence that fishing activities of vessels from several non-members of ICCAT were adversely affecting ICCAT's efforts to conserve bluefin tuna and swordfish.

In 1994, ICCAT adopted the Bluefin Action Plan Resolution. This Plan provides a process for identifying non-members whose vessels are engaged in fishing activities that diminish the effectiveness of ICCAT measures for bluefin tuna. Such non-members are given a year to rectify their fishing practices. If they do not do so, ICCAT can authorise its members to prohibit the importation of bluefin tuna products from the non-members in question.

The very next year, ICCAT identified Belize, Honduras and Panama as non-members whose vessels were fishing in a manner that diminished the effectiveness of ICCAT's bluefin tuna measures. When the governments of these nations failed to rectify the fishing practices of their vessels, ICCAT instructed its members to prohibit the importation of bluefin tuna products from them. These trade embargoes remain in effect.[3]

ICCAT has also adopted a similar approach for dealing with non-member fishing that diminishes the effectiveness of ICCAT's swordfish measures. ICCAT has recently identified the same three states under this procedure, but has not yet imposed trade restrictions.

ICCAT's use of multilateral trade restrictions represents the first time that such measures have been authorised by an international fishery management organisation to ensure co-operation with agreed conservation and management measures. One would expect that other regional fishery organisations will consider similar steps if non-member fishing is not otherwise brought under control.

---

[3]   One of the nations under ICCAT's bluefin tuna trade embargo recently took the step of joining ICCAT, presumably for the purpose of having the trade embargo lifted. Panamanian vessels will henceforth be bound to observe all ICCAT measures.

The other approach, first developed by the NAFO, involves restrictions on landings of fish caught by non-member vessels. Many fish stocks managed by NAFO are in serious trouble. NAFO members have imposed moratoria on fishing for several stocks, causing considerable hardship on those who formerly depended on these harvests for their livelihoods. NAFO enjoys one advantage over ICCAT, however. Because the NAFO Regulatory Area is a relatively compact high seas area, a NAFO joint inspection regime allows for close monitoring of all fishing activity in the Regulatory Area, by members and non-members alike.

In 1997, NAFO adopted a "Scheme to Promote Compliance with the Conservation and Enforcement Measures Established by NAFO." The Scheme sets up a presumption that any non-member vessel that has been observed fishing in the Regulatory Area is undermining the NAFO fishing restrictions. This presumption reflects the fact that all of the valuable groundfish stocks in the Regulatory Area are under moratorium or fully allocated. Even fishing activity for less valuable fish stocks cannot be undertaken without serious, adverse by-catch of depleted fish stocks. If a non-member vessel sighted fishing in the Regulatory Area later enters a port of a NAFO member, the NAFO member may not permit the vessel to land or tranship any fish until the vessel has been inspected. If the inspection shows that the vessel has on board any species regulated by NAFO, landings and transhipments are prohibited unless the vessel can demonstrate that the species were either harvested outside the Regulatory Area or otherwise in a manner that did not undermine NAFO rules.

The Commission for the Conservation of Antarctic Marine Living Resources (CCAMLR) has also adopted a modified version of the NAFO Scheme and is currently considering other related measures, including a catch certification scheme. I am also aware that, for matters closer to Europe, the North-East Atlantic Fisheries Commission is also working to adopt its own programme, which will be based on the NAFO experience.

FAO initiatives

In the spring of 2004, the international community has devoted substantial additional attention to the problem of bad actors. The government of Australia, in particular, is to be commended for its leading role in this endeavour and for coming up with a new acronym – IUU fishing – which stands for "illegal, unauthorised and unregulated" fishing. This phrase, although perhaps not as mellifluous as one might hope, may come as close as the English language permits in capturing the problems posed by the bad actors in a succinct way.

In February, the FAO Committee on Fisheries adopted a far-sighted International Plan of Action to address the problem of overcapacity in many of the world's fisheries. One aspect of that Plan of Action calls upon states to work together in addressing IUU fishing. Two weeks after the COFI meeting, the FAO convened a follow-up ministerial-level meeting on global fisheries issues. At this meeting, the fisheries ministers of the world issued a declaration in which they agreed that the FAO would give priority to develop a full Plan of Action dealing exclusively with IUU fishing, a step that the Commission on Sustainable Development endorsed in March 2004.

Where will these actions take us? It is too soon to tell. One promising development is that policy makers are beginning to think more creatively in approaching the problem of bad actors. For example, within the International Maritime Organisation (IMO), efforts have been underway to control the bad actors of *ocean shipping* – those flag states and vessel owners who do not abide by agreed rules in that area. In light of this, in March 2004 the CSD encouraged the IMO to work with the FAO and the UN itself in dealing with the parallel problems together.

Conclusion

The recent efforts of the international community to deal with the bad actors reflect "a few strong instincts" toward conservation and a heightened need for fair play in ocean fisheries. The international community, on both global and regional bases, is developing "a few plain rules" for the bad actors as well. In time, we may see the plainest rule of all: unless bad actors become good actors, their right to fish will be in jeopardy.

PART II

COMPILING THE EVIDENCE

This session sought to elicit empirical information on IUU fishing, with the key objective of quantifying IUU fishing activities. Discussions focused on the various ways of gathering information and data and assessed their relative efficacy. The session also sought to establish the impact of IUU fishing on resources.

CHAPTER 5

USING TRADE AND MARKET INFORMATION TO ASSESS IUU FISHING ACTIVITIES

Anna Willock, Senior Fisheries Advisor, TRAFFIC International

Introduction

Fisheries commodities generally represent around 25% of the total value of wildlife products in world trade and, after timber, are the most valuable. In the year 2000, fisheries products were estimated to have an export value of USD 55.2 billion (Anon., 2002a). Due to the nature of the activity, reliable global estimates of the value of fisheries products in trade derived from IUU fishing activity are difficult to obtain. However, in relation to general wildlife trade, globally, wildlife smuggling is estimated to be worth USD 6 billion to 10 billion a year, ranking third behind narcotics and arms smuggling (Anon., 2003a).

Analysis of the trade in wildlife products, and in some cases the control of that trade, has long been recognised as a valuable tool contributing to the sustainable use of such resources. The most widely known and well-established regime for the regulation of international trade in wildlife is the Convention on the International Trade in Endangered Species of Wild Fauna and Flora (CITES), which entered into force in July 1975. With 164 current Parties and over 30 000 species listed in the three Appendices to the Convention, CITES represents the most broadly co-ordinated attempt to use international trade as a complement to other management efforts to ensure the sustainability of wildlife. While there are several commercially exploited aquatic species of significance in international trade currently listed in the CITES Appendices, no marine species taken in a large-scale, industrial commercial fishery have yet been listed.

There is also a growing number of documentation and labelling laws and schemes seeking to control and/or identify the source of marine fisheries products in trade, including those concerned with food safety and quarantine. In addition, there has been a growth in eco-labelling schemes underpinned by private organisations, such as the Marine Stewardship Council (MSC), that are designed to enable consumers to identify products from well-managed and sustainable fisheries in the market place. In the case of the MSC, the extent of IUU fishing activity in a fishery seeking certification is recognised as a factor impacting on the health of stocks and taken account of in the decision whether or not to grant certification.

In this respect, moves by regional fisheries management organisations (RFMOs) to implement catch certification and documentation schemes as a complement to other management controls to combat IUU fishing are particularly important in relation to the growth in trade and market-related interventions in fisheries. For the most part, these measures are a response to the inability of traditional management measures and international law to effectively deal with sustainability issues and, in particular, the threat to sustainability of stocks posed by IUU fishing. Trade-related measures introduced by RFMOs are broadly aimed at either gathering information on the source, extent and parties to trade as the basis for other actions to be taken (*e.g.*, the International Commission for the Conservation of Atlantic Tunas' catch certification scheme) or as a direct attempt to prevent product derived from IUU fishing activities from entering trade (*e.g.*, Commission for the Conservation of Antarctic Living Marine Resources' catch documentation scheme).

Given the extent to which fisheries products are present in international trade, knowledge of the trade and the market for those products is almost a prerequisite to good management, with the ability to shed light on issues such as the source of products, extent and nature of demand, and substitute products. In this respect, regardless of whether used as a direct regulatory measure or as a means of gathering information on trade in a fisheries product, trade and market analyses have the potential to make a significant contribution to reducing the threat posed by IUU fishing.

TRAFFIC is the world's largest international wildlife trade monitoring organisation with eight regional offices and 22 national offices. TRAFFIC has carried out a number of analyses of the international trade in and markets for various fisheries products, which have provided valuable information that can be used by governments, nationally, regionally and/or internationally, in developing measures to combat IUU fishing.

This paper:

a) briefly outlines the different methods used to undertake analyses of trade and market information;

b) identifies the range of information on IUU fishing that may arise from trade and market analyses;

c) discusses the key ingredients for trade and market analyses to be able to contribute to assessing IUU fishing activity;

d) provides a number of issues for further consideration including recommendations designed to increase the utility of these forms of analysis in assessing IUU fishing activity.

Methods used in analyses of trade and market information

There are a number of different methods used in the analyses undertaken by TRAFFIC, the main ones being analysis of trade data, market surveys and field research. Such methods must be combined with extensive literature searches and research into any regulatory measures and policies in order to ensure that data derived from trade and market research is placed in its correct context. In applying these methods, some activities may be undertaken that are beyond the normal scope of government, for example, covert market surveys in other countries. Both informal and formal sources of information may be obtained; however, if interventions are to be subsequently made by governments on the basis of these analyses, they must have a strong and objective factual underpinning.

It is extremely important to have the best available information so that certain interpretive decisions can be taken when checking trade data. TRAFFIC is very careful to give a conservative figure when estimating overall trade as there are always inconsistencies when cross-checking export, import and re-export data. For example, when comparing data from different sources it is important to verify that comparisons are being made between the same types of products. Some countries' codes may reflect fish quantities that have been converted to live weight, whereas other sources of data may be for such products as head and gutted, gutted, and fillets. Such data cannot be compared unless this information is known and unless reliable conversion factors are used to convert processed products to live weight equivalents.

In general, statistics, such as those from FAO, underestimate the amount of trade occurring, the quality of this data being dependent on the quality of data its members provide. There are, however, examples where trade statistics at a country or global level may overestimate trade. For example, this occurred in the past with the trade data available for Hong Kong on shark fin imports (Anon., 1996). As shark fins were being imported in to Hong Kong and then re-exported to mainland China for further processing and then re-imported back into Hong Kong, the overall effect was for fins derived from the same animal to be counted twice in imports into Hong Kong. Legitimate industry is often an extremely important advisor in the interpretation of trade information.

Further, even where a country has customs codes for a species it may still be reported under a variety of names – particularly where there may be tariff or tax incentives to do so – therefore care needs to be taken to either use pricing information or intelligence from legitimate industry to correctly identify the species in question or otherwise omit that data from the analyses. When done properly, these forms of analyses will more often provide a minimum estimate of the level of international trade in a species and, in most cases, will be an underestimate.

What useful information can be derived from trade and market analyses?

In providing assessments of a range of different IUU activities, trade and market information can assist in establishing the potential basis for intervention across this range.

Comparison between estimated catch and level of trade

Collating national import, export and re-export data can provide an estimate of the total volume of a particular species in international trade. This may then be compared with the global reported, or estimated, catch of that species. Where the volume of a species in international trade is higher, one of the explanations is that this product has been derived from illegal or unreported fishing activities. Knowledge of the fishery is then likely to indicate whether this is likely to be the case. In situations where a species may be actively managed throughout only part of its range, gaps between trade volume and reported catch may indicate that part of the product comes from an unregulated fishery. While this arguably does not fall within the definition of IUU fishing under the FAO International Plan of Action, it may identify areas where harvest is a matter of concern and so require active management, or where unregulated harvest may undermine trade-related measures for that part of a stock or species that is managed.

The assessment of the international trade in Patagonian Toothfish *Dissostichus eleginoides* undertaken by TRAFFIC in 2001 (Lack and Sant, 2001) is an example of this type of trade analysis. International trade data for Patagonian Toothfish was analysed to determine whether it was possible to use this data to verify the extent of IUU fishing for toothfish and, if so, how the level of international trade compared with estimates of total catch. This analysis, undertaken prior to the implementation of the Commission for the Conservation of Antarctic Marine Living Resources' catch documentation scheme (CDS), showed that IUU fishing may have accounted for half the toothfish in international trade in the year 2000. Comparison of international trade data also indicated that the level of IUU catch may have been four times that estimated by CCAMLR.

In the case of the Patagonian Toothfish trade analysis, catch estimates were available from CCAMLR for other species, however, particularly those harvested from high seas areas not under the mandate of an RFMO; FAO catch estimates may provide the main point of comparison with trade data. For example, in relation to orange roughy *Hoplostethus atlanticus,* a comparison of available international trade information and FAO estimates of global catch indicated that the FAO substantially underestimated the actual global catch of orange roughy (Lack *et al.*, 2003). The FAO has itself recognised that its database underestimates the actual catch of orange roughy (Anon., 2003b), with the trade analysis then confirming that this was indeed likely to be the case and that the underestimate may be as high as 30% in some years. While not solely indicative of the level of IUU fishing activity for orange roughy, such comparisons of global catch and trade provide valuable insights into the potential level of harvest of species and add weight to calls for such stocks to be brought under management arrangements.

Identify discrepancies between export and import figures for a product

Discrepancies between export figures and import data may indicate that fish products are circumventing official trade routes in the country of origin. One of the reasons for this circumvention may be that the product has been illegally obtained.

For example, in the case of the sea cucumber species *Isostichopus fuscus*, harvested mainly in the waters surrounding Ecuador's Galapagos Islands, a comparison between export data from Ecuador and import data from the major import destinations was undertaken. This analysis revealed that the level of exports was likely to significantly underestimate the actual level of trade, with imports of dried sea cucumbers from Ecuador into Hong Kong and Chinese Taipei over the period 1998 to 2002 exceeding the reported exports by at least 10% and in some years by 25% (Willock *et al.*, in press). Of further interest in the trade analyses of *I. fuscus* is the fact that exports from Ecuador were reported during years when the fishery was closed to all commercial harvest. Illegal harvest of the species from the Galapagos is widely recognised by the Ecuadorian Government as the major threat to sustainability of the fishery and the trade comparison contributes data on the extent of the illegal harvest and the need for greater co-ordination between fisheries management and customs authorities as well as with importing countries.

Identify countries engaged in trade in a certain product

Trade analysis can assist in identifying those countries that are engaged in the international trade of a fisheries product and the level of that engagement. RFMOs or national governments can use this information to identify trade flows in a particular fisheries product (and potential IUU products) and ascertain which countries' co-operation is required to effectively manage a species.

CCAMLR, ICCAT and the Commission for the Conservation of Southern Bluefin Tuna (CCSBT) have all used information gathered through trade and market analyses to pinpoint countries from which co-operation is required. In most cases, countries trading in a fisheries product that are not members of the relevant RFMO will be unaware of any issues relating to IUU fishing activity. Therefore, by identifying countries engaged in trading a species where IUU fishing is a problem, it would then be possible to liaise with those countries and seek their co-operation in limiting market access for IUU-caught fish. Invitations to become a party to the relevant RFMO or co-operate in trade-related measures as a co-operating non-party are two types of action that can be taken on the strength of this information. Both ICCAT and CCSBT have also used information on the source of products in trade to identify countries from which their members should not accept imports.

Identify routes/avenues for disposal of IUU products

Gathering information on the export, import and re-export of a particular species can provide information on the routes IUU products take in order to circumvent national management measures, including those relating to trade. This information may provide evidence of the avenues for disposal of products, identify 'hot-spots' (such as porous borders) through which illegally obtained products pass, and provide information on the role of other states in illicit trade as a step towards securing their co-operation to prevent such trade.

The case of the abalone species *Haliotis midae* illustrates this point. *H. midae* is one of three species of abalone endemic to South Africa and is the only species commercially harvested within the country, with over 90% of the catch exported. The main threat to the species, and the future of the fishery based on it, is illegal harvesting (Hauck and Sweijd, 1999). A recent analysis of import data from the major importer, Hong Kong, revealed that imports of the South African endemic abalone came from four other states, including a land-locked country (Willock *et al.*, in press). Given that there is no export of the species into these countries from South Africa, exports from these four countries are likely to consist of abalone smuggled across borders. The South African government is reportedly considering avenues to secure the co-operation of importing countries to stop this illicit trade (Willock *et al.*, in press).

Evidence of adherence to regulatory measures

Market surveys can be useful in obtaining a snapshot of the trade in fisheries products and allow an assessment of the presence or absence of certain forms of IUU product. More detailed surveys over a period of time can provide a more robust assessment of the extent to which IUU products occur in the market place. For example, surveys of major European markets for Swordfish *Xiphias gladius* and Atlantic Bluefin Tuna *Thunnus thynnus* revealed the presence of substantial quantities of undersized specimens of both species, in contravention of ICCAT management measures (Raymakers and Lynham, 1999).

Assessment of information from market surveys can provide independent verification of an enforcement problem and the extent of that problem. Measures can then be developed to respond to these issues.

Main ingredients of robust trade and market analyses

Two factors are essential in ensuring that analyses of trade and market information are sufficiently robust to be used to assess IUU fishing activity, and indeed to be useful in fisheries management in general. These two ingredients, access to data and ability to interpret the data, are similar to other crucial areas of fisheries management, particularly stock assessment. Issues relating to each factor are discussed below.

Access to data

Access to reliable data for analysis is the main barrier to using trade and market information to assess IUU fishing activity. In most cases, species-specific and product-specific customs codes will not be available for the species of interest, with many grouped into generic categories such as 'crustaceans' or 'shark'. Another common practice is to identify certain species, such as 'Bigeye Tuna' and 'Yellowfin Tuna', and then classify all other tuna species under a category 'Other – not Bigeye or Yellowfin'.

Where customs codes are available for a species, these are often only in place in a limited number of the countries potentially engaged in its trade. Fortunately, those countries with detailed customs codes in place are most likely to be the ones most heavily engaged in trade, both as exporters and importers. For example, New Zealand is the major exporter of Antarctic Toothfish *Dissostichus eleginoides* and is one of only two countries with separate export codes for this species and Patagonian Toothfish. The only other country with separate customs codes for the two toothfish species is the U.S., a major importer of these species (Lack, 2001). Under such circumstances, information on trade between the major trading partners can provide at least a minimum estimate of the global trade in a species.

Limited transparency and public availability of trade information and access to markets can also reduce the potential of these tools in assessing IUU fishing activity. Of particular concern is the fact that some of the world's largest importers, exporters and re-exporters have little transparency in their trade figures. For example, China advised CCAMLR that in the first nine months of 2002 it had processed and re-exported nearly 15% of the total global catch of toothfish (Anon., 2003c), yet no official trade data is publicly available.

Although there is reasonable transparency with regard to products in international trade, it is often difficult to access reliable information on domestic trade and consumption. Where IUU-caught fish is traded and consumed domestically, information on which to assess the level of IUU fishing activity may be difficult to obtain. In such cases, market surveys may provide some indications of domestic trade. Where a product is consumed in high volumes and available from a range of sources, however, surveys may not be feasible. In cases where part of the catch landed in a country is consumed locally and the rest exported, trade data will only be available for the exported component, which may assist in providing estimates of local consumption where data on landings is also available.

Ability to interpret data

Access to reliable data is clearly a crucial element in assessing IUU fishing activity. Equally crucial is the ability to correctly interpret that data.

It is essential to marry good information about the relevant fishery from which the product has been derived with trade or market data, as otherwise there is significant potential to misinterpret that data. Factors such as the dynamics of the industry, levels of catch, transhipment and processing practices, and the management measures in place will all potentially affect the interpretation of trade and market data.

IUU fishing activity is often very dynamic, moving areas of operation, points of landing and transit countries, and levels of at-sea transhipment in response to management interventions. Therefore the trade routes for a product may change considerably with little warning. However, the markets for products are less likely to vary in the short-term, particularly high value species (often the target of IUU fishing), which often have limited or specialist market niches. Unless the product is landed directly into the consumer country, import data is likely to exist that will then enable identification of the exporting state.

In this regard, the most effective contributions from trade and market analyses are often achieved where there are strong links with governments, relevant RFMOs and legitimate fishing industry. As noted, close liaison with the latter is particularly useful in assisting in the interpretation of processed product and trade routes.

Another aspect of interpreting trade data in particular is the presence of perverse incentives that may result in illegal trade in a product that does not result from IUU fishing. Many countries have complex import and export taxes and tariffs that do not apply uniformly across all fisheries products, so that some products may be highly taxed while others are not taxed at all. This provides incentives to mis-report trade in certain fisheries products. For example, in relation to the shark-fin trade between Hong Kong and mainland China, although Hong Kong is a duty-free port, mainland China imposes high tariffs on imported shark fins. This resulted in a close match between import and export data on the trade in fins from mainland China to Hong Kong, but large discrepancies in data for trade from Hong Kong to mainland China, with one explanation being that traders sought to under-report imports to mainland China to avoid tariffs (Clarke, in press).

Issues for further consideration

Adoption of species-specific and product-specific customs codes

The Harmonised Commodity Description and Coding System (HS) seeks to co-ordinate customs codes internationally. In relation to fisheries products, an argument often raised against the introduction of detailed codes is that this would be overly cumbersome for national customs authorities given the range of species and products in trade. However, where the sustainability of a species in international trade is threatened by IUU fishing, the introduction of customs codes enabling more accurate assessment of trade could be treated as a priority for action. In the case of orange roughy, for example, concerns about the sustainability of catches from unmanaged stocks, particularly those taken in unregulated high seas areas, have been held for a number of years. With management regimes for unmanaged high seas areas likely to be some years away, the introduction of trade codes for orange roughy by the major trading countries would serve to complement catch reporting to FAO and assist in providing a more accurate estimate of catch.

Improved co-ordination of product-specific codes between countries engaged in the trade of a species would greatly assist in reducing the scope for errors in converting processed weights to live weight. In the case of toothfish, for example, the major exporting country, Chile, has very detailed product codes, whereas its major trading partner, the U.S., has much less detailed codes. While co-ordination of customs codes through the HS is preferable, there is scope for countries to choose to implement more detailed codes for certain products where these do not exist through the HS. Where relevant, RFMOs could provide a useful point of co-ordination for species under their mandate.

Greater transparency in national trade data and that collected under RFMO schemes

As noted, some official trade and market data is difficult or, in some cases, even impossible to obtain. Where such data concerns major trading nations, this significantly limits the value of trade and market information in efforts to assess IUU fishing activity.

Greater transparency is required with regard to trade data and market information, including that collated by RFMOs under catch certification and documentation schemes. Furthermore, such information needs to be made available in sufficient detail to enable comparisons with data compiled from customs agencies.

Increased awareness of trade dynamics by fisheries management agencies especially where IUU fishing is considered to be a threat

For many fisheries, harvest for international trade is the primary driver. This is particularly true for many developing countries where higher valued fish species, such as the larger pelagic tunas, are exported to earn valuable foreign revenue. Despite the importance of trade as a driver for harvest, including by IUU operators, fisheries management agencies usually have poor understanding of the trade demand for fisheries products, with efforts commonly directed at managing the resource from the point of harvest to the wharf. This is because the agency responsible for fisheries management at the national level is almost separate from the agency that manages national exports, imports and re-exports, with limited communication between the two.

Increased awareness of the trade and market dynamics for products from a fishery can assist national authorities in better targeting management resources and may result in the identification of areas where complementary trade-related measures can add value to existing management efforts.

Increased engagement by RFMOs and governments in global fisheries trade issues especially when using trade-related measures as part of their management strategy

Despite the increasing use of trade-related measures in the conservation and management of fisheries, specifically in combating IUU activity, moves to co-ordinate the application of such measures have occurred only recently, through a series of FAO expert consultations. Increased co-ordination and, where appropriate, a higher degree of standardisation between the different schemes is to be encouraged.

Of particular relevance is the interpretation of World Trade Organisation (WTO) rules in respect to fisheries trade-related measures. This is a sensitive issue and one that remains open to debate, with "…interaction between trade measures adopted by RFMOs and WTO rules containing possibilities for both conflict and compatibility" (Tarasofsky, 2003). More concerted efforts should be directed towards ensuring that trade measures implemented in support of the sustainable development and exploitation of fisheries resources are recognised and supported under the WTO.

Increased engagement by legitimate industry

As noted, engagement by legitimate industry greatly contributes to trade and market analyses as it strengthens ability to interpret data and gather intelligence on product movement as well insight into IUU operations. This engagement strengthens the ability of government and other organisations to monitor trade, interpret data and gather reliable data on trade routes, prices and sources of product, which in turn should benefit legitimate industry if such information can be used to reduce or eliminate the threat posed by IUU fishing.

The potential for increased co-ordination between fisheries agencies and CITES

CITES, as the international instrument with the mandate to monitor and regulate international trade in wildlife products, has well-established processes that may readily complement and strengthen broader fisheries management objectives. CITES may provide a range of conservation benefits to marine fish species that are or may be threatened by demand for international trade, particularly where this threat arises from IUU fishing. In broad terms, such benefits can include:

- providing support to national, bilateral and multilateral fisheries management measures;

- providing a tool to combat IUU fishing, where this targets fish that primarily enter international trade;

- providing a standardised global monitoring system for the application of trade-related measures to marine fish (Anon., 2002b).

A number of countries have already sought to use the provisions available under CITES to assist in combating IUU fishing for a particular species. The most recent example is the listing of the sea cucumber species *I. fuscus* in Appendix III of CITES by Ecuador in order to gain international support for its national efforts to combat illegal harvest for international trade.

The increased consideration of trade-related measures also highlights the need for strengthened co-operation between CITES and the FAO, as well as, potentially, between CITES and individual RFMOs.

Limitations of trade and market analyses

While trade and market analyses can contribute to the assessment of IUU fishing, there are a number of limitations to this contribution.

One obvious limitation is that trade and market analyses, by their very nature, only provide data on the valued and retained component of the catch. Therefore the impact of IUU fishing on non-target species and the broader marine environment cannot be directly assessed through trade and market data. Another limitation is that this data does not indicate where the catch was taken and so sheds little light on, for example, particular stocks that may be subject to more intensive IUU fishing activity.

Trade and market information cannot, of itself, identify products derived from IUU fishing unless analysed in conjunction with other information; for example, the presence of products in trade during periods when the corresponding fishery is closed.

Conclusions

Analysis of the trade in wildlife products has long been recognised as a valuable source of information contributing to the sustainable use of natural resources. Such analysis can provide a direct point of intervention as well as guide interventions at other points of the management system.

In the context of IUU fishing, analysis of trade and market information is a potentially powerful tool to assess these activities and so assist efforts to combat them. In broad terms, contributions from trade and market analyses may include:

- increasing the understanding of the nature, scope and extent of IUU activity;

- providing independent verification of the extent of a known IUU problem;

- assessing the effectiveness of an existing trade- and/or market-related measure;

- revealing the existence of a problem that may not have been previously documented, or showing that demand for a species in international trade is a key driver for IUU activity.

As with other data and statistics, including those relating to estimates of catch and fishing effort for example, trade and market information is unlikely to provide absolute results in terms of quantities of a fisheries product in international trade. However, with care taken in its interpretation, such data may form a valuable source of information to assist in assessing IUU fishing and thereby contribute to reducing and eliminating this global threat to sustainable fisheries.

REFERENCES

Anon. (1996), "The World Trade in Sharks: a compendium of TRAFFIC's regional studies", I. TRAFFIC International, Cambridge, UK.

Anon. (2002a), "The State of World Fisheries and Aquaculture 2002", FAO, Rome, Italy.

Anon (2002b), "A CITES priority: Marine fish and the twelfth meeting of the Conference of the Parties to CITES", Santiago, Chile 2002 IUCN, WWF and TRAFFIC briefing document October 2002, TRAFFIC International, Cambridge, UK.

Anon. (2003a), "New Zealand's First Interpol Conference: Wildlife Smuggling" http://www.maf.govt.nz/mafnet/press/131003interpol.htm. Viewed 3 April.

Anon. (2003b), "Deepwater Fisheries: The Challenge for Sustainable Fisheries at the Final Frontier". Background Document. Committee on Fisheries, FAO Rome, Italy.

Anon. (2003c), Report of the Twenty-first Meeting of the Commission, Hobart, Australia, 21 October – 1 November 2002, CCAMLR, Hobart, Australia.

Clarke, S. (in press), "Shark Product Trade in Hong Kong and Mainland China, and Implementation of the CITES Shark Listings", TRAFFIC East Asia, Hong Kong.

Hauck, M. and Sweijd, N.A. (1999), "A Case Study of Abalone Poaching in South Africa and its Impact on Fisheries Management", *ICES Journal of Marine Science* 56:1024-1032.

Lack, M. (2001), "Antarctic Toothfish: An Analysis of Management, Catch and Trade", TRAFFIC Oceania, Sydney, Australia.

Lack, M. and Sant, G. (2001), "Patagonian Toothfish: Are Conservation and Trade Measures Working?" TRAFFIC Bulletin Vol. 19 No. 1, TRAFFIC International, Cambridge, UK.

Lack, M., Short, K. and Willock, A. (2003), "Managing Risk and Uncertainty in Deep-Sea Fisheries: Lessons from Orange Roughy", TRAFFIC Oceania and WWF Endangered Seas Programme.

Oldfield, S. (Ed.) (2003), "The Trade in Wildlife: Regulation for Conservation", *Earthscan Publications Ltd*, London, UK.

Raymakers, C. and Lynham, J. (1999), "Slipping the Net: Spain's Compliance with ICCAT Recommendations for Swordfish and Bluefin Tuna", TRAFFIC Europe, Brussels, Belgium.

Tarasofsky, R.G. (2003), "Regional Fisheries Organisations and the World Trade Organisation: Compatibility or Conflict?" TRAFFIC International, Cambridge, UK.

Willock A., Burgener, M. and Sancho, A. (in press), "First Choice or Fallback? An Examination of Issues Relating to the Application of Appendix III of CITES to Marine Species", TRAFFIC, Cambridge, UK.

CHAPTER 6

FLAGS OF CONVENIENCE, TRANSHIPMENT, RE-SUPPLY AND AT-SEA INFRASTRUCTURE IN RELATION TO IUU FISHING

Matthew Gianni and Walt Simpson, International Oceans Network for WWF

Executive summary

The problem of Illegal, Unreported and Unregulated Fishing on the high seas has been the subject of much discussion and debate at the regional and global level for the past decade or more. Increasing restrictions have been put into place to attempt to deal with the problem of IUU fishing on the high seas. At the same time, the scope of the restrictions have expanded in recognition of a number of important issues: One, that the infrastructure needed to support IUU fishing on the high seas goes well beyond the IUU fishing fleets themselves; two, unless and until the flag of convenience system is eliminated, port states, market states and countries of beneficial ownership will need to employ a suite of measures to combat IUU fishing; and three, regional fisheries management organisations may, in some cases, need to be reformed to ensure that all parties agree to and effectively implement the conservation and management measures adopted by the regional organisation.

In addressing these issues, the focus of this paper is to:

- review recent trends in the numbers of fishing vessels flying Flags of Convenience;

- focus on a key aspect of IUU fishing: the at-sea transhipment and re-supply fleets:

- recommend specific measures to manage at-sea transhipment and re-supply; and,

- place these recommendations within the context of international actions necessary to implement the UN FAO International Plan of Action on IUU Fishing.

The case study approach was chosen to enable a focused assessment of one of the key components of IUU fishing, the infrastructure facilitating at-sea transhipment and re-supply. This report contains specific information on the character of this infrastructure and recommendations to manage at-sea transhipment and re-supply, particularly in high seas tuna fisheries. If effectively implemented, these would provide a significant deterrent to IUU fishing for high valued tuna species. Other key components of IUU fishing include the ports used by IUU vessels, markets for IUU-caught fish, other businesses supporting IUU fishing operations, and loopholes in the international legal regime which allow for the continuance of the flag of convenience system in fisheries. It is hoped that future, collaborative reports containing similarly specific recommendations on these issues will follow.

Introduction

This paper reviews the general trend in the numbers of fishing vessels flying Flags of Convenience (FOC), then focuses on one of the main aspects of the IUU fishing problem – the at-sea transhipment and re-supply fleets. The information on general trends is based primarily on analysis and comparison of information obtained from Lloyd's Register of Shipping. The character and extent of the at-sea transhipment and re-supply fleets is based on a variety of sources of information and a number of assumptions outlined in the paper. At-sea transhipment and re-supply fleets provide an important service to high seas fishing vessels, both legal and IUU, and are an essential component of the global infrastructure associated with high seas fishing. A better understanding of the specific character of this industry will provide governments, regional fisheries management organisations, legitimate fishers and other interested parties a much clearer picture of what can and should be done to prevent, deter and eliminate IUU fishing through regulating this aspect of high seas fisheries.

It must be emphasised that the effective management of high seas fisheries will never be possible until the problem of IUU fishing is largely eliminated. However, the elimination of IUU fishing alone will not guarantee effective fisheries conservation and management. Much more needs to be done, consistent with the conservation provisions of the 1995 UN Fish Stocks Agreement, various provisions of the UN FAO Code of Conduct for Responsible Fisheries and related agreements to put high seas fisheries on a 'sustainable' track.

Recent trends in flags of convenience fisheries

An analysis of information available from Lloyd's Register of Shipping provides some indication of trends in relation to fishing vessels and the flag of convenience system. The data analysed were for the periods 1999, 2001 and 2003. These years were chosen to coincide with the two years preceding and following the adoption of the UN FAO International Plan of Action to Prevent, Deter and Eliminate IUU Fishing. This paper analyses information available on the Lloyd's database on fishing vessels ("fishing vessels", "trawlers" and "fish factory ships") registered to the fourteen countries with open registries listed on Table 5.1.

Table 6.1. Numbers, Average tonnage and Average Age of Fishing Vessels Registered to 14 Countries with Open Registries 1999-2003

Year	Flag State	Total Vessels	Total Tonnage	Average Tonnage	Average Age
1999	Belize	409	348 892	853	23.4
	Bolivia	1	232	232	52
	Cambodia	6	6 547	1 091.2	22.3
	Cyprus	46	103 573	2 251.6	19.1
	Equatorial Guinea	56	30 984	553.3	18.8
	Georgia	29	10 792	372.1	20.9
	Honduras	416	175 387	421.6	25.9
	Marshall Islands	11	18 701	1 700.1	20.2
	Mauritius	22	7591	345	30
	Netherlands Antilles	18	17 481	971.2	25.4
	Panama	224	169 679	757.5	31.6
	St. Vincent	110	81 956	745.1	23.7
	Sierra Leone	34	9 750	286.8	28.7
	Vanuatu	34	50 609	1 488.5	21.9
2001	Belize	455	349 381	767.9	22.8
	Bolivia	11	7 935	721.4	16
	Cambodia	16	17 336	1 083.5	22.6
	Cyprus	51	108 826	2 133.8	19.6
	Equatorial Guinea	51	28 088	550.7	18.4
	Georgia	39	25 338	649.7	23.3
	Honduras	313	125 975	402.5	26.2
	Marshall Islands	11	13 289	1 208.1	19.4
	Mauritius	23	7 860	341.7	30.1
	Netherlands Antilles	24	28 131	1 172.1	20.6
	Panama	198	149 070	752.9	30
	St. Vincent	101	154 787	1 532.5	23.8
	Sierra Leone	30	8953	298.4	28.7
	Vanuatu	46	116 870	2 540.7	15
2003	Belize	279	258 681	933.9	22
	Bolivia	24	21 399	891.6	20
	Cambodia	43	39 224	912.2	20
	Cyprus	41	92 405	2 253.8	18.2
	Equatorial Guinea	41	24 351	593.9	18.5
	Georgia	53	24 080	454.3	18.8
	Honduras	507	178 802	352.7	23.2
	Marshall Islands	14	16 081	1 148.6	13.6
	Mauritius	26	10 676	410.6	28
	Netherlands Antilles	21	18 100	861.9	20.5
	Panama	205	130 512	636.6	27.9
	St. Vincent	86	117 161	1 362.3	23.7
	Sierra Leone	35	10 185	291	26.2
	Vanuatu	64	93 380	1 459.1	7.5

Source: Lloyd's Register of Shipping

The fourteen countries listed on Table 6.1 were chosen on the basis of several factors. Four of the countries – Panama, Belize, Honduras and St Vincent and the Grenadines – consistently top lists of

FOC countries in terms of numbers of registered fishing vessels. They are also the countries most widely identified by regional fisheries management organisations as being the flag states of particular concern in relation to IUU fishing in a survey conducted in 2002.[1] In addition to these four, Bolivia, Georgia, Equatorial Guinea, Sierra Leone, and Cambodia have been subject to import sanctions at one time or another by the International Commission for the Conservation of Atlantic Tunas (ICCAT) because of IUU fishing for tuna in the Atlantic Ocean by vessels flying their flags. The remaining five were chosen from the list of FOC countries identified by the International Transport Workers' Federation (ITF) and the report of the UN Secretary General's Consultative Group on Flag State Implementation[2] as having the highest number of fishing vessels on their registries in addition to the nine countries mentioned above.

In fact the list of countries on Table 6.1 could be much longer. The International Transport Workers' Federation identifies 28 countries as operating flags of convenience, including fishing and merchant vessels.[3] A UN FAO report published in 2002 lists 32 states as operating flags of convenience or open registries and having registered fishing vessels within recent years.[4]

To be clear, not every vessel flagged to the 14 countries listed above is necessarily engaged in IUU fishing. Twenty-one vessels flagged to Panama, for example, are listed on the ICCAT 'white list' of fishing vessels as authorised by Panama to fish in the Atlantic Ocean. The ICCAT list of 3 176 vessels authorised by contracting or co-operating parties to fish for tunas and tuna like species in the Atlantic, Caribbean, and Mediterranean Sea, also contains another twenty vessels combined flagged to Panama, St Vincent and the Grenadines, Honduras, and Belize as well as Bolivia, Vanuatu, and Sierra Leone. Most of these vessels are authorised to fish by Brazil.[5] The Indian Ocean Tuna Commission (IOTC) does not list any vessels flagged to these 14 countries as being amongst the 2 030 vessels authorised by contracting or co-operating parties to fish tuna and tuna-like species in the Indian

---

[1]   Swann, J., "Fishing Vessels Operating under Open Registers and the Exercise of Flag State Responsibilities: Information and Options", FAO Fisheries Circular No. 980, Rome 2002.

[2]   Consultative Group on Flag State Implementation, Advance, unedited text, Oceans and the law of the sea. United Nations, 5 March 2004.

[3]   Antigua and Barbuda, Bahamas, Barbados, Belize, Bermuda, Bolivia, Burma/Myanmar, Cambodia, Cayman Islands, Comoros, Cyprus, Equatorial Guinea, Germany (second register), Gibraltar, Honduras, Jamaica, Lebanon, Liberia, Malta, Marshall Islands, Mauritius, Netherlands Antilles, Panama, Sao Tome e Principe, Sri Lanka, St Vincent and the Grenadines, Tonga, Vanuatu. The primary criteria the ITF uses in making such a designation is the extent to which there is a genuine link between the flag state and the owners of the vessels on its registry; that is, the extent to which vessels on the registry are foreign-owned. In classifying states as flag of convenience countries, the ITF also takes into consideration a state's ability and/or willingness to enforce international minimum social standards on its vessels, including respect for basic human and trade union rights, freedom of association and the right to collective bargaining with *bona fide* trade unions; its social record as determined by the degree of ratification and enforcement of ILO Conventions and Recommendations; and safety and environmental record as revealed by the ratification and enforcement of IMO Conventions and revealed by port state control inspections, deficiencies and detentions. *Source:* International Transport Workers' Federation *Steering the Right Course: Towards an era of responsible flag states and effective international governance of oceans and seas.* June 2003. http://www.itf.org.uk/english/fisheries/pdfs/steeringrightcourse.pdf.

[4]   Swann, J., "Fishing Vessels Operating under Open Registers and the Exercise of Flag State Responsibilities: Information and Options", FAO Fisheries Circular No. 980, Rome 2002. Appendix I.

[5]   ICCAT record of vessels as per the 2002 Recommendation by ICCAT Concerning the Establishment of an ICCAT Record of Vessels over 24 m Authorised to Operate in the Convention Area. http://www.iccat.org/vessel2/vessels.aspx (accessed 29 March 2004).

Ocean.[6] The Inter-American Tropical Tuna Commission (IATTC) lists fifty-two Panamanian-flagged longline vessels and nineteen purse seiners (the flag and status of two are under dispute) authorised by Panama to fish in the Eastern Pacific Ocean. Honduras, Belize, Bolivia, and Vanuatu combined have an additional 18 vessels on the IATTC list of purse seine vessels.[7] Unfortunately, the authors were unable to review the South Pacific Forum Fisheries Agency's Regional Register of Fishing Vessels to determine whether vessels flagged to these fourteen countries are on the list of vessels in good standing.

Given that many of the vessels flagged to the fourteen countries on Table 6.1 are longline vessels targeting tuna and other highly migratory species, this begs an important question: aside from the relatively small percentage authorised to fish as indicated above, where do these vessels fish? Taking Honduras as an example, it had 507 vessels over 24 metres registered in 2003. The website for the Honduras ships Registry states that, as a condition for obtaining the Honduran flag, "...fishing vessels have to submit an affidavit which states, according to the Resolution issued by the International Commission for the Conservation of Atlantic Tunas, that there is to be no tuna fishing. If this document is not presented, a clause which prohibits such activity will be placed on the back of the Certificate of Registration."[8]

On the ICCAT list, there are four Honduran-flagged vessels authorised by Brazil to fish in the ICCAT area under charter arrangements with Brazilian companies. An additional two tuna purse-seine vessels are authorised to fish in the Eastern Pacific in the IATTC area. No Honduran-flagged vessels are listed as authorised to fish for tuna in the Indian Ocean. Of the remaining 501 large-scale fishing vessels on the Honduran registry, many, if not most, are likely to be tuna fishing vessels. If not the Atlantic, Indian Ocean or Eastern Pacific tuna fisheries, where are the remaining longline vessels authorised to fish?

In addition to the vessels registered to the fourteen countries listed on Table 1, the unknown category contains at least some vessels registered to flags of convenience as well. For example, in a random selection of thirty vessels on the 2003 Lloyd's database listed as flag "unknown", the authors determined the flags of thirteen of these by using data from other sources including Lloyds Marine Information Group, the International Telecommunications Union, INMARSAT and various national agencies responsible for the IMO programme of Port State Control. Of these thirteen, eight were flagged to one of the 14 FOC countries, another 4 were flagged in countries not listed on Table 1, and one vessel was found to have been scrapped.

Trends

With these caveats in mind, a number of interesting trends emerge from the information on the Lloyd's database.

---

[6]   IOTC Record of vessels over 24 metres authorised to operate in the IOTC area (updated 2004-03-29). http://www.iotc.org/English/record/search.php

[7]   List of authorised large longline vessels, IATTC Vessel database. Inter-American Tropical Tuna Commission. http://www.iattc.org/vessellistopen/ALLLVList.aspx (accessed 1 April 2004). Active purse-seine capacity lists, IATTC Vessel database. 1 March 2004. http://www.iattc.org/PDFFiles2/ActivePurseSeineCapacityList03012004.pdf

[8]   http://www.marinamercante.hn/registry2.html fishing.

Top four flag of convenience countries

Belize, Panama, Honduras, and St Vincent and the Grenadines collectively have had over 1100 fishing vessels registered to fly their flags in each of the three years. Over the period 1999-2003, although the number of vessels flagged to Belize declined by approximately 30% while the number flagged to Honduras increased by some 20%, all four countries remained at the top of the list of FOC countries in terms of the numbers of fishing vessels on their registries.

A number of measures have been adopted over the past several years by ICCAT, CCAMLR, IOTC and other regional fisheries management organisations, including, in some cases, trade measures and import bans directed specifically at all four countries. While these measures apparently have resulted in some deregistration of fishing vessels from the registries of one or more countries (*e.g.* Panama) they have not prevented any of these states from continuing to maintain large numbers of fishing vessels on their registries if the Lloyd's information is at all correct. Nor have the measures adopted by the regional fisheries management organisations discouraged large numbers of ship owners interested in flying FOCs from continuing to register their ships to Panama, Belize, Honduras, and St. Vincent and the Grenadines.

Up and coming FOCs/others

Amongst the other countries on the list, Georgia, Cambodia, Vanuatu and Bolivia appear to be 'up and coming' flags of convenience for fishing vessels. The number of fishing vessels flagged to each of these four countries rose markedly between 1999 and 2003, with an increase from 70 to 184 fishing vessels registered to all four countries combined. Of the 64 vessels flagged to Vanuatu, twenty have been built in the last three years.

Cyprus continues to maintain over 40 fishing vessels on its registry despite becoming a member of the European Union in May 2004 and the commitments made by the European Union to crack down on IUU fishing. Finally, while the number of vessels flagged to Honduras declined between 1999 and 2001, the number jumped from 313 vessels to over 500 vessels in 2003. In general terms, this dramatic change in the numbers of fishing vessels on the Honduran registry would appear to be an ongoing indication of the relative ease with which fishing vessels are able to 'hop' from flag to flag.

Effectiveness of UN FAO IPOA

One of the most obvious trends is that the number of fishing vessels on the Lloyd's Register database registered to these fourteen flag of convenience countries combined has declined only slightly, even two years after the adoption of the UN FAO IPOA on IUU fishing. Moreover, the number of vessels listed as flag "unknown" on the database has increased over the same period. As indicated earlier, eight vessels of a random sample of 30 vessels listed as flag "unknown" on the Lloyd's database were found to be registered to FOC countries, suggesting that substantial numbers of vessels on this list may in fact be registered to FOC countries. Further investigation into the vessels registered to flags of convenience in the "unknown" category, and the reasons why these and others vessels are listed as such on the Lloyd's database, would be useful in providing a clearer picture of trends in the flagging of fishing vessels over the past several years. Nonetheless, assuming the information on the Lloyd's database is reasonably indicative of overall trends in the flag of convenience registries, from a global perspective the adoption of the UN FAO IPOA on IUU fishing and the efforts of regional fisheries management organisations and some states to combat IUU fishing have so far had limited effect.

Table 6.2. Summary of trends, average tonnage and average length of fishing vessels

(Fishing Vessels, Trawlers and Fish Factory Ships) registered to the 14 countries listed in Table 6.1, 1999-2003, compared to all fishing vessels ≥ 24 metres in length)

	Country of Registration	Number of Vessels	% of total Vessels	Average Length	Average Gross Tonnage	Total Gross Tonnage	% of Total G. T.	Average Age
1999	All	19 581		42.13	546.4	10 698 619		25.3
	FOC (14 countries)	1 449	7.4%	50.41	780.8	1 131 449	10.6%	25.2
	Unknown	1 108	5.7%	42.17	353.5	391 732	3.7%	33
2001	All Countries	19 206		42.38	543.6	10 441 289		25
	FOC (14 countries)	1 340	7.0%	50.35	845.1	1 132 447	10.8%	24.4
	Unknown	1 248	6.5%	43.46	429.4	535 878	5.1%	30.1
2003	All	19 905		42.40	548.7	10 922 794		24
	FOC (14 countries)	1 279	6.4%	48.51	806	1 030 883	9.4%	22.4
	Unknown	1 485	7.5%	42.66	416.5	618 490	5.7%	28.4

Source: Lloyd's Maritime Service

New vessel construction

Another trend that emerges is the fact that some 14% of large-scale fishing vessels built within the past three years were flying flags of convenience by the end of 2003. This represents a real problem in that a significant portion of new vessels appear to be built with a view to engaging in IUU fishing.

Most of these vessels are built in Chinese Taipei (see Table 6.4). In fact, of the 51 fishing vessels over 24 metres built in Chinese Taipei over the past three years, 50 were flagged in FOC countries by the end of 2003, while only one was flagged in Chinese Taipei. It would be worth further investigation to determine whether any of the companies in Chinese Taipei involved in building new vessels have benefited from funds for the joint Japan/Chinese Taipei programme designed to decommission large-scale tuna longline vessels. Further, given the status of Chinese Taipei as a "Co-operating Party, Entity or Fishing Entity" of ICCAT, the government should be encouraged to ensure that no vessels built in Chinese Taipei shipyards are allowed to register to flag of convenience countries.

Table 6.3. Summary: New Fishing Vessel Construction 2001, 2002, 2003

	Fishing Vessels > 24m built in 2001, 2002, 2003	
	Number of Vessels Built	Total Gross Tonnage
Registered in All Countries	478	263 354
Registered FOC or Unknown	58	36 985
FOC and Unknown Vessels as a Percentage of Total Tonnage	12%	
FOC Flag		
Belize	11	3 644
Bolivia	5	4 159
Cambodia	1	2 495
Cyprus	0	0
Equatorial Guinea	0	0
Georgia	6	3 289
Honduras	0	0
Marshall Islands	1	1 152
Mauritius	0	0
Netherlands Antilles	1	393
Panama	9	2 744
St. Vincent	1	635
Sierra Leone	0	0
Vanuatu	20	17 631
Unknown	3	843

Table 6.4. Names of Fishing Vessels Flagged to FOCs and Unknown, built in 2001, 2002 and 2003

Vessel Name	Registered Owner	Residence of Registered Owner	Nationality of Builder	Length	Gross Tonnage
Belize					
Ruey Tay	Ruey Yih Fishery	Belize	Chinese Taipei	29.9	119
San Jose	Sedamanos Arevalo	Ecuador	Ecuador	29.9	131
Southern Star No. 888	Grace Marine	Chinese Taipei	Chinese Taipei	56.5	520
Wang Jia Men	Owner Unknown	Unknown	Chinese Taipei	29.8	140
Yu Long	Owner Unknown	Unknown	Chinese Taipei	29.9	125
Yu Long No. 10	Owner Unknown	Unknown	Chinese Taipei	29.9	125
Yu Long No. 2	Owner Unknown	Unknown	Chinese Taipei	29.9	125
Yu Long No. 6	Owner Unknown	Unknown	Chinese Taipei	29.9	125
Zee Chun Tsai No. 22	Wu Lai Ming	Chinese Taipei	Chinese Taipei	29.9	119
Zee Chun Tsai No. 23	Owner Unknown	Unknown	Chinese Taipei	29.9	119
Zhou Shan 18	Zhoushan Putuo	China	China	86.2	1 996
Average				**37.4**	**331.3**
Bolivia					
Champion	Sun Hope Investment	Chinese Taipei	Chinese Taipei	54.6	647
Georgia	Georgia Fishery	Chinese Taipei	Chinese Taipei	62.6	878
Hunter	Hunter Fishery	Chinese Taipei	Chinese Taipei	62.6	878
Isabel	Isabel Fishery	Chinese Taipei	Chinese Taipei	62.6	878
Jackson	Jackson Fishery	Chinese Taipei	Chinese Taipei	62.6	878
Average				**60.99**	**831.8**
Cambodia					
Shin Ho Chun No. 102	Lubmain Shipping		Chinese Taipei	**85.2**	**2 495**
Georgia					
Chen Chieh No. 31	Pi Ching Fishery	Chinese Taipei	Chinese Taipei	24.0	101
Chen Chieh No. 32	Pi Ching Fishery	Chinese Taipei	Chinese Taipei	24.0	101
Kiev	Kiev Fishery	Chinese Taipei	Chinese Taipei	54.6	647
Monas	Monas Fishery	Chinese Taipei	Chinese Taipei	63.2	1 105
Nantai	Nantai Fishery	Chinese Taipei	Chinese Taipei	63.2	1 105
Shang Jyi	Shine-Year Maritime	Singapore	Chinese Taipei	24.0	230
Average				**42.1**	**548.2**
Netherlands Antilles					
Patudo	Overseas Tuna	Spain	Spain	**44.5**	**393**

Table 6.4. (cont.) Names of Fishing Vessels Flagged to FOCs and Unknown, built in 2001, 2002 and 2003

Vessel Name	Registered Owner	Residence of Registered Owner	Nationality of Builder	Length	Gross Tonnage
Panama					
Chung Kuo No. 81	Genesis Ocean	Panama	Chinese Taipei	32.0	179
Chung Kuo No. 85	Genesis Ocean	Panama	Chinese Taipei	32.0	179
Chung Kuo No. 86	Genesis Ocean	Panama	Chinese Taipei	32.0	179
Chung Kuo No. 91	Genesis Ocean	Panama	Chinese Taipei	32.0	179
Chung Kuo No. 95	Genesis Ocean	Panama	Chinese Taipei	32.0	179
Chung Kuo No. 96	Gilontas Ocean	Panama	Chinese Taipei	32.0	179
Marine 303	Tuna Globe	Chinese Taipei	Chinese Taipei	50.8	420
Pesca Rica No. 2	Rica Panama	Chinese Taipei	Chinese Taipei	59.2	625
Pesca Rica No. 6	Grande Panama	Chinese Taipei	Chinese Taipei	59.2	625
Average				**40.1**	**304.9**
St. Vincent & The Grenadines					
Tuna Bras No. 216	Tunabras Int.	British Virgin Isl.	China	**57.4**	**635**
Vanuatu					
Chin Chun No. 12	Sheng Sheng Fishery	Vanuatu	Chinese Taipei	61.0	637
Fair Victory 707	Fair Victory International	Vanuatu	Chinese Taipei	70.6	1,180
Fong Seong 168	Trans-Global Int.	Vanuatu	Chinese Taipei	90.0	2,380
Fong Seong 196	Trans-Global Int.	Vanuatu	Chinese Taipei	90.0	2,386
Fu Chun No. 126	Fu Chun Fishery	Vanuatu	Chinese Taipei	61.0	637
Heng Chang No. 168	Ever Fortune Fishery	Vanuatu	Chinese Taipei	61.0	637
Hf No. 88	Hf Fishery	Vanuatu	Chinese Taipei	69.0	1,150
Hsiang Sheng No. 6	Hsiang Sheng Fishery	Vanuatu	Chinese Taipei	70.6	1,280
Hsiang Shun	Hsiang Chan Fishery	Vanuatu	China	52.7	560
Jin Hong No. 308	Jin Hong Ocean Ent.	Vanuatu	Chinese Taipei	60.1	625
Jui Der No. 36	Jui Fu Fishery	Vanuatu	China	61.5	558
Jupiter No. 1	Jupiter Fishery	Vanuatu	Chinese Taipei	61.5	699
Ming Man No. 2	Ming Shun Fishery	Chinese Taipei	Chinese Taipei	61.5	660
Mitra No. 888	Ryh Chun Fishery	Vanuatu	Chinese Taipei	61.5	660
More Rich	Sun Rise Fishery	Vanuatu	Chinese Taipei	59.2	625
Ocean Harvest	Ocean Harvest Fishery	Vanuatu	Chinese Taipei	50.1	490
Pacific Tracker No. 116	Melanesia Marine	Vanuatu	Chinese Taipei	40.0	327
Shun Fa No. 8	Shun Fa Fishery	Vanuatu	Chinese Taipei	69.0	1,150
To Chan No. 2	Sun Rise Fishery	Vanuatu	China	45.0	492
Tunago No. 62	Tunago Fishery	Vanuatu	Chinese Taipei	45.0	498
Average				**62.0**	**881.6**
Unknown					
Brave	Bravotime	Hong Kong	Chinese Taipei	33.0	227
Great Ocean I	Southern Cross	Vanuatu	Chinese Taipei	34.6	296
Seta 70	Owner Unknown		Chinese Taipei	46.0	320
			Average	**37.9**	**281.0**

Average size of FOC flagged vessels:

Finally, it is worth noting that the average length and tonnage of the vessels registered to the fourteen countries listed are substantially higher than the averages for all fishing vessels combined (flying all flags) greater than or equal to 24 metres on the Lloyd's database (Table 6.2). For 2003, while the number of fishing vessels flying the flag of one of the fourteen FOC countries is only about 6.4% of the total, this fleet represents close to 10% of the capacity of all 'large-scale' fishing vessels on the Lloyd's database as measured in Gross Tonnage.

At-sea transhipment, tankers and re-supply fleets

The viability of IUU fishing, like legal fishing, requires infrastructure and support services as well as access to market. A number of the provisions of the UN FAO International Plan of Action on IUU fishing recognise this fact. Paragraphs 73 and 74 of the IPOA call upon states to deter importers, transhippers, buyers, consumers, equipment suppliers, bankers, insurers and other services suppliers within their jurisdiction from doing business with vessels engaged in IUU fishing, including adopting laws to make such business illegal.

One of the major elements of the supporting infrastructure for distant water fleet fishing on the high seas consists of at-sea transhipment and re-supply vessels. Many high seas distant water fishing vessels stay at sea for long periods of time, transhipping their catches, refuelling, rotating crews, and re-supplying bait, food, and water through transhipment and re-supply vessels servicing the fishing fleets at sea. Aware of the essential role played by at-sea transhipment and re-supply vessels in the operation of IUU fleets, the IPOA further elaborates on the subject of transhipment and re-supply at sea and, in paragraphs 48 and 49 states:

> *"48. Flag States should ensure that their fishing, transport and support vessels do not support or engage in IUU fishing. To this end, flag States should ensure that none of their vessels re-supply fishing vessels engaged in such activities or tranship fish to or from these vessels. This paragraph is without prejudice to the taking of appropriate action, as necessary, for humanitarian purposes, including the safety of crew members.*
>
> *49. Flag States should ensure that, to the greatest extent possible, all of their fishing, transport and support vessels involved in transhipment at sea have a prior authorisation to tranship issued by the flag State..."*

Transhipment: Fish transport vessels ("Reefers")

At-sea transhipment of the catch of fishing fleets targeting high value species of tuna such as Bigeye and Bluefin tuna operating in the Atlantic and Indian Oceans is a major component of the infrastructure supporting longline tuna fishing on the high seas. While there is no published list of transhipment vessels as far as the authors are aware, Table 6.5 contains a sample list of refrigerated cargo vessels that are likely to be transhipping high-grade tuna in the Atlantic, Indian Ocean and Pacific Oceans.

Table 6.5. Sample List of Refrigerated Cargo Vessels Delivering Sashimi Grade Tuna to Japan

Vessel Name	Flag	Owner/Manager	Nationality of Owner/ Manager	Country of Financial Benefit	Principal Areas of Operation
Amagi	Panama	Kyoei Kaiun Kaisha	Japan	Japan	Pacific-Indian
Asian Rex	Panama	Azia Sekki	Japan	Japan	Atlantic-Indian
Chikuma	Panama	Hakko Marine	Japan	Japan	Med-Indian-Atlantic
Corona Reefer	Japan	Tachibana Kaiun	Japan	Japan	Atlantic-Indian-Med.
Eita Maru	Panama	Toei Reefer Line	Japan	Japan	Atlantic
Fortuna Reefer	St. Vincent	Habitat International	Chinese Taipei	Chinese Taipei	Pacific
Fuji	Bahamas	Kasuga Kaiun	Japan	Japan	Indian - Atlantic
Golden Express	Panama	Dongwon Industries	Korea	Korea	Pacific-Indian
Gouta	Panama	Chin Fu Fishery	Chinese Taipei	Japan	Atlantic
Harima 2	Panama	Hakko Marine	Japan	Japan	Atlantic-Indian
Haru	Panama	Chuo Kisen	Japan	Japan	Atlantic-Indian
Hatsukari	Panama	Atlas Marine	Japan	Japan	Atlantic-Pacific
Honai Maru	Panama	Kyoei Kaiun Kaisha	Japan	Japan	Pacific-Indian
Kyung Il No.7	Korea	Yung Il Shipping	Korea	Korea	Pacific
Luo Hua	St. Vincent	Luoda Shipping	China	China	Pacific-Indian
Meita Maru	Panama	Toei Reefer Line	Japan	Japan	Atlantic-Pacific
New Prosperity	Panama	Nisshin Kisen	Japan	Japan	Indian-Pacific-Atlantic
Reifu	Liberia	Korea Marine	Korea	Japan	Atlantic-Indian-Pacific
Ryoma	Panama	Chuo Kisen	Japan	Japan	Atlantic-Indian
Sagami 1	Panama	Wakoh Kisen	Japan	Japan	Indian-Pacific-Atlantic
Satsuma 1	Panama	Tachibana Kaiun	Japan	Japan	Pacific-Indian-Atlantic
Seita Maru	Panama	Toei Reefer Line	Japan	Japan	Indian-Pacific
Shin Izu	Panama	Kyoei Kaiun Kaisha	Japan	Japan	Indian-Pacific
Shofu	Liberia	Korea Marine	Korea	Korea	Atlantic-Pacific
Tenho Maru	Panama	Hayama Senpaku	Japan	Japan	Indian-Atlantic-Pacific
Tuna Queen	Panama	Alavanca	Japan	Japan	Mediterranean
Tunabridge	Japan	Shinko Senpaku	Japan	Japan	Atlantic-Indian-Pacific
Tunastates	Panama	Shinko Senpaku	Japan	Japan	Indian-Atlantic
Yamato 2	Panama	Wakoh Kisen	Japan	Japan	Atlantic-Indian
Yurishima	Panama	Alavanca	Japan	Japan	Pacific

Methodology

This list was compiled on the basis of the following method and criteria: The major market for sashimi grade tuna is Japan, and the major ports of entry for transhipped tuna into Japan were determined to be Shimizu and Yokosuka. Using the Lloyds Seasearcher database, a list of reefers regularly unloading in these ports was drawn up. The voyages of each of these reefers was then analysed, looking for frequent transits through known tuna fishing areas and to ports known to be transhipment points for tuna, and for ships that spent significantly longer at sea in the tuna fishing areas than would normally be required for a typical transit. Once a likely candidate was identified, we then looked at other vessels owned or managed by the same company to see if any followed a similar trading pattern. This research yielded a list of over 150 reefers. We then investigated each vessel using the internet and various databases held by government and commercial organisations to narrow down the list to those most likely to be transhipping tuna at sea. The results of this procedure gave a provisional list of 66 reefers likely to be regularly picking up tuna from fishing vessels and delivering it to market in Japan. However, more research would be needed to determine the level of accuracy of the list. A representative sample of these vessels is listed in Table 6.4. Annex 6.A. lays out the port visits and itineraries of several of these vessels over the period 2001-2003.

Table 6.6. Numbers and Frequency of Reefers Likely to be Delivering Transhipped Tuna to Shimizu and Yokosuka Ports in Japan

Ship Port Visits	2001	2002	2003	Average Visits per Year
Shimizu	285	346	329	320
Yokosuka	38	145	139	141
Different Ships	**2001**	**2002**	**2003**	**Average Ships per Year**
Shimizu	64	69	65	66
Yokosuka	50	8	45	48

The case of the M/V Hatsukari, a vessel documented by Greenpeace International as transhipping sashimi grade tuna in the South Atlantic from both IUU and legal longline vessels in May 2000 in the international waters in the South Atlantic, provides a practical illustration of the typical operation of a vessel involved in at-sea transhipment of high grade tuna destined for market in Japan (see Box 6.1).[9]

---

[9] Bours H., M. Gianni, D. Mather, *Pirate Fishing Plundering the Oceans*, Greenpeace International February 2001.

Box 6.1. Case Study: M/V Hatsukari

On the 3rd of March, 2000, the M/V Hatsukari sailed from her home port of Shimizu in Japan. The Hatsukari is a Japanese-owned and Panama-flagged refrigerated cargo ship, 94 metres long, displacing 3,029 tons, with a crew of Japanese officers and Filipino sailors. After stopping in Busan, South Korea on the 12th and 13th of March and in Kaoshiong, Chinese Taipei on the 16th and 17th of March where she most likely took on supplies for Korean and Chinese Taipei fishing vessels to add to those already on board for the Japanese fleet, she sailed toward Singapore to take on fuel.

The Hatsukari departed Singapore on the 24th of March for the 5,700 mile voyage to Cape Town. This voyage would normally take about 18 days, but the Hatsukari arrived in Cape Town on the 26th of April, 33 days after leaving Singapore. Given this passage time, it is likely that she made several rendezvous with vessels fishing in the western Indian Ocean to take on board their catch of frozen tuna. After servicing this fleet, the Hatsukari proceeded on to Cape Town where more supplies and spare parts were loaded for the longline fleets fishing for Bigeye tuna in the Atlantic Ocean off the African coast.

Companies that own or manage the longline tuna fishing vessels working the Eastern Atlantic Ocean had pre-arranged with the owners of the Hatsukari to have their catch picked up at sea and delivered to markets in Japan. Contact by radio was made between the Hatsukari and the fishing vessels, and a position and time for the rendezvous was arranged. As the Hatsukari entered the area, the longline fishing vessels pulled up their gear and one by one came alongside the Hatsukari to discharge their cargo of frozen tuna and to pick up food, supplies and spare parts.

On the 6th of May near position 9° 00 S - 5° 00 W, several hundred kilometres off the coast of Angola, the Greenpeace vessel M/V Greenpeace encountered the Hatsukari. The Hatsukari was observed meeting the Chien Chun No. 8, a Belize flag longliner, and began transferring bait and receiving frozen tuna from the longline vessel. Soon afterward, two more Belize flagged vessels, the Jeffrey 816 and Jackie 11 came alongside the Hatsukari. Later the same day, the Cambodian flagged Benny No. 87 and two Chinese Taipei vessels, Yu I Hsiang and Jiln Horng 206, also took their turns.

Almost a month after leaving Cape Town, on the 25th of May, the Hatsukari made a brief stop at St. Vincent in the Cape Verde Islands. The Hatsukari arrived back in Cape Town on the 20th of June where it reportedly offloaded seventy-two tons of tuna of indeterminate species. She departed Cape Town on the 21st of June for the return voyage to Japan via Singapore. Again, this voyage, which would normally take approximately 18 days, took over a month due most likely to stops to service fishing vessels at-sea in the Indian Ocean. The Hatsukari arrived in Singapore on the 26th of July, departing the 29th to sail back to Japan. The Hatsukari arrived in Shimuzu on the 8th August where the transhipped cargo of high grade tuna was offloaded for market.

The M/V Hatsukari is one of a fleet of refrigerated cargo vessels or "reefers" that regularly travel from the ports of Shimuzu and Yokosuka in Japan, stopping at Busan, South Korea, Kaoshiong, Chinese Taipei and Singapore, then continuing to the Indian and Atlantic Oceans, with stops at Cape Town, South Africa, Las Palmas in the Canary Islands of Spain and occasionally other Atlantic or Indian Ocean ports. These vessels spend relatively long periods of time at sea, transhipping sashimi grade tuna and re-supplying high seas tuna longline fleets. The sample of reefers and their itineraries in Annex I follow similar patterns.

The Hatsukari was transhipping fish on the high seas from IUU fishing vessels as well as legal vessels fishing for tuna. Similarly, Greenpeace documented an attempted transhipment from a Belize flagged tuna longline vessel to the reefer M/V Toyou in the same area on 12 May 2000.[10] Like the

---

[10]   *Ibid* Greenpeace.

Hatsukari, at least some portion of the transhipment fleet is likely to be servicing both IUU and legal tuna longline fishing vessels operating on the high seas. Although not impossible, it seems unlikely that a fleet of transhipment vessels would service IUU fishing vessels only.

Observers aboard transhipment vessels

In the same way that ICCAT, IOTC and the IATTC have developed lists of vessels authorised to fish in their respective areas of competence, the authors would argue that these and other RFMOs should require that all transhipment vessels operating in the area of competence of the organisation have an authorisation to tranship at sea and that a list be compiled of such vessels. Furthermore, we would argue that relevant Regional Fisheries Management Organisations should agree to establish an observer programme on board all transhipment vessels to monitor and report on all transhipments in fisheries regulated by the RFMO at sea. The programme should be operated under the authority or auspices of the RFMO, in co-operation with, but independent of, the flag states of the transhipment vessels (similar to the observer programme on fishing vessels run by the IATTC). The failure of a tuna transhipment vessel to co-operate in the programme should be made grounds for denial of port access (in other than emergency situations) and the imposition of other sanctions by the member countries of the RFMO, and others where possible.

Some of the practicalities of establishing an observer programme emerge in reviewing the information on this list. All but seven of the sixty-six vessels on the provisional list of reefers we identified as being involved in at-sea transhipping of high grade tuna are flagged to contracting parties of ICCAT, with most flagged to Panama and Japan. All but a handful are owned or managed by companies based in Japan and Korea. The co-operation of these three states: the flag states, market states and and/or countries of beneficial ownership of most of the transhipment fleet should be relatively straightforward – all are contracting parties of ICCAT and have committed to the IPOA on IUU fishing as well as similar resolutions on transhipment adopted by ICCAT.[11]

A similar situation applies for the fisheries in the IATTC area. Assuming that either or both ICCAT and the IATTC were to establish such an observer programme involving Panamanian-flagged transhipment vessels and others, it should not be difficult to do the same for the Indian Ocean fisheries. Both Japan and Korea are members of the IOTC and it would be reasonable to assume that Panama could be persuaded to co-operate in such a programme even though it is not currently a member of the IOTC. However, in addition to establishing observer programmes, RFMOS should adopt measures to require that all transhipment vessels should be flagged to contracting parties or co-operating parties/entities of the RFMO, with sanctions applied to vessels (*e.g.* denial of port access) and countries (import restriction/bans) in contravention of the measures.

Tankers and re-supply vessels

Fleets of vessels that refuel and re-supply high seas fishing vessels are also an essential element of the infrastructure necessary to maintain IUU fishing as well as fishing by legal operators. In Table

---

[11]   For example Recommendation 02-23 adopted by ICCAT in 2002: Recommendation by ICCAT to Establish a List of Vessels presumed to have carried out Illegal, Unreported and Unregulated Fishing Activities in the ICCAT Convention Area - Paragraph 9 "Contracting Parties and Co-operating non-Contracting Parties, Entities or Fishing Entities shall take all necessary measures, under their applicable legislation: e) To prohibit the imports, or landing and/or transhipment, of tuna and tuna-like species from vessels included in the IUU list". See also ICCAT Resolution 01-18: Scope of IUU Fishing. Adopted by ICCAT in 2001.

6.6, the authors attempted to put together a sample list of vessels most likely to be servicing distant water fishing vessels operating on the high seas and, in some cases within other countries' EEZs.

Methodology

The methodology used in this case was as follows:

- an internet search yielding several companies that specialise in refuelling (bunkering) vessels at sea,

- investigating tankers belonging to these companies, producing a profile of the vessels engaged in this type of work,

- finding tankers fitting this profile using the Lloyds Register database,

- reviewing the voyage history of each tanker to find those making regular voyages into areas known to be frequented by tuna fishing vessels and spending significantly longer at sea than would have been required for a routine transit.

This research produced a list of over 100 tankers, which was then narrowed down to 54 that, for at least part of the year, are engaged in refuelling and re-supplying fishing vessels at sea. Again, this list is provisional and would require further research to verify that all of these vessels are involved, or highly likely to be involved, in refuelling and reprovisioning distant water fishing vessels at sea. A sample of 30 of these vessels is included in Table 6.7.

While the ownership and registered flags of these vessels involves a greater number of countries than do the high value tuna transhipment fleets, at least some the companies that own or manage tanker vessels are involved in a variety of other at-sea services. For example, ADDAX Bunkering Services owns or charters a fleet of 10-12 tankers that re-supply fishing vessels in the Atlantic and Indian Oceans. This fleet also supplies offshore mining operations, oil platforms and seismic survey vessels. Amongst the services it supplies are fuel, provisions and fresh water. ADDAX is a subsidiary of the Geneva based transnational, ADDAX & ORYX group.[12] Another company, SK Shipping operates a fleet of over 20 tankers supplying fuel and supplies to fishing fleets, worldwide. According to their website, SK provides "...port bunkering and bunker-trading services in the North and South Pacific, the Atlantic Ocean, the Indian Ocean, PNG, Guam, and the Arafura Sea. We have also diversified our business to offer comprehensive fishing-vessel services that include crew repatriation, spare parts, and bait. In addition, we bring integrated logistics services to the fishing industry, including reefer service and fish trading". SK is a subsidiary of SK Group, the 3rd largest conglomerate in Korea.[13]

---

[12]   http://www.addax-oryx.com/media/pdf/bunkers.pdf

[13]   http://www.skshipping.com/jsp/eng/company/overview.jsp

Table 6.7. Tankers and Re-supply Vessels Servicing Fishing Vessels at Sea - Provisional List

Tanker Name	Flag	Owner/Manager	Nationality of Owner/Manager	Principal Area of Operation
Arsenyev	Russia	Primorsk Shipping	Russia	Atlantic
Atom 7	Panama	Sekwang Shipping	Korea	Pacific
B.Cupid	Singapore	Aceline Ship Mngt.	Singapore	Atlantic
Dae Yong	Korea	Cosmos Shipping	Korea	Pacific
Dalnerechensk	Cyprus	Primorsk Shipping	Russia	Atlantic
Hai Gong You 302	China	China National Fisheries	China	Atlantic
Hai Soon 16	Singapore	Hai Soon	Singapore	Indian
Hai Soon Ii	Singapore	Hai Soon	Singapore	Indian
Hai Soon Ix	Singapore	Hai Soon	Singapore	Pacific
Hai Soon Xv	Singapore	Hai Soon	Singapore	Atlantic
Hl Tauras	Singapore	Hong Lam Marine	Singapore	Pacific
Hobi Maru	Ecuador	Toko Kaiun	Japan	Pacific
Hosei Maru	Japan	Toko Kaiun	Japan	Indian
Hozen Maru	Japan	Toko Kaiun	Japan	Pacific
Japan Tuna No.3	Panama	Japan Tuna Co-Op	Japan	Pacific-Indian
Katie	Liberia	Aquasips	Latvia	Atlantic
Kosiam	Singapore	Kosiam Trading	Singapore	Pacific
L. Star	Singapore	Sekwang Shipping	Singapore	Indian
Sea Pearl	Seychelles	Al Dawood		Atlantic
Mighty 7	Panama	Sekwang Shipping	Korea	Ind-Pacific
Nagayevo	Cyprus	Primorsk Shipping Corp.	Russia	Atlantic
New Kopex	Korea	Sekwang Shipping	Korea	Pacific
Nipayia	Panama	Lotus Shipping	Greece	Indian
Oriental Bluebird	Panama	New Shipping Kaisha	Japan	Pacific
Shin Co-Op Maru	Panama	Kumazawa	Japan	Pacific
Smile No.3	Korea	Sekwang Shipping	Korea	Pacific
Soyang	Korea	Sekwang Shipping	Korea	Pacific
Star Tuna	Panama	Korea Ship Managers	Korea	Pacific
Starry	Singapore	Honglam Shipping	Singapore	Pacific-Indian
Vesta 7	Panama	Sekwang Shipping	Korea	Pacific

Finally, some companies are involved in both transhipment of fish and re-supply. Sunmar Shipping, for example, services international fleets operating in the Russian Far East. According to its website, the company operates 20 vessels which tranship "frozen fish and fish meal products" at sea and delivers the fish to markets in Europe, the United States, China, Korea, Japan and elsewhere. Sunmar also delivers provisions and supplies directly to the fishing fleets.[14]

It is difficult to understate the importance of tankers and re-supply vessels to the operations of high seas IUU fishing fleets. Given the size, scope, visibility and the diversity of the operations of major companies involved in the business, RFMOs should engage these companies as they may be amenable to co-operating in international efforts to prevent, deter and eliminate IUU fishing, whether through observer programmes, bringing company policies and business practices into line with RFMO recommendations, and/or by other means. Integrating tankers and re-supply vessels and the companies that own, manage or charter these vessels into regional efforts to ensure effective compliance with RFMO measures are a necessary and potentially very effective means of combating IUU fishing.

Recommendations/discussion

The following recommendations are drawn from the above research into recent trends in the use of flags of convenience fisheries and the role and character of the at-sea transhipment, refuelling and re-supply fleets in supporting the operations of high value tuna longline fleets and other fishing fleets on the high seas.

The recommendations are as follows:

1. Further investigation into the numbers of vessels registered to flags of convenience in the "unknown" category, and the reasons why these and others vessels are listed as such on the Lloyd's database, would be useful in providing a clearer picture of trends in the flagging of fishing vessels over the past several years.

2. It would be worth further investigation to determine whether any of the companies in Chinese Taipei involved in building new fishing vessels over the past three years, virtually all of which have been flagged to FOC countries, have benefited from funds for the joint Japan/Chinese Taipei programme designed to decommission large-scale tuna longline vessels.

3. Given the status of Chinese Taipei as a "Co-operating Party, Entity or Fishing Entity" of ICCAT, the government should be encouraged to ensure that no vessels built in Chinese Taipei shipyards are allowed to register to flag of convenience countries.

4. RFMOs should require that all transhipment vessels operating in the area of competence of the organisation have an authorisation to tranship at sea and that a list be compiled of such vessels.

5. RFMOs should agree to establish an observer programme on board all transhipment vessels to monitor and report on all transhipment at sea. The programme should be operated under the authority or auspices of the RFMO, in co-operation with, but independent of, the flag states of the transhipment vessels concerned.

6. RFMOS should adopt measures to require that all transhipment vessels should be flagged to contracting parties or co-operating parties/entities of the RFMO, with sanctions applied to vessels (*e.g.* denial of port access) and countries (import restriction/bans) in contravention of the measures.

7. RFMOs should engage companies that own, manage or charter tankers and re-supply vessels servicing fishing vessels on the high seas to co-operate in international efforts to prevent,

---

[14]   http://www.sunmar.com/ssi/default.htm

deter and eliminate IUU fishing, whether through observer programmes, bringing company policies and business practices into line with RFMO recommendations, and/or by other means.

As mentioned in the executive summary, in addition to the above, a number of other aspects of the infrastructure support and facilitate IUU fisheries worldwide. It is clear from the Lloyd's data that the number of fishing vessels flying flags of convenience remains high in spite of the adoption of the UN FAO International Plan of Action on IUU fishing and the many efforts of regional fisheries management organisations over the past several years.

In the absence of (or, in effect, as a substitute for) effective flag state control, responsible nations will continue to incur the cost of deterring IUU fishing. These costs are essentially twofold: one, the cost of monitoring control and enforcement, whether at sea, in port, regulating imports or investigating and prosecuting nationals or companies within their jurisdiction involved in IUU fishing; two, the cost to responsible fishing nations in terms of research, conservation and management, and the loss of actual or potential revenue to IUU fishing.

As was discussed in a paper prepared by Gianni for WWF for the June 2003 meeting Ministerial level OECD Round Table on Sustainable Development related to fisheries, the financial benefit derived by Flag of Convenience states in registering fishing vessels are relatively small. By some estimates, the top four flag of convenience countries may derive only a few million US dollars per year in revenues from the flagging of over 1000 fishing vessels combined. By comparison, the cost to the international community of the failure of these states to exercise control over the activities of their fishing vessels is likely to be far greater.

It would be well worth considering a means or method to document and/or reasonably estimate the types of costs incurred by responsible flag states as a result of FOC fishing. Then, on this basis, seek compensation through international arbitration mechanisms available from specific states operating open registries whose vessels are fishing in a region in contravention of the measures established by a relevant fisheries management organisation to the detriment of responsible flag states' fleets. Whether or not there is a genuine economic link between the flag state and the IUU fishing vessels or fleets flying its flag, the flag state bears the ultimate responsibility for the activities of the vessels. If an FOC state is faced with the prospect of paying substantial sums in compensation to other states for its failure to regulate its fishing fleets, this could prove a significant and cost-effective deterrent to IUU fishing in ways which port state controls, market restrictions, and enhanced monitoring, control and surveillance have so far been unable to accomplish.

The authors hope to further develop this line of inquiry as part of a larger project involving further research into the variety of components of the international infrastructure supporting IUU fishing on the high seas.

ANNEX 6.A.

Table 6.A1.1. Sample of Port Visits and Itineraries of Refrigerated Cargo Vessels Transhipping High Value Tuna At Sea for Delivery to Japan, 2001-2003

	JAN	FEB	MAR	APR	MAY	JUN	JUL	AUG	SEP	OCT	NOV	DEC

M/V ASIAN REX

2001: CP-JPN | SH · YK-SH-KA | SI-CP | E&W ATL LA-ALG-COL | PC-JPN | YK-TO-SH-YK-KA | SI-CP | E & W. ATLANTIC | PC-YK

2002: YK-SH-YK-SH-KA | SI-CP | EAST ATLANTIC (CP) | CP-SI | SH-YK-SH-YK-KA | SI-CP | EAST ATLANTIC (LAS PALMAS) | CP-SI

2003: SH-BU-SU-KA | SI-CP | E. ATLANTIC (LAS PALMAS) | CP-SI | SH-YK-SH-SU-KA | EAST ATLANTIC (CAPE VERDE, ISLANDS) | CP-SI

M/V CHIKUMA

2001: MED. | MED-JPN | SH-TO-YK | JPN-ESP | MEDITERRANEAN | MED-AUS PL | AUS-JPN | TO-SH-BU-KA | SI-CP | MEDITERRANEAN

2002: MED-(CRO) | MED-JPN | TO-SH | SI-SZC | MEDITERRANEAN | SZC-SI | TO-SH-BU | KOR-MED | MEDITERRANEAN

2003: MED-JPN | TO-SH | JPN-MED | MEDITERRANEAN | MED-JPN | SH-TO-SH-KA | JPN-MED | MEDITERRANEAN

M/V EITA MARU

2001: SH-YK-SH-YK-SH-YK-KA | INDIAN OCEAN | IO-JPN | SH-YK-SH-YK-SH-KA | JPN-IO | INDIAN OCEAN | IO-JPN

2002: SH · KA | SI-CP | E-W. ATLANTIC | PC-JPN | SH-YK-SH-YK-KA | JPN-TRN | WEST-EAST ATLANTIC | CP-SI

2003: SH-YK-SH-YK-KA | SI-CP | EAST ATLANTIC | CP-SI | S.E.A. | SH-YK-SH-YK-SH-YK-KA

Table 6.A1.1. Sample of Port Visits and Itineraries of Refrigerated Cargo Vessels Transshipping High Value Tuna At Sea for Delivery to Japan, 2001-2003 (cont.)

M/V HARIMA 2

Year	JAN	FEB	MAR	APR	MAY	JUN	JUL	AUG	SEP	OCT	NOV	DEC
2001	EAST ATLANTIC		CP-SH	SH-YK-SH-YK-KA-BU		SOUTH CHINA SEA-INDONESIA			SH-YK-SH-YK-KA-BU			S. CHINA S.-IND
2002		SH	BU-KA	S. CHINA SEA-IND		SOUTH CHINA SEA-INDONESIA			YK-SH-YK-SH-BU-KA			W.PAC
2003	WEST PAC.(PA)	SH-YK-SH-YK-KA		TAI-PC	W-E.ATLANTIC(LAS)		CP-SI	SH-YK-SH-YK-SH-YK-KA		S. CHINA SEA-IND		SH

M/V HARU

Year	JAN	FEB	MAR	APR	MAY	JUN	JUL	AUG	SEP	OCT	NOV	DEC
2001			KA-BU-KA-BU-SH	JPN-PC	W.& E.ATLANTIC. (TRN-CP)		CP-SI	SH-YK-SH-YK-BU-KA		S. CHINA SEA-IND		SH
2002	YK-SH-KA		INDIAN OCEAN (MA)		SH-YK-TO-KA		INDIAN O. (SEYCHELLES)	SH-YK-KA		MEDITERRANEAN		SH-KA
2003	SI-CP		EAST-WEST ATLANTIC (CP-LA)	CP-SI	SH-YK-SH-BU-KA		SI-CP	EAST ATLANTIC	CP-SI	SH-TO-TO-KA	SI-SY	INDIAN OCEAN

M/V HATSUKARI

Year	JAN	FEB	MAR	APR	MAY	JUN	JUL	AUG	SEP	OCT	NOV	DEC
2001	SH-YK-SH-YK-KA-SH			WEST PACIFIC (PA)		BU-SH-YK-KA-BU		WEST PACIFIC		SH-BU-SH-YK-SH-HA		
2002		JPN-PC	WEST ATLANTIC (LA-CVI)		PC-JPN	BU-SH		SI-CP	EAST ATLANTIC (CP-SVI)	CP-SI		SH-YK
2003	SH-YK		WEST-EAST PACIFIC			SH-YK-TO-SH-BU-KA		EAST PACIFIC (CA)		YK-SH-TO-BU-KA		

M/V MEITA MARU

Year	JAN	FEB	MAR	APR	MAY	JUN	JUL	AUG	SEP	OCT	NOV	DEC
2001	SH-HA	SI-CP	EAST ATLANTIC (LAS PALMAS)	CP-SI		SH-YK-SH-YK-SH-YK		JPN-PER	W.&E. PACIFIC (CA)		PER-JPN	SH-BU-SH-HA-BU
2002	SI-CP		E. ATL. (LAS PALMAS)	CP-SI	SH-BU-SH-KA		EAST ATLANTIC (CAPE VERDE)		CP-SI	SH-YK-SH-SU-KA		SI-CP
2003	EAST ATLANTIC (LAS PALMAS)		CP-SI		SH	YK-HA-KA	SI-CP	E & W. ATLANTIC-LA-PC-E PAC		PC-JPN		SH-YK-SH-TO-SH

Table 6.A1.1. Sample of Port Visits and Itineraries of Refrigerated Cargo Vessels Transhipping High Value Tuna At Sea for Delivery to Japan, 2001-2003 (cont.)

Months: JAN | FEB | MAR | APR | MAY | JUN | JUL | AUG | SEP | OCT | NOV | DEC

M/V NEW PROSPERITY

Year	Itinerary (JAN → DEC)
2001	YK-SH-YK-SH-TO-KE · JPN-PER · EAST PACIFIC (CA) · PER-JPN · SH-BU-TO-YK-BU-SH · SI-CP · EAST ATLANTIC (CP-WB) · CP-SI
2002	SH-YK-SH-BU-KA · W.PAC (SU-PA) · YK-SH-TO-YK-SH-YK-SH · JPN-PER · E.PACIFIC (PERU) · PER-JAP · YK-SH-KA
2003	SI-CP · EAST ATLANTIC (CP) · CP-SI · SH · BU-KA · W.PAC. (SV) · YK-TO-BU-KA · INDIAN OCEAN (SY)

M/V SHOFU

Year	Itinerary (JAN → DEC)
2001	EAST ATLANTIC (LAS PALMAS) · CP-SI · SH · YK-BU-KA · WEST PAC (PA) · SH · BU-KA-BU-SH
2002	WEST PACIFIC (PAPEETE) · SH-BU-SH-YK-BU-KA · SI-CP · EAST ATLANTIC (CAPE TOWN) · CP-SI · BU-SH · BU-SH-BU-KA · WPAC
2003	WEST PACIFIC · YK-SH · SH · SI-MED · E.ATL-MED (LA-SP) · MED-SI · SH-TO-BU-KA · SI-PC · WEST & EAST ATLANTIC (LA)

M/V TENHO MARU

Year	Itinerary (JAN → DEC)
2001	E.PACIFIC · SH-YK-SH · SI-SZC · MED-E.ATLANTIC (LA-ESP) · SZC-SI · SH · YK-SH-YK · EPAC (MA) · SH
2002	WPAC · SH-YK-SH-TO-SH-KA · SI-SZC · ME.E.ATLANTIC (LA-ESP) · SZ-SI · BU-TO-SH-KA-BU-SH · W.PACIFIC (PAPEETE)
2003	SH-BU-SU-SH-KA-BU · W.PACIFIC (PAPEETE) · SH · SH-BU-KA · SI SY · INDIAN OCEANS (SYCHELLES) · SY-INO · SH-BU-KA

M/V TUNASTATES

Year	Itinerary (JAN → DEC)
2001	EAST ATLANTIC (CAPE TOWN) · CP-SI · SI-CP · SH-TO-SH-BU-KA · EAST ATLANTIC (CP) · CP-SI · SH · BU-KA-SU
2002	INDIAN OCEAN (SY) · SH-BU-KA · SI-LA · E.ATLANTIC · LA-COL-PC-JPN · BU-SH-TO-SH-SU-KA · INDIAN O. (SY)
2003	IND · YK-SH-BU · WEST PACIFIC (SUVA) · SH-TO-YK-BU-KA · INDIAN OCEAN · YK-TO-SU-KA · SI-CP

100

Table 6.A1.1. Sample of Port Visits and Itineraries of Refrigerated Cargo Vessels Transhipping High Value Tuna At Sea for Delivery to Japan, 2001-2003 (cont.)

M/V YAMATO 2

	JAN	FEB	MAR	APR	MAY	JUN	JUL	AUG	SEP	OCT	NOV	DEC
2001	SH-KA	SI-CP	EAST ATLANTIC		CP-SI	SH-YK-SH-BU-KA			S. CHINA SEA - INDONESIA		SH-YK-SH-BU-KA	
2002	S. CHINA SEA - INDONESIA	S. CHINA SEA - IND.	SH-YK-SH-YK-SH-KA		S. CHINA SEA - INDONESIA		SH-YK-SH-BU-KA		SI-CP	SOUTH ATLANTIC	CP-SI	SH
2003	YK-SH-KA	S. CHINA SEA - IND.		SH-YK-SH-BU-KA		S. CHINA SEA - IND.		SH-TO-SH-KA-SH		JPN-TRN	WEST AND EAST ATLANTIC	

= LOADING FISH AT SEA OR IN PORT

= VESSEL IN TRANSIT

= VESSEL IN PORT DISCHARGING TUNA

ATL	-	ATLANTIC OCEAN	MED	- MEDITERRANEAN
KA	-	KAOSHIUNG, CHINESE TAIPEI	TAI	- CHINESE TAIPEI
SH	-	SHIMUZU, JAPAN	DU	- DURBAN, SOUTH AFRICA
BU	-	BUSAN, S. KOREA	MN	- MANTA, ECUADOR
KE	-	KESENNUMA, JAPAN	TO	- TOKYO, JAPAN
SI	-	SINGAPORE	ESP	- SPAIN
CA	-	CALLAO, PERU	PA	- PAPEETE, TAHITI
KOR	-	KOREA	TUN	- TUNESIA
SU	-	SUAO, CHINESE TAIPEI	FR	- FREMANTLE, AUSTRALIA
CO	-	COLUMBIA	PAC	- PACIFIC OCEAN
LA	-	LAS PALMAS, CANARY ISLANDS	WB	- WALVIS BAY, NAMIBIA
SV	-	SUVA, FIJI	HA	- HAHINOHE, JAPAN
CP	-	CAPE TOWN, SOUTH AFRICA	PC	- PANAMA CANAL
MA	-	MAURITIUS	YK	- YOKOSUKA, JAPAN
SY	-	SEYCHELLES	IN	- INDONESIA
CRO	-	CROATIA	PER	- PERU
MAL	-	MALTA	INO	- INDIAN OCEAN
SZC	-	SUEZ CANAL	PL	- PORT LINCOLN, AUSTRALIA
CVI	-	CAPE VERDE ISLANDS		

Table 6.A1.2. Sample Itineraries of Tankers Refuelling Fishing Vessels At Sea, 2001-2003

JAN	FEB	MAR	APR	MAY	JUN	JUL	AUG	SEP	OCT	NOV	DEC

M/T STAR TUNA

2001: HI N.E. PACIFIC | HI EAST PACIFIC | HI NORTH WEST PACIFIC | HI WEST PACIFIC | SH

2002: UL-BU WEST PACIFIC | HI EAST PACIFIC | HI EAST PAC | BU WEST PACIFIC | HI EAST PACIFIC | HI

2003: WEST PACIFIC | BU | SI NORTH PACIFIC | HI EAST PACIFIC | HI EAST PACIFIC

M/T B. CUPID

2001: EATL LP ETL EATL LP EATL AB EATL AB E. ATLANTIC | LP EAST ATLANTIC | LP EAST ATLANTIC | AB

2002: EATL LP EAST ATLANTIC | AB EATL TE AB EAST ATLANTIC | TE E. ATLANTIC | LO E ATL | LO EATL | LO E. ATLANTIC | LO E ATL

2003: TE TE EAST ATLANTIC | LP EAST ATLANTIC | LP EATL TE EATL LP EAST ATLANTIC | TE

M/T ATOM 7

2001: WEST PACIFIC | VO YO WPAC | YO-AK PACIFIC | YO PACIFIC

2002: PAC VO WEST PACIFIC | GU | NA

2003: UL W PACIFIC | UL PACIFIC | BA EPAC BA EAST PACIFIC | BA EAST PACIFIC | BA EAST PACIFIC | BA EAST PACIFIC | BA

Table 6.A1.2. Sample Itineraries of Tankers Refuelling Fishing Vessels At Sea, 2001-2003 (cont.)

	JAN	FEB	MAR	APR	MAY	JUN	JUL	AUG	SEP	OCT	NOV	DEC

M/T VESTA 7

- 2001: CU CARIBBEAN — CU CARIB — CU CARIBBEAN — BA PACIFIC — BU PAC — UL WPAC — BU
- 2002: WPAC — BU WEST PACIFIC — UL W. PACIFIC — UL WEAT PACIFIC — UL WEST PACIFIC — GU WEST PACIFIC — BU-BL
- 2003: UL WEST PACIFIC — UL WEST PACIFIC — GU WEST PACIFIC — DA WEST PACIFIC

M/T SHIN CO-OP MARU

- 2001: HI PACIFIC — HI E PACIFIC — BA-CA E. PACIFIC — HI EAST PACIFIC — BA EAST PACIFIC — HI EAST PACIFIC — B
- 2002: HI PACIFIC — HI PACIFIC — HI PACIFIC — HI PACIFIC — HI WEST PACIFIC — UL
- 2003: HI PACIFIC — HI EAST PACIFIC — CA EPAC CA — HI E PACIFIC — EAST PACIFIC — BA EAST PAC — HI

M/T JAPAN TUNA NO.3

- 2001: SI S.W.PAC NC — SI S.W.PAC — NC S.W. PACIFIC — AK S.W.PACIFIC — GE S. AUSTRALIA — SI E.INDIAN O. — FR E.INDIAN OCEAN — SI INDIAN O. — SI
- 2002: INDIAN OCEAN — SI S.W. PACIFIC — AK S.W. PACIFIC — AK SOUTH AUSTRALIA — GE INDIAN O. — SI INDIAN O. — FR INDIAN OCEAN — SI WEST PACIFIC
- 2003: HI N.PAC. — HI S.W.PACIFIC — AK S.W. PACIFIC — AK S.W. PACIFIC — AK S.W. PACIFIC — AK S.W. PACIFIC — SI WESTERN PACIFIC — HI

M/T SMILE NO. 3

- 2001: GU W. PACIFIC — GU
- 2002: WEST PACIFIC — GU WEST PACIFIC — GU WEST PACIFIC — GU NORTH PACIFIC — GU N. PAC — HI N.PAC — HI W. PAC — GU
- 2003: GU WPAC GU — WPAC — GU WEST PACIFIC — BU WEST PACIFIC — GU W. PAC — GU WEST PACIFIC — GU WEST PACIFIC

Table 6.A1.2. Sample Itineraries of Tankers Refuelling Fishing Vessels At Sea, 2001-2003 (cont.)

	JAN	FEB	MAR	APR	MAY	JUN	JUL	AUG	SEP	OCT	NOV	DEC

M T KOSTAM

2001: N.PACIFIC | N.PACIFIC | HI | N.PACIFIC | HI | NORTH PACIFIC | HI | N.PACIFIC | HI | N.PACIFIC | HI | N.PACIFIC | HI | NORTH PACIFIC | HI | N.PACIFIC | HI | N.PACIFIC | HI

2002: N.PACIFIC | HI | N.PAC | HI | N.PACIFIC | HI | N.PACIFIC | HI | N.PACIFIC | N.PACIFIC | HI | N.PACIFIC | HI

2003: N.PACIFIC | WEST PACIFIC | BU | WEST PACIFIC | HI | NORTH PACIFIC | HI | N.PACIFIC | HI | N.PACIFIC | HI | NORTH PACIFIC | HI | N.PACIFIC | HI | N.PAC | HI

■ = TANKERS AT SEA - IN TRANSIT, OR SERVICING FISHING VESSELS.

▨ = TANKERS IN PORT REFUELING AND LOADING SUPPLIES

AB   ABIDJAN
AK   ALASKA
BA   BALBOA
BU   BUSAN
CU   CURACAO
GU   GUAM
HI -   HAWAII
LO   LOME
LP   LAS PALMAS
NA   NAGOYA
SI   SIAPAN
TE   TEMA
UL   ULSAN
YO   YOKOHAMA

CHAPTER 7

PATAGONIAN TOOTHFISH - THE STORM GATHERS

Dr. Denzil G.M. Miller, CCAMLR, Tasmania, Australia [1]

Abstract

This paper documents the experiences of the Commission for the Conservation of Antarctic Marine Living Resources (CCAMLR) in managing marine living resources in the waters (*i.e.* south of about 45°S) for which it is responsible. Emphasis is given to legal and institutional aspects, particularly sovereignty issues and jurisdictional controls. Recent high levels of Illegal, Unreported and Unregulated (IUU) fishing for Toothfish (Dissostichus sp.) in the CCAMLR area are used to illustrate the management and enforcement measures taken by this particular organisation to combat such fishing. While it is concluded that these measures have relied heavily on national (particularly coastal state) enforcement to be effective, their clear affinity with other recent fisheries agreements is highlighted. Various factors are identified for further consideration.

Introduction

It has been stated that:

"An old spectre haunts fisheries management today: governance without government". [2]

Although provocative, this statement clearly demonstrates that much appears to have gone horribly wrong with humankind's efforts to manage fishing on the high seas. These efforts are perceived to have failed miserably despite expectations to the contrary flowing from general customary international law. Such expectations, first outlined in Principle 21 of the 1972 Stockholm

---

[1]   Email: denzil@ccamlr.org. The opinions expressed in this paper are those of the author and do not reflect the collective, or official, views of CCAMLR.

[2]   See p. 157 in O.S. Stokke, "Governance of high seas fisheries: The role of regime linkages", in D. Vidas and W. Østreng (eds.), *Order for the Oceans at the Turn of the Century*. (Kluwer Law International, The Hague, 1999), pp. 157-172.

Declaration[3] and embodied in the 1982 United Nations Convention on the Law of the Sea (LOSC)[4] were subsumed into Principle 2 of the 1992 Rio Declaration.[5] They clearly intimate that there is a general obligation on all states to ensure that "activities within their jurisdiction or control do not cause damage to the environment of other States or of areas beyond the limits of national jurisdiction".

In substantiating the Rio interpretation, Freestone[6] asserts that the above obligation, although minimal, assumes generality when applied to the global commons of the high seas. However, he maintains that the extent to which it represents a clear invocation to avoid environmental damage not only applies to activities confined within state territory, "but also to activities under State jurisdiction (including State registered vessels)". Arguably, therefore, the Rio interpretation is relevant to the extent that protection of the environment and certain activities are linked in the context of being subject to state jurisdiction [including over nationals (*i.e.* legal and natural individuals)].

The dichotomy between the opening quotation's "realism" and Freestone's "idealism" has become alarmingly evident over the past decade. As more and more fisheries are affected by heavy exploitation, the search for new resources increases.[7] Irresponsible operators have taken advantage of prevailing circumstances to optimise their own economic advantages, often to the detriment of the stocks concerned and at the expense of their more responsible competitors. While the serious consequences of such behaviour have been clearly recognised by the international community[8,9,10,11,12,13,14], the extent of fishing activity violating applicable laws and regulations

---

3   Declaration of the United Nations Conference on the Human Environment. (United Nations Environment Programme, 1972). 5 pp.

4   *United Nations Convention on the Law of the Sea, 1982.* (United Nations, New York, 1983). 224 pp.

5   *Rio Declaration on Environment and Development.* (United Nations Environment Programme, 1992). 4 pp.

6   See p. 104 in D. Freestone, "The Conservation of Marine Ecosystems under International Law", in C Redgewell and M. Bowman (eds.), *International Law and the Conservation of Biodiversity.* (Kluwer Law International, 1995), p. 91-107.

7   *The State of World Fisheries and Aquaculture, 2002.* (Food and Agriculture Organization of the United Nations, Rome, 2002), 150 pp.

8   There are a number of international instruments that set out provisions to address irresponsible fishing practices. These include the *LOSC*[4], the *1993 Agreement to Promote Compliance with International Conservation and Management Measures by Fishing Vessels on the High Seas*[9] (the "*FAO Compliance Agreement*"), the *1995 Agreement for the Implementation of the Provisions of the United Nations Convention on the Law of the Sea of 10 December 1982 Relating to the Conservation and Management of Straddling Fish Stocks and Highly Migratory Fish Stocks*[10] (the "*United Nations Fish Stocks Agreement - UNFSA*") and the *1995 Code of Conduct for Responsible Fisheries*[11] (the "*FAO Code of Conduct*"). It must be emphasised that the *Code* was formulated as a practical framework to be applied in conformity with the other instruments listed and in light of, *inter alia*, the *1992 Declaration of Cancun*[12] and the *1992 Rio Declaration on Environment and Development*[13], in particular Chapter 17 of Agenda 21.[14]

9   *Agreement to Promote Compliance with International Conservation and Management Measures by Fishing Vessels on the High Seas, 1993.* (Food and Agricultural Organization of the United Nations, Division for Ocean Affairs and the Law of the Sea, United Nations, New York, 1998), p. 41-49. The *Agreement* entered into force on 4 April 2003.

10   *Agreement for the Implementation of the Provisions of the United Nations Convention on the Law of the Sea of 10 December 1982 Relating to the Conservation and Management of Straddling Fish Stocks and Highly Migratory Fish Stocks, 1995.* (Food and Agricultural Organization of the United Nations, Division

continues to increase dramatically. Such activity is essentially "irresponsible", as it fails the acceptable standards of most international measures aimed at improving ocean governance and at ensuring sustainable management of living resources contained therein.

The Food and Agricultural Organization of the United Nations (FAO)[15] has emphasised that irresponsible harvesting directly undermines effective management of marine fisheries. It impedes efforts to ensure stock sustainability and is "unfair", carrying, as it does, a heightened risk for lost economic and social opportunities. The potential for such losses has serious implications, in both the long- and short-term, since it increases the risk of diminishing future food security.

Consequently, the recent proliferation of pernicious and potentially environmentally damaging fishing practices globally, particularly on the high seas, has come to preoccupy many regional fishery management organisations (RFMOs). This concern has prompted the development of new terminology to describe fishing activities carried out in such a way as to circumvent regulatory controls. Having applied the term in the early 1990s, in 1997 the Commission for the Conservation of Antarctic Marine Living Resources (CCAMLR)[16] became the first RFMO to formally designate these activities as "Illegal, Unreported and Unregulated" (IUU) fishing.[17]

Soon thereafter, the FAO Committee on Fisheries (COFI) took up the matter[18] in 1999. COFI initiated a process to formally define the terminology (Box 7.1) and to combat the problem through an International Plan of Action to Prevent, Deter and Eliminate Illegal, Unreported and Unregulated

for Ocean Affairs and the Law of the Sea, United Nations, New York, 1998), p. 7-40. The *Agreement* entered into force on 11 December 2001.

[11]   *Code of Conduct for Responsible Fisheries, 1995.* (Food and Agricultural Organization of the United Nations, Division for Ocean Affairs and the Law of the Sea, United Nations, New York, 1998), p. 56-78.

[12]   *Cancun Declaration on Responsible Fishing, 1992.*
http://www.oceanlaw.net/txts/summaries/cancun/htm.

[13]   See *Rio Declaration on Environment and Development, 1992, op.cit,* n. 5.

[14]   "Protection of the oceans, all kinds of seas, including enclosed and semi-enclosed seas, and coastal areas and their protection, rational use and development of their living resources", in *Report of the United Nations Conference on Environment and Development, Chapter 17,* (United Nations, New York, A/CONF. 151/26 Vol. II, 1992).

[1515] FAO, *Implementation of the International Plan of Action to Prevent, Deter and Eliminate Illegal, Unreported and Unregulated Fishing. FAO Technical Guidelines for Responsible Fisheries No. 9,* (Food and Agriculture Organization of the United Nations, Rome, 2002), 122 pp.

[16]   The Commission established under Article VII of the *Convention for the Conservation of Antarctic Marine Living Resources, 1980 (CAMLR Convention).* p. 7 of the *Basic Documents,* (CCAMLR, Hobart Australia, 2002), 129 pp. Some *Contracting Parties* (often termed "*Acceding States*") are not *Commission Members* as they do not qualify for such under the conditions outlined in Article VII. These States do not take part in the *Commission's* decision-making under Article XII.

[17]   Letter from the Executive Secretary of *CCAMLR* to *FAO* [Ref. 4.2.1.(l)] as cited by G. Lutgen, *A review of measures taken by Regional Marine Fishery Bodies to address contemporary fishery issues,* Footnote 135 on p. 35, *FAO Fisheries Circular No. 940,* (Food and Agriculture Organization of the United Nations, Rome, 1999), 97 pp.

[18]   COFI, *Report of the Twenty-Third Meeting of the Committee on Fisheries. FAO Fisheries Report No. 595,* (Food and Agriculture Organization of the United Nations, Rome, 1999), 70 pp.

Fishing (IPOA-IUU).[19] The attached Implementation Plan[20] provided various practical suggestions on actions aimed at ensuring the IPOA-IUU's overall success. Nevertheless, and notwithstanding the definitions in Box 7.1, some unregulated fishing may still occur without violating international law and/or may not require application of measures envisaged under the IPOA-IUU. This fishing would be apart from that addressed by the final provision in Box 7.1.

Like many regional bodies responsible for fisheries management (amongst other responsibilities[21]), CCAMLR has been particularly affected by IUU fishing for Patagonian Toothfish (Dissostichus eleginoides) since the mid-1990s. In this paper, I use CCAMLR's experiences to illustrate some of the organisation's successes, and failures, in combating IUU Toothfish fishing. A brief history of the Toothfish IUU problem is provided. Some of CCAMLR's measures to combat the problem are documented, as are the organisation's efforts to develop, and ascribe to, international "best practice". Possible future action(s) are suggested.

The CCAMLR Convention

The boundaries of the CAMLR Convention Area (Figure 7.1) are confined within the Antarctic Polar Front[22] (APF) to the north and the Antarctic continental margin to the south (*i.e.* a major part of the "Southern Ocean"). Assignation of the APF as the Convention's northern boundary confines CCAMLR's area of responsibility within a hydrographic domain on which the underlying biogeography of the many marine species confined therein depends. For instance, the presence of deep-ocean basins south of the APF induces a high degree of species endemism, particularly for fish that inhabit the shallower Antarctic Continental shelf or areas close to the many oceanic islands that are a common feature of the Southern Ocean.[23] As highlighted by Fischer and Hureau,[24] endemism is comparably less for species inhabiting deeper water, although they still may be encountered in areas of high hydrographic variability such as immediately north and south of the APF.

With its entry into force on 7 April 1982, the CAMLR Convention was, and remains, one of the first, and only, regional marine agreements to explicitly balance conservation with rational (*i.e.* "sustainable") use. This is achieved through the implementation of a precautionary and holistic

---

[19]   FAO, *International Plan of Action to Prevent, Deter and Eliminate Illegal, Unreported and Unregulated Fishing,* (Food and Agriculture Organization of the United Nations, Rome, 2001), 24 pp.

[20]   FAO, *op cit.*, n. 15.

[21]   Currently, there is considerable debate concerning *CCAMLR's* exact mandate and role. This is attributable to the fact that Article II of the *Convention* requires *CCAMLR* to manage *both* harvested species and the Antarctic marine ecosystem as a whole. Nevertheless, *CCAMLR's* fishery regulation functions do not differ from those of many other marine fishery bodies with competency to manage fishing in the areas for which they are responsible. Therefore, for the purposes of this paper *CCAMLR* will be regarded as a *RFMO*.

[22]   The Antarctic Polar Front (*APF*) is the zone where colder, less saline waters flowing north from the Antarctic meet warmer, more saline waters flowing south in the Atlantic, Indian and Pacific Oceans. The term has effectively replaced that previously in common use - "the Antarctic Convergence". The latter term was used during negotiation of the *CAMLR Convention* and is referred to in Article I of the *Convention* (*op. cit.* n. 16). The mean position of the *APF* is between 45 and 60°S depending on longitude.

[23]   K.-H. Kock, "*Antarctic Fish and Fisheries*". (Cambridge University Press, Cambridge, 1992), 359 pp.

[24]   W. Fischer and J.-C. Hureau (eds.), "*FAO Species Identification Sheets for Fishery Purposes, Southern Ocean (CCAMLR Convention Fishing Areas 48, 58 and 88), Vol. II*", (Food and Agriculture Organization of the United Nations, Rome, 1985), 232 pp.

approach based on managing exploitation from an ecosystem[25] perspective (Box 7.2). In jurisdictional terms, CCAMLR has had to account for mixed sovereignty, and jurisdictional, imperatives,[26] to ensure that regulation, monitoring, reporting and enforcement of fishing regulatory measures are coherent within the whole Convention Area. The Area itself comprises the high seas as well as areas under some form of national jurisdiction. South of 60°S, application of the Convention is subject to the sovereignty considerations of the Antarctic Treaty.[27]

Article IX of the CAMLR Convention outlines CCAMLR's functions.[28] Paragraph 1 empowers the Commission to collect data, facilitate research and develop measures necessary to ensure effective management of Antarctic marine living resources and the attached ecosystem. Such activities include the need to establish scientific procedures to estimate the yield of harvested stock(s). Article IX, paragraph 2 comprehensively lists management ("conservation") measures that could be applied. These include, *inter alia*, the setting of catch limits, designation of fishing areas and season, designation of protected species and various other input/output controls (*e.g.* effort limits, size limits etc.).

CCAMLR builds on the provisions of Article IX, paragraph 1 through the activities of its Scientific Committee[29] and associated specialist groups. It has instituted model-based procedures to estimate the sustainable yield of harvested stocks along with associated catch limits. The procedures themselves attempt to account for the life history characteristics, as well as the age/size distribution, of the species being harvested so as to provide realistic projections of stock status. They also attempt to allow for uncertainty in either the input data or estimation procedures.[30]

Like LOSC Article 63, the CCAMLR Convention also applies to the management of so-called "transboundary stocks".[31] Patagonian Toothfish is perceived as such a stock since it is distributed throughout, and occurs within, most of the waters falling under national jurisdiction inside the Convention Area. The species also occurs to the north of the APF on the high seas and in the maritime zones of a number of coastal states adjacent to the area, particularly around the southern tip of South America. CCAMLR has been long aware of the difficulties associated with managing transboundary stocks. In 1993, a Resolution was adopted to address management of stocks occurring both within and

---

[25] For example see E.J. Molenaar, "CCAMLR and Southern Ocean Fisheries" (2001), *International Journal of Marine and Coastal Law* 16.(3): 465-499.

[26] See discussion in C. Joyner, "Maritime zones in the Southern Ocean: Problems concerning correspondence of natural and legal regimes" (1990), *Applied Geography* 10: 307-325, and the *Chairman's Statement* attached to the *CAMLR Convention* (CCAMLR *op. cit.*, n. 16, p. 23-24).

[27] See Article III of the *CAMLR Convention* in CCAMLR, *op. cit.*, n. 16, p. 5.

[28] See Article IX of the *CAMLR Convention* in CCAMLR, *op. cit.*, n. 16, p. 8-10.

[29] Articles XIV and XV of the *CAMLR Convention* respectively establish a *Scientific Committee* to advise the *Commission* and outline the kinds of activities which the *Committee* will conduct at the direction of the *Commission* pursuant to the *Convention's* objectives (CCAMLR, *op. cit.*, n. 16, p. 12-14).

[30] See A.J. Constable, W.K. de la Mare, D.J. Agnew, I. Everson and D.G.M Miller, "Managing fisheries to conserve the Antarctic marine ecosystem: Practical implementation of the Convention on the Conservation of Antarctic Marine Living Resources (CCAMLR)" (2000), *ICES Journal of Marine Science* 57: 778-791.

[31] The *FAO Fisheries Glossary* defines "transboundary stocks" as those "stocks of fish that migrate across international borders", (Food and Agriculture Organization of the United Nations, Rome, 2002). From the *FAO* Website: http://www.fao.org/fi/glossary/default.asp; LOSC, *op. cit.*, n. 4.

outside the Convention Area.[32] Interestingly, this Resolution foreshadowed many similar UNFSA provisions.

The Patagonian toothfish fishery

Exploratory fishing for Patagonian toothfish began north of the APF in about 1955.[33] The development of deepwater longlining in the early 1980s allowed a commercial fishery for the species to develop in Chilean waters, where annual catches between 5 000 and 10 000 tonnes have been taken since about 1985.[34] During the same period, and until the early 1990s, toothfish catches were trawled in Argentine and Falkland (Malvinas) Island waters. Thereafter, both trawling and longlining were employed.[35]

In both the CCAMLR Area and closely adjacent waters, toothfish have been the target of a trawl fishery around the French Kerguelen Islands since the mid-1980s.[36] The species has also been taken as a by-catch around South Georgia since the late 1970s.[37] However, it was not until the Soviet Union developed a longline fishery in the South Georgia region in 1988/89, followed by Chile in 1991/92, that large-scale commercial harvesting of toothfish in CCAMLR waters developed. The fishery expanded in 1996/97 with nationally sanctioned fisheries in the South African Exclusive Economic Zone (EEZ) at the Prince Edward Islands and in the Australian Fishing Zone (FZ) around Heard and McDonald Islands. Toothfish catches at various locations within the Convention Area are illustrated in Figure 7.2.

History of IUU fishing for toothfish

The emergence and development of IUU fishing for toothfish has been well documented[38] for the Southern Ocean in general, and for the CCAMLR Area in particular (Figure 7.3). Consequently, I

---

[32]   *CCAMLR Resolution 10/XII* (adopted in 1993) addresses "Harvesting of Stocks Occurring both within and outside the Convention Area". p. 121 of *CCAMLR Schedule of Conservation Measures in Force 2000/04*, (*CCAMLR*, Hobart Australia, 2003), 156 pp. The *Resolution* "reaffirms that *Members* should ensure that their flag vessels conduct harvesting of any stock or stocks of associated species to which the *Convention* applies in areas adjacent to the *Convention Area* responsibly and with due respect for *Conservation Measures* adopted under the *Convention*". It also pre-dated more detailed *UNFSA* provisions (especially Article 19) (*op. cit.* n. 8 and 9).

[33]   D.J. Agnew, "The illegal and unregulated fishery for toothfish in the Southern Ocean, and the CCAMLR catch documentation scheme" (2000), *Marine Policy* 24: 361-374.

[34]   Table 1 in D.J. Agnew, *op. cit.* n. 33.

[35]   From "*FIFD, Fishery Department Fishery Statistics, Vol. 3 (1989-1998).* (Falkland Islands Government, Stanley, Falkland Islands, 1999) and "*Report of the Workshop on Methods for the Assessment of Dissostichus eleginoides*", SC-CAMLR-XIV, (*CCAMLR*, Hobart, Australia, 1995), Annex. 5, Appendix E: 387-417.

[36]   G. Duhamel, "Biologie et exploitation de *Dissostichus eleginoides* autour des Iles Kerguelen (Division 58.5.1)", *CCAMLR Selected Scientific Papers, Vol. SC-CAMLR-SSP/8*, (*CCAMLR*, Hobart, Australia, 1991), p. 85-106.

[37]   *CCAMLR Statistical Bulletin, Vols 1 and 2 (1970-1979 and 1980-1989)*, (*CCAMLR*, Hobart, Australia, 1990).

[38]   Various publications deal with toothfish *IUU*. Reference is limited to: D.J. Agnew, *op. cit.* n. 33; K. Dodds, "Geopolitics, Patagonian Toothfish and living resource regulation in the Southern Ocean" (2000), *Third World Quarterly* 21.(2): 229-246.; J.A. Green and D.J. Agnew, "Catch Documentation Schemes to combat Illegal, Unreported and Unregulated fishing: CCAMLR's experience with the

have only provided a brief summary here with focus being given to CCAMLR estimates of IUU catch levels.

Prior to 1996, CCAMLR used sightings of unlicensed fishing vessels in the Convention Area to determine IUU activities and attendant catch levels. However, with the expansion of legitimate fishing activities alluded to above, along with the simultaneous expansion of the IUU fleet, CCAMLR developed a standard methodology to assess IUU catches based on a variety of information (Box 7.3). Essentially, and as explained by Sabourenkov and Miller,[39] CCAMLR calculates the IUU catch per vessel as a function of daily catch rate for the days fished per fishing voyage summed over the number of voyages per year. The calculation uses catch rate information from the geographically closest legitimate fishery. The total IUU catch per year is then summed over all the vessels identified.

Toothfish IUU catch estimates are reviewed annually by the CCAMLR Working Group on Fish Stock Assessment (WG-FSA) to estimate total removals for stock assessment purposes. Account is taken of any new information on IUU fishing derived from both catch and trade data. This information usually comes from the CCAMLR Toothfish Catch Documentation Scheme (CDS)[40] (see following section, CCAMLR's management of toothfish IUU fishing). Figure 7.4 illustrates CCAMLR's estimates of annual IUU catch compared with legitimate catches during the period 1996/97 to 2002/03. The estimated value of these catches is illustrated in Figure 7.5. It can be seen that cumulative financial losses arising from IUU fishing (USD 518 million) in the Convention Area are likely to be substantive, and at least in the order of benefits enjoyed by legitimate operators (USD 486 million).

Nevertheless, many recent publications on IUU fishing in the Convention Area have emphasised the high levels of uncertainty attached to such estimates when these are compared with regulated catch levels. The situation is complicated by the fact that CCAMLR estimates have undergone many revisions in light of new information at hand. As Sabourenkov and Miller[41] indicate, estimates derived from trade statistics are often noticeably higher than direct CCAMLR estimates[42] using the procedures outlined in Box 7.3. This is probably attributable to "double accounting" where reported trade levels for some countries may include both fish imported for processing and exported quantities of processed product(s). Further bias may arise from transhipments in port areas being recorded as imports or

---

Southern Ocean Toothfish", (2002), *Ocean Yearbook* 16: 171-194.; G.P. Kirkwood and D.J. Agnew, "Deterring IUU Fishing" in A.I.L Payne, C.M. O'Brien and S.I. Rogers (eds.), *Management of Shared Fish Stocks.* (Blackwell, Oxford, 2004): 1-22; G. Lutgen, "The Rise and Fall of the Patagonian Toothfish - Food for Thought" (1997), *Environmental Policy and Law* 27 (5): 401-407, and E.N. Sabourenkov and D.G.M. Miller, "The Management of Transboundary Stocks of Toothfish, *Dissostichus* spp., under the Convention on the Conservation of Antarctic Marine Living Resources" in A.I.L Payne, C.M. O'Brien and S.I. Rogers (eds.), *Management of Shared Fish Stocks.* (Blackwell, Oxford, 2004): 68-94.

[39]   Derived from E.N. Sabourenkov and D.G.M Miller, *op. cit.* n. 38.

[40]   Table 2 in SC-CAMLR, "*Report of the Twenty-First Meeting of the Scientific Committee for the Conservation of Antarctic Marine Living Resources - SC-CAMLR XXI*". (CCAMLR, Hobart, Australia, 2002), 524 pp. It should also be noted that the statistics compiled by *CCAMLR* on IUU Toothfish catches pool catches of both toothfish species (*Dissostichus eleginoides* and *D. mawsoni*) found in the Convention Area, especially when these are compiled from *CDS* information [see Section 5.(b)].

[41]   E.N. Sabourenkov and D.G.M Miller, *op. cit.* n. 38.

[42]   M. Lack and G. Sant, "Patagonian Toothfish: Are conservation and trade measures working?" (2001), *TRAFFIC Bulletin,* 19(1): 18 pp; E.N. Sabourenkov and D.G.M Miller, *op. cit.* n. 38.

exports. Finally, there may be misclassification of other fish species (*i.e.* bass or sea bass) that resemble toothfish or carry similar trade classifications.

The catch figures derived *via* the above procedures are likely to be incomplete as they are heavily dependent on the assumptions underlying the supporting analyses. Consequently, CCAMLR has recognised that estimates of IUU-caught toothfish in the Convention Area are both coarse and probably only represent a crude limit approximation on the potential extent of such catches.[43]

Compared with initial levels, there has been a noticeable decrease in the overall estimated IUU toothfish catch over the past four seasons (Figure 7.4). Although the underlying reasons for this trend are not entirely clear,[44] there is some suggestion that any decrease in the level of IUU catch could be attributed to CCAMLR's introduction of measures to better identify fishing location(s) and to monitor toothfish trade (see below). Thus the combined effects of CCAMLR measures with those of individual states, particularly coastal states, may have worked in concert to deter IUU fishing through increasing costs attached to "doing business" in the face of more effective enforcement action and/or improved intelligence on IUU operations as a whole.[45] In particular, the latter has allowed CCAMLR and its Members to focus better, and more directly, on the most persistent IUU vessels, their flags and their beneficial owners.

Based on CCAMLR's experience, the task of effectively bringing IUU fishing in one area under control has been complicated greatly by the fishery's ability to relocate elsewhere. Translocation is often accompanied by a change of flag, vessel name and/or ownership. The potential for obfuscation is compounded by the eastward progression of IUU fishing from the Atlantic Ocean sector of the Convention Area (CCAMLR Statistical Area 48) into the Indian Ocean (Area 58) since 1996/97. The fishery moved initially from the South African Prince Edward Islands to the French Crozets and Kerguelen Islands, and finally to the Australian Heard and McDonald Islands[46] (Figure 7.3). A similar trend is evident from CCAMLR area estimates of IUU catch over the past six seasons (Figure 7.6). Since about 2000, the IUU fishery has probably penetrated into the higher latitudes of the Indian Ocean, most notably around Ob and Lena Banks (see Figure 7.3), and possibly farther south into Prydz Bay.

CCAMLR's management of toothfish IUU fishing

General

CCAMLR has long endorsed the notion that IUU fishing compromises sustainability of toothfish stocks in the Convention Area. In turn, this seriously undermines the effectiveness of the organisation's management measures.[47] There is deep concern that continued high levels of IUU fishing would also compromise CCAMLR's long-standing objective to reduce incidental seabird by-catch during longlining operations [Section 5(b)]. In CCAMLR's view, the catching of seabirds by

---

[43]   SC-CAMLR, "*Report of the Eighteenth Meeting of the Scientific Committee for the Conservation of Antarctic Marine Living Resources - SC-CAMLR XVIII*", (CCAMLR, Hobart, Australia, 1999), p. 1-107.

[44]   D.J. Agnew, *op. cit.* n. 33; E.N. Sabourenkov and D.G.M Miller, *op. cit.* n. 38.

[45]   G.P. Kirkwood and D.J. Agnew, *op cit.* n. 38; E.N. Sabourenkov and D.G.M Miller, *op. cit.* n. 38.

[46]   D.J. Agnew, *op. cit.* n. 33.

[47]   CCAMLR, "*Report of the Sixteenth Meeting of the Commission for the Conservation of Antarctic Marine Living Resources - CCAMLR XVI*", (CCAMLR, Hobart, Australia, 1997), p. 8-12 and 24-28.

IUU longliners exerts an unacceptable and negative effect on many threatened seabird species of conservation concern.[48]

Let us now look at the tools that CCAMLR has in its armoury, or has employed, to combat toothfish IUU fishing in the Convention Area as a whole.

System of Inspection

CCAMLR's progressive development of fishery control measures provided for the collection of standard fisheries data as well as information on fish biology, ecology, demography and productivity. Such information is crucial to monitoring fishing activity and in assessing the status of various stocks.

In 1989, CCAMLR implemented a system of inspection to formalise procedures for the at-sea inspection of Contracting Party vessels fishing in the Convention Area by designated inspectors from CCAMLR Member States. Details of the CCAMLR System of Inspection are provided in the CCAMLR Basic Documents.[49] The System is nationally operated with inspectors being appointed by national authorities that in turn report *via* the Member State concerned to CCAMLR. Inspections may be carried out from vessels of the designating Member, or from on board vessels being inspected.[50] Arrangements for scheduling inspections are a matter between the Flag and Designating State.[51] However, inspectors are permitted to board fishing, or fisheries research, vessels in the Convention Area at will on the proviso that such vessels are flagged to CCAMLR Contracting Parties.[52] The System also provides for reporting sightings of Non Contracting Party (NCP) flagged vessels fishing in the CCAMLR Area. While the total number of at-sea inspections undertaken annually in the CCAMLR Area is relatively small, inspection efforts have tended to concentrate on areas of most intensive fishing activity. The outcomes of such inspections have been comprehensively summarised elsewhere.[53]

Scheme of International Scientific Observation

In 1992, the CCAMLR Scheme of International Scientific Observation augmented the System of Inspection.[54] Under this Scheme, observers are taken aboard vessels engaged in fisheries research or

---

[48]   K.-H. Kock, "The direct influence of fishing and fishery-related activities on non-target species in the Southern Ocean with particular emphasis on longline fishing and its impacts on albatrosses and petrels - A review", (2001), *Reviews in Fish Biology and Fisheries*, 11: 31-56; CCAMLR, "*Report of the Twentieth Meeting of the Commission for the Conservation of Antarctic Marine Living Resources - CCAMLR XX.*", (CCAMLR, Hobart, Australia, 2001), p. 1-69.

[49]   CCAMLR, "System of Inspection", *op. cit.* n. 16, p. 105-112; *CCAMLR Basic Documents*, CCAMLR, *op. cit.* n. 16.

[50]   Article III of the *CCAMLR* System of Inspection, *op. cit.* n. 49.

[51]   Article III (c) of the *CCAMLR* System of Inspection, *op. cit.* n. 49.

[52]   CCAMLR, "*Report of the Fourteenth Meeting of the Commission for the Conservation of Antarctic Marine Living Resources - CCAMLR XIV*", (CCAMLR, Hobart, Australia, 1995), paragraph 7.25, p. 25. This particular paragraph should be read in conjunction with paragraph 7.26 which provides for the addition of a new Article (Article IX) to be added to the System of Inspection to provide a definition of activities assumed to comprise scienific research on, or harvesting of, marine living resources in the *Convention Area*.

[53]   D.J. Agnew, *op. cit.* n. 33; E.N. Sabourenkov and D.G.M Miller, *op. cit.* n. 38.

[54]   See p. 115-119 ("CCAMLR Scheme of International Scientific Observation"), CCAMLR *op. cit.* n. 16.

commercial fishing in the Convention Area. This is arranged bilaterally between the Designating Member (i.e. the Member wishing to place an observer aboard a vessel) and the Receiving Member (i.e. the Flag State of the vessel concerned).[55] The observer's primary task is to collect essential scientific data and to promote the Convention's objectives. To ensure scientific impartiality, observers designated under the Scheme are confined to the nationals of a CCAMLR Member other than the Flag State of the vessel on which the observer serves. A recent requirement has directed observers to provide factual data on sightings of activities by vessels other than those on which they are deployed.[56] Application of the CCAMLR Observation Scheme is mandated for all CCAMLR-sanctioned toothfish fisheries, particularly in areas outside national jurisdiction.

Management ("Conservation") measures

As indicated, the initial increase in IUU fishing for toothfish in the Convention Area coincided with the expansion of legitimate fishing activity sanctioned either by CCAMLR or by coastal states in the Indian Ocean. The level of IUU fishing was unprecedented, with more than 40 IUU fishing vessels being sighted within the South African EEZ at the Prince Edward Islands[57] alone during the 1997/98 season. Since then, CCAMLR has been constantly developing and revising its management ("conservation") measures[58] in an effort to eliminate IUU fishing (Box 7.4). Briefly, these measures promote co-operation between CCAMLR Contracting Parties to improve compliance, implement at-sea inspections of Contracting Party vessels, ensure marking of all vessels and fishing gear, and introduce satellite-based vessel monitoring systems (VMS) to verify catch location. Additional measures address mandatory Port State inspections by Contracting Parties of their vessels licensed to fish in the Convention Area and further aim to develop ties with NCPs involved in toothfish fishing or trade. As already highlighted, scientific observers have been tasked with collecting and reporting factual information on fishing vessel sightings. Most recently, CCAMLR has established a vessel database to facilitate information exchange between Members on vessels known to have fished in contravention of the organisation's Conservation Measures These Measures (Conservation Measures 10-06, 10-07 and Resolution 19/XXI) respectively set in place procedures to list Contracting Party and NCP vessels that have engaged in IUU fishing in the CCAMLR Area as well to take measures against vessels flying the flags of states deemed not to be complying with such measures.[59]

---

[55]   Section B of the *CCAMLR* Scheme of International Scientific Observation, *op. cit.* n. 54.

[56]   CCAMLR, "*Report of the Seventeenth Meeting of the Commission for the Conservation of Antarctic Marine Living Resources - CCAMLR XVII*". (CCAMLR, Hobart, Australia, 1998), p. 12-22.

[57]   D.J. Agnew, *op. cit.* n. 33.

[58]   CCAMLR ("*Schedule of Conservation Measures in Force, 2003/04*"), *op. cit.* n. 32. *Conservation Measures* are binding on all *Commission Members* (*op. cit.* n. 28). While one body of opinion does not accept that *Conservation Measures* are binding on all *CCAMLR Contracting Parties*, *Convention* Article XXI (1) mandates *each Contracting Party* to take appropriate measures within its competence to ensure compliance with the *Convention's* provisions and with *Conservation Measures* adopted by the Commission to which the *Party* is bound under Articles IX. In contrast to *Conservation Measures*, *CCAMLR Resolutions* are not legally binding. The *Schedule* may be found on the CCAMLR Website: http://www.ccamlr.org/pu/e/pubs/cm.drt.htm.

[59]   See also E.N. Sabourenkov and D.G.M Miller, *op. cit.* n. 38.; CCAMLR, *op. cit.* n. 32 and 58. It should be noted that, unlike the numbering system for *CCAMLR Conservation Measures* that for *Resolutions* was *not* changed in 2002.

CCAMLR Toothfish Catch Documentation Scheme (CDS)

Toothfish IUU fishing not only undermines CCAMLR's Conservation Measures, it also violates the principles of UNFSA Articles addressing Flag State duties (Article 18), the obligations of Non-Members, or Non-Participants, in regional fisheries arrangements (Article 17) and LOSC Articles 116-119. Given its relatively high economic value, the demand for toothfish continues to attract significant prices internationally. As fishable stocks occur both within and outside the CCAMLR Area, IUU-caught fish in the Area have been difficult to trace through the trade cycle. This has resulted in a level of undetermined and non-restricted access to international markets by IUU fishing operators.[60]

In 1998, CCAMLR began developing trade-based measures to monitor landings, and the access to international markets, of toothfish caught in the Convention Area by its Members, as well as in waters under their jurisdiction.[61] At the time, other international initiatives to trace trade in specific fish species had been negotiated, or were being refined. The most prominent of these was the Bluefin Tuna Statistical Document (BTSD) introduced by the International Commission for the Conservation of Atlantic Tuna (ICCAT) in 1992.[62] The BTSD monitors trade in fresh and frozen tuna. A subsequent measure requires that ICCAT Members deny landings in their ports of tuna caught outside ICCAT measures or in the absence of a BTSD.

In contrast to ICCAT-type systems, CCAMLR toothfish trade-related measures introduce a number of new and important elements. Agnew[63] has considered CCAMLR's development of the CDS in some detail. He, and others,[64] stress that the design, adoption and implementation of the Scheme by far constitutes CCAMLR's most significant attempt to combat IUU fishing in the Convention Area.

While a number of unique principles underpin the CDS (Box 7.5), it must be stressed that the CDS was never seen as a stand-alone measure but rather as an integral component in a suite of CCAMLR measures to combat IUU fishing. Thus, its two main objectives are best summarised as:

- To track global landings of, and trade in, toothfish caught both within and outside the Convention Area, and

- To restrict access to international markets for toothfish from IUU fishing in the Convention Area.

As a CCAMLR Conservation Measure, the CDS tracks toothfish landings and requires both identification and verification of catch origin. This enables CCAMLR, through either landing or transhipment records, to identify the origin of toothfish entering the markets of all CDS Parties. It also facilitates determination of whether toothfish in the Convention Area have been caught in a manner consistent with CCAMLR Conservation Measures.

---

60   D.J. Agnew, *op. cit.* n. 33; E.N. Sabourenkov and D.G.M Miller, *op. cit.* n. 38.

61   D.J. Agnew, *op. cit.* n. 33; J. A. Green and D.J. Agnew, *op. cit.* n. 38.

62   ICCAT, "Recommendations Adopted by the Commission at its Eighth Meeting -Report for Biennial Period, 1992-1993, Part 1", (ICCAT, Madrid, Spain, 1993). Resolutions 92-1 and 92-3.

63   D.J. Agnew, *op. cit.* n. 33.

64   E.N. Sabourenkov and D.G.M Miller, *op. cit.* n. 38.

With the CDS' entry into force on 7 May 2000, CCAMLR was able to implement a comparatively robust mechanism to collect toothfish data from areas both within, and adjacent to, the Convention Area. Such data are vital for estimating "total" toothfish removals; a key input parameter to improve stock assessment and provide clearer insights into global catch levels and associated market forces.[65]

Other considerations

The various measures outlined in Section 5(a)(iii) are fully consistent with the provisions of LOSC[66] Articles 116 to 119, UNFSA[67] Articles 21 to 23 and Articles III to VIII of the Compliance Agreement.[68] In reaction to UNFSA Articles 8 (particularly paragraphs 3 and 4) and 17, CCAMLR encourages its Members to accept and promote the entry into force of UNFSA[69] as well as the Compliance Agreement. Acceptance of the FAO Code of Conduct[70] has also been encouraged. Furthermore, CCAMLR has frequently acknowledged that both the UNFSA's and the Compliance Agreement's recent entries into force are likely to contribute significantly to the reduction, and ultimately elimination, of IUU fishing in the Convention Area.[71]

Many CCAMLR Members actively contribute to the FAO's work in implementing the above agreements. Most notably, both CCAMLR and its Members promoted development of the 1999 FAO International Plan of Action for Reducing Incidental Catch of Seabirds in Longline Fisheries[72] and the IPOA-IUU.[73] CCAMLR participates as an institutional observer at the biennial meetings of COFI and its attached sub-committees.

Institutionally, CCAMLR also co-operates with various other regional fisheries organisations, especially those managing fisheries in waters adjacent to the Convention Area (*e.g.* ICCAT, the Indian Ocean Tuna Commission [IOTC], the Commission for the Conservation of Southern Bluefin Tuna [CCSBT] and the recently formed South East Atlantic Fisheries Commission).[74] This includes, *inter alia,* the exchange of information on IUU fishing on the high seas and efforts to combat such fishing.

---

[65]   E.N. Sabourenkov and D.G.M Miller, *op. cit.* n. 38.

[66]   LOSC, *op. cit.* n. 4.

[67]   UNFSA, *op. cit.* n. 8.

[68]   Compliance Agreement, *op. cit.* n. 8.

[69]   *UNFSA* entered into force when the necessary 30 ratifications had been deposited (*op. cit.* n. 10). CCAMLR, *op.cit.* n. 47.

[70]   FAO Code of Conduct, *op. cit.* n. 8 and 11; CCAMLR, *op. cit.* n. 47.

[71]   CCAMLR, *op. cit.* n. 47.

[72]   FAO, *International Plan of Action for Reducing Incidental Catch of Seabirds in Longline Fisheries.* (Food and Agriculture Organization of the United Nations, Rome, 1999), 26 pp.

[73]   IPOA-IUU, *op. cit.* n. 19.

[74]   The annual *CCAMLR* meeting considers its co-operation with other international organisations as a standing agenda item. It also considers such co-operation under other agenda items where appropriate, including during various discussions by the *Commission's* subsidiary bodies, particularly the *Scientific Committee.*

CCAMLR and the IPOA-IUU

The IPOA-IUU's major purpose is to provide a comprehensive and integrated global approach to combat IUU fishing through prevention, deterrence and elimination.[75] In so doing, the IPOA-IUU strives to address various key principles and strategies (Box 7.6).

The various steps already, or to be, taken by CCAMLR to address IUU fishing for toothfish (previous section entitled CCAMLR's management of toothfish IUU fishing) are assessible in the context of the following statement:[76]

- Providing all CCAMLR Contracting Parties with comprehensive, effective and transparent measures to combat IUU fishing within the Convention Area and for fish stocks for which CCAMLR is responsible.

Pursuant to the IPOA-IUU's general principles shown in Box 7.6, as well as the more practical steps outlined in the IUU Implementation Plan,[77] CCAMLR has already implemented most of the Plan's necessary steps through its various Conservation Measures. From available information, it is clear that CCAMLR has developed a cohesive framework of measures to combat IUU toothfish fishing that is fully compatible with international "best practice" as identified by the IPOA-IUU.[78] To illustrate the point, it is worth working through an example.

The IPOA-IUU Implementation Plan prescribes 14 items to deal with - "Actions to Prevent, Deter and Eliminate IUU Fishing". All 14 have been directly addressed by CCAMLR. For example a suite of CCAMLR measures have focused on developing, implementing and maintaining records of vessels fishing in the Convention Area. These are clearly subject to Convention Articles XX and XXI[79] and provide specifically for the marking of vessels in the Convention Area (CCAMLR Conservation Measure 10-01), a requirement to license fishing vessels (Conservation Measure 10-02), the promotion of compliance with measures by Contracting Party vessels (Conservation Measure 10-06), the promotion of compliance with measures by NCP vessels (Conservation Measure 10-07) and the taking of measures in relation to flags of non-compliance (Resolution 19/XXI). Space does not permit the inclusion here of similar details for other CCAMLR measures consistent with the activity categories addressed by the Implementation Plan. However, based on information presented elsewhere,[80] these measures are equally comprehensive and substantive.

However, a possible CCAMLR failing has been that its measures to combat IUU fishing have evolved piecemeal and consequently have not necessarily been developed according to any plan or determined timetable. This shortcoming has been recognised by the Commission, which has recently

---

[75]   See paragraphs 8 and 9 of the *IPOA-IUU*, *op. cit.* n. 19.

[76]   D.G.M. Miller, E. Sabourenkov and N. Slicer, "Unregulated Fishing and the Toothfish Experience" in D. Vidas (ed.), *Antarctica 2000 and Beyond.* (Kluwer, In Press).

[77]   IPOA-IUU, *op. cit.* n. 15.

[78]   D.G.M. Miller, E. Sabourenkov and N. Slicer, *op. cit.* n. 76.

[79]   CCAMLR, *op. cit.* n. 16.

[80]   D.G.M. Miller, E. Sabourenkov and N. Slicer, *op. cit.* n. 76.

initiated development of an organisational IUU implementation plan[81] within the prescriptions of the IPOA-IUU Implementation Plan.[82]

Some ancillary thoughts

When illustrating CCAMLR's effectiveness in combating IUU fishing in the Convention Area, it is necessary to highlight a few additional considerations. These relate as much to the organisation's successes as they do to its shortcomings.

International

CCAMLR Article IV specifically binds its Parties to the sovereignty provisions of Antarctic Treaty Articles IV and VI. There is an added complication, however. By including all waters south of the APF, CCAMLR raises sovereignty issues that cannot be dealt with directly by the Treaty. As a result, a special statement made by the Chairman of the Conference on the Conservation of Antarctic Marine Living Resources was attached to the Convention.[83] This sets out the conditions for the Convention's application in waters adjacent to any land (*i.e.* islands) where existence of sovereignty is recognised by all CCAMLR Contracting Parties.

The above arrangement provides for coastal state enforcement within national waters inside the Convention Area in conformity with CCAMLR's needs. On balance, this has been the case[84] with most affected CCAMLR Members having endeavoured to ensure harmonisation between national and CCAMLR measures. In this context, it is interesting to note that no CCAMLR Member has ever voiced a reservation under the Chairman's Statement to significant measures aimed at combating IUU fishing.[85] These include Conservation Measures 10-04 [mandating deployment of Vessel monitoring systems (VMS), 10-05 (the CDS), 10-06 (promoting compliance by Contracting Party vessels) and 10-07 (promoting compliance by NCP vessels)].

By implication, therefore, it could be argued that the CCAMLR Members most likely to be affected by application of the Chairman's Statement view IUU fishing not only as a CCAMLR issue, but also as a priority concern for coastal states with sovereign waters in the Convention Area. With the exception of South Africa (largely for technical reasons associated with a lack of enforcement capability[86]), the depth of this concern has been visibly manifest for the CCAMLR coastal states most affected. France and Australia, in particular, have devoted considerable time, effort and money to protect their waters from IUU activities. Despite their efforts, IUU fishing has impacted toothfish

---

[81]   CCAMLR, "*Report of the Twenty-First Meeting of the Commission for the Conservation of Antarctic Marine Living Resources - CCAMLR XXI*", (CCAMLR, Hobart, Australia, 2002), paragraph 8.15, p. 32.

[82]   See particularly paragraphs 80-82 in FAO, *op. cit.* n. 15, p. 101-102.

[83]   CCAMLR, *op. cit.* n. 16.

[84]   D.G. M. Miller, "The International Framework for the Management of Fishing in the Southern Ocean". *Paper Presented at the Outlook 2004 Conference,* (ABARE, Canberra, Australia – 2-3 March 2004)

[85]   D.G. M. Miller, *op. cit.* n. 84.

[86]   A.D. Brandao, A., D.S. Butterworth, B.P. Watkins and D.G.M. Miller, "A first attempt at an assessment of the Patagonian Toothfish (*Dissostichus eleginoides*) resource in the Prince Edward Islands EEZ". (2004), *CCAMLR Science.* 9: 11-32.

stocks in the Indian Ocean,[87] most notably around the Prince Edward Islands, where the future sustainability of D. eleginoides has been seriously compromised.[88]

An ancillary consideration is the extent to which the Convention's provisions (particularly Conservation Measures) can be effectively applied on the high seas within the CCAMLR Area.[89] The situation is exacerbated by the Area's geographic extent (*ca.* 35 x 10^6 sq. km) and by the remoteness of many fishing grounds.[90] This tends to favour fishing outside CCAMLR's regulatory control, particularly by vessels flying the flags of CCAMLR NCPs.[91] While the list of specific Conservation Measures dealing with CCAMLR NCPs systematically grows, there is still a need to balance the implied regulatory provisions of such Measures with the rights of *all* States (CCAMLR Contracting and NCPs alike) to fish the high seas under LOSC Article 116.[92]

However, it needs to be recognised that when LOSC Article 116 is read in conjunction with Articles 117 to 119,[93] there is a clear obligation on *all* states to co-operate in the conservation and management of marine living resources on the high seas and to take appropriate measures to ensure that this occurs. Together with the FAO Compliance Agreement[94] and UNFSA Articles 8, 19 to 23,[95] these general provisions obligate states fishing on the high seas in the CCAMLR Convention Area to do so in cognisance of measures aimed at ensuring stock sustainability and in a manner not discharging them from co-operating with CCAMLR in the conservation and management of relevant fisheries resources.

Despite these positive associations and inferences, there is still scope to explore how effectively LOSC provisions, and especially those of UNFSA, can be aligned with CCAMLR's efforts to combat toothfish IUU fishing[96] in the Convention Area and closely adjacent areas. The development of a

---

[87]   See discussion in paragraph 5.4 of CCAMLR, "*Report of the Eighteenth Meeting of the Commission for the Conservation of Antarctic Marine Living Resources - CCAMLR XVIII*", (CCAMLR, Hobart, Australia, 1999). This states - "The Scientific Committee drew the attention of the Commission to the potential similarities between the implications for future sustainability of *Dissostichus* spp. stocks as a consequence of *IUU* fishing and the collapse of *Notothenia rossii* stocks due to overfishing in the late 1970s".

[88]   A.D. Brandao, A., D.S. Butterworth, B.P. Watkins and D.G.M. Miller, *op cit.* n. 86.

[89]   See C.C. Joyner, "The Antarctic Treaty System and the Law of the Sea: Competing regimes in the Southern Ocean" (1995), *International Journal of Marine and Coastal Law*, 10(2):301-331 and M. Levy "The enforcement of Antarctic marine living resources claims" (1997), *Duke Development Clinic/Adcock*. 155 pp.

[90]   G. P. Kirkwood and D. J. Agnew *op. cit.* n. 38

[91]   D. J. Agnew *op. cit.* n. 33

[92]   LOSC, *op. cit.* n. 4.

[93]   LOSC, *op. cit.* n. 4.

[94]   FAO, *op. cit.* n. 9.

[95]   UNFSA, *op. cit.* n. 10.

[96]   K. Dodds, *op. cit.* n. 38.

CCAMLR institutional plan to provide regional focus for the IPOA-IUU[97] is obviously a step in the right direction to address this particular problem.[98]

The toothfish saga revisited

Like whaling, finfish fishing in the Southern Ocean has been characterised by "boom and bust" cycles,[99] with successive discovery, exploitation and depletion of each new target stock taking place over progressively shorter time scales. In this context, we have seen that the cumulative value (Figure 7.5) of the IUU fishery for toothfish in the CCAMLR Area over the past eight years is close to that for the legitimate fishery. Figure 7.5 also illustrates that the profits enjoyed by IUU operators were nearly twice those of the legitimate fishery until about 1998/99, when a drop-off in IUU catches is observable. While considerable uncertainties are associated with estimating early IUU-catch levels,[100] Kirkwood and Agnew[101] suggest that a decline in IUU operations in 1998/99 may have occurred as a result of the CDS negotiations nearing finality. It is therefore difficult to say whether the observed reduction in IUU activities resulted from operators reducing fishing or whether they made efforts to legitimise their operations. Equally, stocks may have become so depleted as to defy profitable exploitation, even for IUU operators.

In contrast to its more modest success in combating the IUU problem directly, CCAMLR has had considerable success (Figure 7.7) in reducing bird mortality associated with toothfish longlining in the Convention Area through promulgation of measures specifically aimed at minimising incidental by-catch.[102] However, the take of seabirds by the IUU fishery in the CCAMLR Area and by longline vessels fishing on the feeding grounds of particular bird species farther north still raises considerable cause for concern[103] and is likely to be unsustainable for most of the species affected[104] despite CCAMLR's efforts to the contrary.

The CDS

Initial evaluation of the CDS is encouraging.[105] Not only is the Scheme unique in its scope and application, but it also became fully operational relatively quickly (within less than two years). It has also drawn in a number of CCAMLR NCPs and its overall coverage extends to more than 90% of the global world trade in toothfish (Figure 7.8).

The advent of the CDS has led to the Scheme's Parties denying toothfish landings and/or shipments in the absence of the required documents. The absence of such documentation provides a

---

[97]   See FAO, *op. cit.* n. 15 and the preceding section.

[98]   See Paragraph 8.15 in CCAMLR, *"Report of the Twenty-First Meeting of the Commission for the Conservation of Antarctic Marine Living Resources – CCAMLR-XXI"*, (CCAMLR, Hobart, Australia, 2002), 205 pp.

[99]   K.-H. Kock, *op. cit* n. 23.

[100]   E. N. Sabourenkov and D. G. M. Miller, *op. cit.* n. 38.

[101]   G.P. Kirkwood and D.J. Agnew, *op. cit.* n. 38.

[102]   K.-H. Kock, *op. cit.* n. 48; IFF, "Second International Fishers Forum" (2002), *SPC Fisheries Newsletter*, No. 103: 32 pp.; D.G. Miller *et al. op. cit.* n. 76;

[103]   K.-H. Kock, *op. cit.* n. 48

[104]   *Conservation Measure* 25-02 in CCAMLR, *op. cit.* n. 32 and 98.

[105]   E.N. Sabourenkov and D. G. M. Miller, *op. cit.* n. 38.

rebuttable presumption that triggers enforcement action. It has also improved appreciation of toothfish global catch levels and focused on incidents of malpractice or fraud. With evidence that the introduction of the CDS has made trading in IUU-caught fish less profitable, it is notable that the Scheme also seems to restrict unfettered market access to IUU-caught products.[106] While some of the improvements to the CDS suggested by Sabourenkov and Miller[107] are likely to make it even more effective in combating IUU fishing in the Convention Area, it is still worth asking:

"What would the consequences have been in the absence of the CDS?"

Based on current levels of IUU fishing for toothfish in the CCAMLR and closely adjacent areas, the answer appears obvious – the situation would have been much worse, because the CDS has had a noticeable impact on accessibility to global markets (particularly in the United States and Japan) thereby deterring IUU operators.[108] A key illustration of such deterrence is the fact that IUU-caught fish fetch a significantly lower price ($\pm 20\%$) than fish with attached CDS accreditation.[109]

With the CDS as a significant step, CCAMLR is able to promote multilateral co-operation to combat toothfish IUU fishing. In contrast to other CCAMLR Conservation Measures that are limited to the Convention Area and to CCAMLR Members, the CDS is applicable globally. Furthermore, its implementation remains consistent with many of the provisions of UNFSA Articles 7, 8 and 17.[110] As the CDS is generally aimed at minimising any national bias,[111] there is every expectation that its effectiveness will benefit from enhanced international co-operation. In this respect, and following a 2002 proposal to list toothfish under Appendix II of the Convention on Trade of Endangered Species (CITES), the recent decisions by both CCAMLR and the Twelfth Conference of CITES Parties (COP-12)[112] to improve co-operation and the exchange of information between the two organisations is a gratifying development. It should also broaden the CDS' application globally. As highlighted by Miller et al.,[113] this should serve to reduce possible World Trade Organization (WTO) scrutiny arising from the perception that relatively few parties participate in the Scheme. Consequently, the CDS would better qualify as a "multilateral solution based on international co-operation and consensus" aimed at combating a transboundary environmental problem, or one of a global nature – a status favoured by the WTO's Committee on Trade and Environment (CTE).

---

[106]   G. P. Kirkwood and D. J. Agnew, *op. cit.* n. 38.

[107]   E.N. Sabourenkov and D. G. M. Miller, *op. cit.* n. 38.

[108]   E.N. Sabourenkov and D. G. M. Miller, *op. cit.* n. 38.

[109]   E.N. Sabourenkov and D. G. M. Miller, *op. cit.* n. 38.

[110]   UNFSA, *op. cit.* n. 10.

[111]   K. Larson, "Fishing for a compatible solution: Toothfish conservation and the World Trade Organization" (2000), *The Enivronmental Lawyer,* 7(3): 123-158.

[112]   Need for co-operation between *CCAMLR* and *CITES* was addressed in paras 10.72 to 10.75 of CCAMLR, *op. cit.* n. 98 and by *CITES COP-12* Conference Resolution 12.4 and Decisions 12.57 to 12.59 CITES, "*Report of the Twelfth Conference of Parties*", (CITES, Geneva), Website: http://www.cites.org.

[113]   D. G. M. Miller, E. N. Sabourenkov and D. Ramm, "*CCAMLR's* approach to managing Antarctic Marine Living Resources, (In Press), *Deep Sea 2003 Conference Proceedings.* (Food and Agriculture Organization of the United Nations, Rome).

Finally, it is notable that Article 30 of the Vienna Convention on the Law of Treaties[114] addresses the application of successive treaties relating to the same, or similar, subject matter. In these terms the competency of relevant international law arrangements such as LOSC, UNFSA and CCAMLR need to be carefully considered in relation to the potential, and added, involvement of such instruments as CITES in their day-to-day affairs. Every effort needs to be made to ensure that essential provisions/competencies are not undermined or overridden. This clearly implies that initiatives to harmonise the application of more than one international instrument (say in response to IUU fishing) must not violate the rights, obligations and duties of any Party under any other instrument to which it is specifically contracted.

National enforcement

Apart from the CDS, it is probably true to say that deterrence of toothfish IUU fishing in the Convention Area has been most effectively addressed by coastal state action rather than by the direct application of specific CCAMLR Conservation Measures alone.[115] There appear to be two primary reasons for this. First, the levels of punitive fines imposed (in some cases in excess of USD 1 million) for IUU fishing within sovereign waters inside the CCAMLR Area (combined with the seizure of vessels, and/or catch and increased risk of apprehension) by coastal states have undoubtedly contributed to enhancing deterrence. A clear example of this is the recent ruling by the International Tribunal for the Law of the Sea (ITLOS)[116] on Australia's prosecution of the Russian flagged Volga for fishing in its FZ around Heard and McDonald Islands[117] (Table 7.1). Second, combined with recent strong statements by the Australian government on deployment of armed patrols, enhanced co-operation between Australia, South Africa and France, and the building of specially-designated patrol vessels by both South Africa and Australia, there appears to be growing political will to combat IUU fishing in the CCAMLR Area. Such developments are clearly evident in a number of recent, and successful, prosecutions of IUU fishing vessels in the CCAMLR Area, particularly by coastal states in the Indian Ocean (Table 7.1).

The comparability or equivalence of imposed sanctions[118] is another issue closely linked to effective deterrence. This is a complex matter that depends on factors such as the equivalence of judicial, or regulatory, procedures between states as well as currency exchange conversion rates. In its broadest interpretation, Article XI of the CAMLR Convention may be seen as implying that any harmonisation of conservation measures for species occurring in both the Convention Area and in adjacent areas under national jurisdiction could also include consideration of equivalence in the imposition of sanctions. However, CCAMLR has never specifically discussed the matter and there may be some merit in pursuing a similar course of action to that outlined in Article 8.4.(b) of the Southern African Development Community (SADC) Protocol on Fisheries where SADC Parties are urged to co-operate in:

---

[114] Vienna Convention on the Law of Treaties, (1969). Website: http://www.un.org/law/ilc/texts/treaties.

[115] G. P. Kirkwood and D. J. Agnew, *op. cit.* n. 38.

[116] *ITLOS* ruled on 23 December 2002 that Australia should release the *Lena* on the posting of a bond of AUD 1 920 000. For details, see Website - http://www.itlos.org.

[117] G. P. Kirkwood and D. J. Agnew, *op. cit.* n. 38.

[118] C.C. Joyner, "Compliance and enforcement in new fisheries law" (1998), *Temple International and Comparative Law Journal* 10(2): 301-331.

"Establishing region-wide comparable levels of penalties imposed for illegal fishing by non-SADC vessels and with respect to illegal fishing by SADC vessels in the waters of other State Parties".[119]

It is not difficult to envisage the potential benefits of such an approach being applied consistently by CCAMLR Contracting Parties.

From this discussion, it should be clear that any significant reduction in (*i.e.* deterrence of) IUU fishing is the key to assessing the effectiveness of any attached enforcement action.[120] Clearly, the absence of severe penalties, combined with limited enforcement (for whatever reason) only serves to enhance the lucrative rewards of IUU fishing with profits outweighing penalties. Fishing thus becomes more cost-effective.[121] It follows, therefore, that effective enforcement action must take account of where, and by whom, IUU fishing benefits are being enjoyed. However, as highlighted by Rayfuse,[122] certain potential shortcomings inherent in flag state enforcement need to be effectively addressed as a first step, particularly the use of "flags of convenience".[123] Inadequate flag state enforcement is compounded by the apparent unwillingness, or inability, of many national authorities to focus enforcement action on individuals (*i.e.* nationals) or companies[124] that benefit from the proceeds of IUU fishing. Such considerations become even more important in the face of general reluctance to extend state jurisdiction through additional application of coastal state rights to the high seas.[125] Given that RFMOs like CCAMLR are generally recognised as being responsible for fisheries governance at a regional level, then establishing specific multilateral arrangements to boost enforcement certainly appears worthy of consideration, This would have implications not only for the daily business of RFMOs, but also for exploring the application of non-flag state enforcement powers in the event that the primacy of flag state responsibilities are not being fulfilled.

While it may be argued that references to "nationals" in the LOSC[126] are perfunctory rather than obligatory, there is growing appreciation that some control is necessary over natural and legal persons to facilitate fulfilment by states of their obligations to co-operate in taking the necessary measures for the conservation of high seas living resources. Clear evidence of this intent can be found in LOSC Articles 117-118, UNFSA Article 10.(1)[127] and in various initiatives by states to exert direct control

---

[119]   SADC Fisheries Protocol, Website: http://www.sadc.int/english/protocol.

[120]   G. P. Kirkwood and D. J. Agnew, *op. cit.* n. 38.

[121]   M. Levy, *op. cit.* n. 89.

[122]   R. Rayfuse, "Enforcement of high seas fisheries agreements: Observation and inspection under the Convention on the Conservation of Antarctic Marine Living Resources" (1998), *International Journal of Marine and Coastal Law* 13(4): 579-605.

[123]   B. Vukas and D. Vidas, "Flags of Convenience and High Seas Fishing: The Emergence of a Legal Framework", in O.S. Stokke (ed.), *Governing High Seas Fisheries: The Interplay of Regional Regimes.* (Oxford University Press, Oxford, 2001). 53-90.

[124]   D. G. M. Miller, E. N. Sabourenkov and D. Ramm, *op. cit.* n. 113.

[125]   G. P. Kirkwood and D. J. Agnew, *op. cit.* n. 38.

[126]   Various *LOSC* Articles make reference to the obligations of "nationals" to comply with, or co-operate in, the implementation of conservation measures governing marine living resource utilisation. The most prescriptive of these include Articles 62.(4) and 117. LOSC, *op. cit.* n. 3.

[127]   See *UNFSA* Article 11.(1) which states - "ensure the full cooperation of their relevant national agencies and industries on implementing the recommendations and decisions of the organization of arrangement". UNFSA, *op. cit.* n. 10.

over the activities of their nationals to enhance compliance with third party and international fisheries management measures.[128]

Bearing these considerations in mind, there is little doubt that control of "nationals" is a question worth exploring in any agenda or global effort to combat IUU fishing. Furthermore, and following 11 September 2001, globally heightened sensitivity to transnational crime provides an opportunity to address contrary behaviour by natural persons in the international arena. In these terms, the environmentally, as well as economically, damaging practice of IUU fishing is likely to be viewed as contrary behaviour, even if the generally perceived criminal intent is often seen as relatively minor compared with other criminal acts.

Discussion

General

Apart from the CAMLR Convention, other international agreements outside the Antarctic Treaty System are relevant to the ongoing, and environmentally sustainable, management of Antarctic marine living resources. The most recent and noticable of these is the 1992 Convention on Biological Diversity (CBD).[129] With its attached 1994 Jakarta Mandate, the CBD may be linked to relevant marine management institutions. However, the details of its potential interactions with CCAMLR in particular, remain unclear.

Probably more relevant, Article XIII of the recent Agreement on the Conservation of Albatrosses and Petrels[130] explicitly references the rights and obligations of its Parties under the CAMLR Convention. This clearly, and directly, links the common subject matter of the two agreements insofar that the species subject to the former are also directly of concern to the latter (particularly in terms of their incidental mortality in the toothfish longline fishery).

---

[128]   Various States have introduced regulatory provisions to ensure that their nationals comply with international conservation and management measures inside or outside national waters. Notable examples include Australia under the *Fisheries Management Act, 1991* (Act No. 162 of 1991); New Zealand subject to Part 6A of the *New Zealand Fisheries Act, 1996*; Norway in application of *Article 6 of the Regulations Relating to Fishing and Hunting Operations by Foreign Nationals in the Economic Zone of Norway, 1977*; South Africa in application of the *Marine Living Resources Act, 1998* (Act. No. 18 of 1988 - *South Government Gazette Notice No. 189630 of 27 May 1998*) and Spain under *Directive 1134/2001 of 31 October 2002*. A recent and interesting development has been the indictment by United States authorities of a number of South African citizens and joint South African-United States nationals under the United States *Lacey Act* on 21 counts for various offences, including alleged illegal harvesting of South Coast Rock Lobster and Patagonian Toothfish, in defiance of South African statutes and *CCAMLR Conservation Measures*. See *"Conspiracy to Violate the Lacey Act and to Commit Smuggling"* (2003), United States District Court, Southern District of New York. *Indictment S1 03 Crim. 308 (LAK)*: 36 pp. The principals charged in this case have recently pleaded guilty and stand to forfeit at least USD 5 million worth of assets.

[129]   CBD, "Convention on Biodiversity, 1992", Website: http://www.biodiv.org/default.aspx. The *Convention* aims to develop and implement strategies for the sustainable use, and protection, of biodiversity. The *Jakarta Mandate* specifically applies this objective to marine and coastal biodiversity. Article 22 of the *Convention* makes general reference to, and recognises, "rights and obligations" under other international agreements. The *CBD* entered into force on 29 December 1993.

[130]   "Agreement on the Conservation of Albatrosses and Petrels, 2001". The *Agreement* entered into force on 1st February 2004, Website: http://www.aad.gov.au/default.asp?casid=13504.

On a different matter, it is premature to assess the extent to which, in combination with other related CCAMLR measures, the CDS - *a)* will prove indispensable in the battle against toothfish IUU fishing, or whether *b)* it is particularly effective in managing the exploitation of transboundary stocks within, and outside, the Convention Area. For this to be so, all international toothfish trade should be limited only to fish taken legally, or in a regulated manner compatible with CCAMLR's approach. Consequently, IUU-taken fish should not enter world markets. This is something with which the CDS has had considerable circumstantial success, but which remains to be universally realised.[131] In these terms, the question arises of how effectively RFMOs like CCAMLR uphold the long-held legal precedent of 'flag state control'. As the issue presents itself, it provides motivation to consider how such control could be enhanced by utilising more widely focused and/or trade-based agreements, such as CITES, and associated measures under the WTO. Given the interesting ancillary questions proposed, and as already highlighted, the issue is unlikely to be easily, or quickly, resolved. A key consideration remains the definition of boundaries between organisational competencies in terms of designating common standards across organisations addressing similar matters but subject to different international arrangements.

Equally, CCAMLR should continue to expand the role of 'Port' and 'Market' States to discourage IUU-caught toothfish trade. Without diminishing flag state responsibilities, CCAMLR's recent efforts have brought into focus the need for NCPs to take more responsibility for discouraging the trade of toothfish caught in a manner that undermines CCAMLR Conservation Measures. The question of NCP co-operation remains at the heart of improving CCAMLR's ability to combat IUU fishing. To be effective, such co-operation needs to be fully consistent with the obligations set out in UNFSA Articles 20, 21 and 23.

Any trade-based regime like the CDS should remain dynamic, so that it can respond appropriately to changing circumstances. Thus, the CDS must undergo periodic and regular review. Consequently, every effort should be applied to the comparable tightening of associated measures to ensure the successful realisation of CCAMLR's overall objectives in, and approach to, combating IUU fishing.[132] Not only should such review be transparent, it is essential that worthwhile incentives are provided to economically-disempowered developing states, where these may perceive greater economic benefits from being linked to IUU operations, either as flags, or ports, of convenience. In particular, there is a need for future, and further, consideration of the attendant economic insecurities experienced by some developing countries (such as Kenya, Mozambique and Mauritius[133]) that have become involved in the trade of IUU-caught toothfish. Therefore, any effort to improve the application of relevant LOSC Article 140 and UNFSA Articles 24-26 should be boosted, with the particular aim of providing these countries with alternative incentives to counteract the economic benefits accrued from IUU fishing and to enhance their commitment to responsible fishing practices.

While there is little doubt that the CDS is a vital component in CCAMLR's "toolbox" of regulatory measures, it cannot be implemented and evaluated in isolation.[134] This is clearly recognised by CCAMLR through its development of a wide variety of Conservation Measures (*e.g.* Measures 10-

131   Larson, *op. cit.* n. 111; G. P. Kirkwood and D. J. Agnew, *op. cit.* n. 38.

132   M. Lack and G. Sant, *op. cit.* n. 42.

133   G. Mills, "Insecurity and the Developing World", in G. Mills (ed.), *Maritime Policy for Developing Nations.* (SAIIA, Johannesburg, 1995), pp. 12-37.

134   D. G. M. Miller, E. N. Sabourenkov and D. Ramm, *op. cit.* n. 113.

02, 10-03, 10-04, 10-06, 10-07) and Resolutions (14/XIX, 15/XXIII, 16/XIX and 17/XX) augmenting the CDS's application and efficacy.[135]

It is noticeable that, in keeping with the Antarctic Treaty's key provisions, international co-operation has been carried over to the CAMLR Convention.[136] In practice, CCAMLR has done much to advance co-operation, again in the form of the CDS and its growing involvement with various organs of the FAO. Furthermore, various CCAMLR Conservation Measures are dependent on institutionalising international co-operation at a global level[137] to combat IUU fishing in the Convention Area. Therefore, with UNFSA's recent entry into force,[138] there is every expectation that CCAMLR will benefit from enhanced international co-operation and that its capacity to meet the Convention's objectives will be improved.[139]

CCAMLR has frequently acknowledged that both UNFSA and the FAO Compliance Agreement[140] are likely to contribute significantly to the Commission's work in general and to reducing, and hopefully eliminating, IUU fishing in the Convention Area in particular.[141] Again involvement of both CCAMLR and its Members in the FAO's work is important and should be encouraged.

To summarise, and as matters now stand, in common with many other fisheries-related instruments, the effective application of the CAMLR Convention on the high seas (*i.e.* outside national territorial jurisdiction) is confounded by insufficient flag state control (UNFSA Articles 18 and 19)[142] over IUU vessels. The situation is further compounded by the deliberate use of flags of convenience to circumvent fisheries management measures.[143]

Conclusions

With the exception of the CDS, we have seen that the enforcement of CCAMLR Toothfish Conservation Measures has generally met with limited success outside areas where national jurisdiction is vigorously applied. Consequently, much still needs to be done to ensure compatibility

---

[135]   CCAMLR, *op. cit.* n. 32.

[136]   Article XXII of the *CAMLR Convention* (CCAMLR, *op. cit.* n. 16) strives to build co-operative relationships between *CCAMLR* and relevant inter-governmental and non-governmental organisations. Article XXIII specifically mandates co-operation with other elements of the *Antarctic Treaty System* and the Scientific Committee for Antarctic Research (*SCAR*). F.O. Vicuna, "Antarctic conflict and international cooperation," in *Antarctic Treaty System: An Assessment.* (Polar Research Board, National Academy of Press, Washington, 1986). 55-64.

[137]   G. Lutgen, "A review of measures taken by Regional Fishery Bodies to address contemporary issues" (1999), *FAO Fisheries Circular* 940: 97 pp; G. Lutgen, "Cooperation and regional fisheries management" (2000), *Environmental Policy and Law* 30/5: 251-257.

[138]   *UNFSA* Part III (Articles 8 to 16) (UNFSA, *op. cit.* n. 10) outlines various mechanisms for international co-operation in the management of the resources concerned. These complement similar sentiments implicit in *LOSC* Articles 61, 63, 64 and 117-119 (LOSC *op. cit.* n. 4).

[139]   K. Dodds, *op. cit.* n. 38.

[140]   FAO, *op. cit.* n. 9.

[141]   For example see Paragraphs 5.11 and 5.32 in CCAMLR *op. cit.* n. 47.

[142]   R. Rayfuse, *op. cit.* n. 122.

[143]   B. Vukas and D. Vidas, *op. cit.* n. 123.

between various relevant legal instruments in order to provide for more effective management of Antarctic marine living resources in the broadest sense. Obvious topics for consideration include:

- Improving the enforcement of regulatory measures to protect the environment in which Antarctic marine living resources are found (*i.e.* facilitate effective implementation of the CAMLR Convention Article II elements in particular).

- Developing legal mechanisms to ensure compatibility between national and international instruments applicable to Antarctic marine living resources issues (*e.g.* sovereignty/jurisdictional disputes must be resolved to minimise potential political, legal and administrative conflicts). Steps should also be taken to harmonise the application of regulatory measures in areas under national jurisdiction and on the high seas.

- Improving co-operation on issues related to enforcement and the sanction of perceived transgressions. This implies a need to reinforce international co-operation and information exchange to promote responsible fishing activity. Instruments such as the FAO Code of Conduct and the Compliance Agreement[144] go some way to formalising these responsibilities. Implementation of the FAO IPOA-IUU should be particularly encouraged, and it appears worthwhile exploring how the provisions of instruments like CITES and the CBD may be used to augment current CCAMLR management initiatives such as the CDS.

- Giving additional, and serious, consideration to the role of NCPs in RFMO arrangements. In this regard the CDS is an especially welcome initiative, as is the UNFSA's entry into force (especially the provisions of Article 17 which do not discharge non-RFMO participants from their obligations to co-operate in the conservation and management of relevant straddling fish stocks and highly migratory fish stocks).

- Elaborating operational definitions, and practical application, of certain key LOSC provisions. Particular attention should be given to further developing co-operative management and conservation regimes on the high seas in accordance with LOSC Article 116-119 and to improving flag state controls through the establishment of genuine links between fishing vessels and their flags.[145] The responsibilities/obligations of nationals may be best suited for examination in this light.

Taken together, the above considerations imply a need for a robust, and collective, political will aimed at promoting:[146]

- A steadfast commitment to combating IUU fishing;

- International engagement to take strong action in all relevant forums;

- Continued strengthening and testing of international law;

---

[144] FAO, *op. cit.* n. 11 and 9.

[145] B. Vukas and D. Vidas, *op. cit.* n. 123.

[146] From Senator the Hon. Ian Macdonald (Australian Minister for Fisheries, Forestry and Conservation), "Statement to the Australian Press Club", (Canberra, 19 August 2003). Website:

http://www.affa.gov.au/ministers/macdonald/speeches/2003/pressclubfishing.html.

- Building co-operative alliances between "like-minded" countries; and

- Maintaining effective on-the-water patrols.

Mills[147] has emphasised that the co-operative elements of "political will" are the key to promoting economically fair and sustainable use of any resource, insofar as they reduce regional economic insecurity arising from irresponsible fishing practices.[148] In Freestone's words,[149] the CAMLR Convention has been described as "a model of the ecological approach". While this paper, on balance, judges CCAMLR to have notably faced up to its obligations, only time will show how successful and effective it has been.

Acknowledgements

I wish to thank the organisers of the OECD Workshop on Illegal, Unreported and Unregulated Fishing Activities for inviting and sponsoring me to attend this important event. I also thank my colleague, Dr. Eugene Sabourenkov, for his assistance with the figures and his sage advice. This paper is dedicated to all responsible fishers.

---

[147] G. Mills, *op. cit.* n. 132.

[148] G. Mills, *op. cit.* n. 132.

[149] D. Freestone, *op. cit.* n. 6.

ANNEX 7.A.

Box 7.1. FAO IPOA-IUU[150] Definition of Illegal, Unreported and Unregulated Fishing

---

- **ILLEGAL FISHING**

Activities conducted by national or foreign vessels in waters under the jurisdiction of a State, without the permission of that State, or in contravention of its laws and regulations;

Activities conducted by vessels flying the flag of States that are parties to a relevant regional fisheries management organization but operate in contravention of the conservation and management measures adopted by that organization and by which the States are bound, or relevant provisions of the applicable international law; or

Activities conducted in violation of national laws or international obligations, including those undertaken by co-operating States to a relevant regional fisheries management organization.

- **UNREPORTED FISHING**

Fishing activities that have not been reported, or have been misreported, to the relevant national authority, in contravention of national laws and regulations; or

Fishing activities undertaken in the area of competence of a relevant regional fisheries management organization that have not been reported or have been misreported, in contravention of the reporting procedures of that organization.

- **UNREGULATED FISHING**

Fishing activities carried out in area of application of a relevant regional fisheries management organization by vessels without nationality, or by those flying the flag of a State not party to that organization, or by a fishing entity, in a manner that is not consistent with or contravenes the conservation and management measures of that organization; or

Fishing activities carried out in areas or for fish stocks in relation to which there are no applicable conservation or management measures and where such fishing activities are conducted in a manner inconsistent with State responsibilities for the conservation of living marine resources under international law.

---

[150]   Paragraph 3 of the IPOA-IUU – FAO, *op. cit.*. n. 19.

Box 7.2. Summary of the General Provisions of CAMLR Convention Article II[1]

- **CONVENTION OBJECTIVE**

Conserve Antarctic Marine Living Resources

- **CONSERVATION AND RATIONAL USE**

Conservation includes rational use

- **CONSERVATION PRINCIPLES**

Harvesting and associated activities according to conservation principles below:

- **HARVESTED SPECIES**

Prevent decrease of harvested population to levels below those ensuring stable recruitment (*i.e.* not below level close to that ensuring greatest net annual increment)

- **ECOSYSTEM CONSIDERATIONS**

Maintain ecological relationships between harvested, dependent and related species restore depleted populations

- **PRECAUTIONARY APPROACH**

Minimise risks of change not reversible in 20-30yrs

- **Take Account Of**

Harvesting Effects (Direct/Indirect)
Alien Introduction
Effects of Associated Activities
Effects of Environmental Change

---

[1]   See Article II of the *CAMLR Convention* in CCAMLR, *op. cit.*, n. 16, p. 4-5.

Box 7.3. Information Used by CCAMLR to Estimate IUU Toothfish Fishing Activities[2]

- **CCAMLR LICENSED VESSELS**

Type, size, catch, fishing effort and fishing trip duration

- **IUU VESSELS SIGHTED FISHING**

Number, type and size

- **RECOVERED LONGLINE GEAR FROM ILLEGAL FISHING**

- **TOOTHFISH LANDINGS**

CCAMLR Members' Ports
Other States' Ports (where known)

- **CATCH & EFFORT INFORMATION**

Vessels apprehended for IUU Fishing by Coastal States in Convention Area

- **VERIFIED INFORMATION FROM THE INTERNATIONAL MEDIA**

- **CATCH & TRADE STATISTICS**

Various sources (e.g. Published Trade Information, Customs Declarations)

---

[2] D.J. Agnew, *op. cit.* n. 33 and E.N. Sabourenkov and D.G.M Miller, *op. cit.* n. 38 in particular.

Box 7.4. CCAMLR Toothfish Conservation Measures (CM) Aimed at Eliminating IUU Fishing in the Convention Area

Measures have been developed since 1996/97 and are referenced as CMs currently in force[3]

Measure	Conservation Measure
Fishery Regulatory Measures	
Prohibition of directed toothfish fishing in the Convention Area except in accordance with CMs	CM 32-09
Advance notification of new fisheries.	CM 21-01
Advance notification and conduct of exploratory toothfish fisheries, including data collection and research plans	CMs 21-02 & 41-01
Reporting catch and effort, and biological data, including reporting of fine-scale data	CMs 23-01, 23-02, 23-03, 23-04 & 23-05
Placement of international scientific observers on vessels targeting toothfish	CM 41-01
	Various area-specific measures
Reducing seabird mortality during longline and trawl fishing	CMs 25-02 & 25-03
Flag State Measures	
Contracting Party licensing and inspection obligations for fishing vessels under their flag operating in the Convention Area	CM 10-02
At-sea inspections of Contracting Party fishing vessels	System of Inspection
Marking of fishing vessels and fishing gear	CM 10-01
Compulsory deployment of satellite-based VMS on all vessels (except the krill fishery) licensed by CCAMLR Members to fish in the Convention Area	CM 10-04
Toothfish Catch Documentation Scheme	CM 10-05
Port State Measures	
Port inspections of vessels intending to land toothfish to ensure compliance with CCAMLR conservation measures	CM 10-03
Scheme to promote compliance by Contracting Party vessels with CCAMLR conservation measures	CM 10-06
Scheme to promote compliance by Non-Contracting Party vessels with CCAMLR conservation measures	CM-10-07

---

[3]   CCAMLR, *op. cit.* n. 32 & 58; E.N. Sabourenkov and D.G.M Miller, *op. cit.* n. 38.

Box 7.4. CCAMLR Toothfish Conservation Measures (CM) aimed at eliminating IUU Fishing in the Convention Area. (Cont.)

Measure	Conservation Measure
Resolutions	
Harvesting stocks occurring both within, and outside, the Convention Area, paying due respect to CCAMLR CMs	Resolution 10/XII
Implementation of the Catch Documentation Scheme by Acceding States and Non-Contracting Parties	Resolution 14/XIX
Use of ports not implementing Toothfish Catch Documentation Scheme	Resolution 15/XIX
Application of VMS in Catch Documentation Scheme	Resolution 16/XIX
Use of VMS and other measures to verify CDS catch data outside the Convention Area, especially in FAO Statistical Area 51	Resolution 17/XX
Harvesting of Patagonian toothfish outside areas of Coastal State jurisdiction adjacent to the Convention Area in FAO Statistical Areas 51 and 57	Resolution 18/XXI
Flags of Non-Compliance	Resolution 19/XXI

Box 7.5. Key Principles Underpinning the Toothfish CDS[4]

- Ascertain Catch Origin for all Toothfish Transhipped/Landed/Imported/Exported

- Require Authorization to Fish for Toothfish

- Apply to IUU Fishing by both CCAMLR Contracting and Non-Contracting Parties

- Aim to Prohibit Toothfish Entering World Markets without Valid/Verified Catch Documents

- Non-Discriminatory, Fair and Transparent

- Practical and Capable of Easy/Rapid Implementation

- Applies to fishing within and outside the CCAMLR Area (*e.g.* Recognition Given to "Transboundary" Nature of Toothfish Distribution)

- Conducive to CCAMLR Non-Contracting Party Participation

- Includes Validation & Verification Procedures to Ensure Confidence in Information Produced

- Indicates Responsibilities and/or Obligation of All Participants

---

[4] G.P. Kirkwood and D.J. Agnew, *op cit.* n. 38; K. Larson, *op. cit.* n. 111; E.N. Sabourenkov and D.G.M Miller, *op. cit.* n. 38.

Box 7.6. FAO IPOA-IUU's Key Principles and Strategies[5]

- *PARTICIPATION & CO-ORDINATION*

IPOA-IUU implemented directly by all states or in co-operation with other states, or indirectly through RFMOs or through FAO/other appropriate international organisations. Close co-operation and full stakeholder participation (*e.g.* by the fishing industry, non-governmental organisations and other interested parties) are important to the plan's successful implementation.

- *PHASED IMPLEMENTATION*

Measures to prevent, deter and eliminate IUU fishing to be based on urgent and phased approach taking account of national as well as regional and global actions in accordance with IPOA-IUU.

- *COMPREHENSIVE AND INTEGRATED APPROACH*

Measures to prevent, deter and eliminate IUU fishing should address factors affecting all capture fisheries. Approach taken should build on flag state responsibility and use all available jurisdiction consistent with international law. Latter includes port state measures, coastal state measures, market-related measures and measures to ensure nationals do not support, or engage in, IUU fishing.

States encouraged to use all IUU-directed measures where appropriate and to co-operate to ensure that these are applied in coherent and integrated manner. IPOA-IUU should address all economic, social and environmental impacts of IUU Fishing.

- *CONSERVATION*

Measures to prevent, deter and eliminate IUU fishing to be consistent with conservation and long-term sustainable use of fish stocks and protection of the environment.

- *TRANSPARENCY*

IPOA-IUU to be implemented in transparent manner in accordance with Article 6.13 of Code of Conduct.

- *NON-DISCRIMINATION*

IPOA-IUU to be developed and applied without discrimination in form or in fact against any State or its fishing vessels.

---

[5]   See paragraphs 9.1 to 9.6 of the *IPOA-IUU, op. cit.* n. 19.

Table 7.1. Recent Action Against IUU Toothfish Fishing

(HIMI - Heard and McDonald Islands; FZ – Fishing Zone; ITLOS – International Tribunal for the Law of the Sea; t - tonnes; RSA – Republic of South Africa; AFMA – Australian Fisheries Management Act, 1991; MLRA – South African Marine Living Resources Act, 1998; UK – United Kingdom; USA – United States)[6]

VESSEL/ COMPANY	FLAG/ NATIONALITY	ACTION	OUTCOME(S)
SouthTomi	Togo	*March 2001* Illegal Fishing HIMI FZ >100t Toothfish Australian Arrest off RSA Coast RSA Assistance	AUD 136 000 Fined under AFMA Largest Fine to Date Catch/Vessel Confiscated Failure Secure Release Bond Vessel to be Sunk Winter 2004
Volga	Russian Federation	*February 2002* Illegal Fishing HIMI FZ 126t Toothfish Australian Arrest in FZ	Prosecuted under AFMA Vessel/Catch Confiscated ITLOS Bond AUD2 million Bond Close Commercial Value Bond not Paid Vessel Dispatched for Scuttling 14/4/2003
Lena	Russian Federation	*February 2002* Illegal Fishing HIMI FZ/CCAMLR 80t Toothfish Previously Sighted HIMI Area Australian Arrest	Prosecuted under AFMA 3 Crew Fined AUD 100 000 each Catch/Vessel Confiscated Vessel Scuttled 19/11/2003
Viarsa	Uruguay	*August 2003* Illegal Fishing HIMI FZ 85t Toothfish Australian Arrest Mid-Atlantic 3900 n. ml. (21-day) Hot Pursuit RSA/UK Assistance	Catch/Vessel Confiscated AUD 5 m Bond All Crew Charged Legal Process Ongoing
Maya V	Uruguay	*January 2004* Illegal Fishing HIMI FZ 202t Toothfish Australian Arrest	Charged under AFMA Legal Action Pending AUD 550 k Charge All Crew Charged Catch/Vessel Confiscated

---

[6]   From various sources.

Table 7.1. Recent Action Against IUU Toothfish Fishing (cont.)

VESSEL/ COMPANY	FLAG/ NATIONALITY	ACTION	OUTCOME(S)
Hout Bay Fishing	South Africa	*June 2001* Illegal/Possession/Trade Toothfish RSA *June 2003* Smuggling Conspiracy USA	Prosecuted under MLRA Fined R 40 m (AUD 8 m) Licenses Revoked Closed down Indicted US Lacey Act 21 Counts Charges pending Fines to USD 250 k /Count Asset Forfeiture USD 11.5 mil Possible Jail Time 5 Years/Count *March 2004* Key Defendants Plead Guilty USD 5 m Asset Forfeiture

Figure 7.1. The CCAMLR Area

Statistical Areas, Sub-areas and Divisions are shown

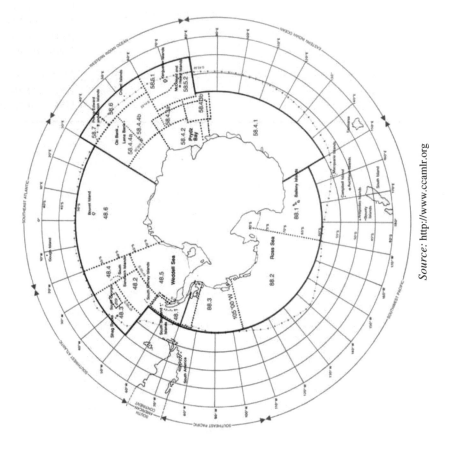

Source: http://www.ccamlr.org

138

Figure 7.2. Toothfish (Dissostichus spp.) Catches in CCAMLR Statistical Subarea Divisions.

Catches reported by split-year, beginning 1 July one year and ending 30 June the next (*e.g.* 1988/89 split year). Statistical Areas are "48" – Southwest Atlantic Ocean; "58" – Indian Ocean; "88" – Pacific Ocean – Ross Sea (Predominantly *D. mawsoni* catches).

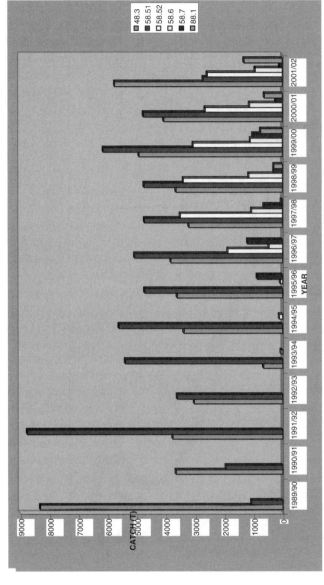

Source: All data from CCAMLR Statistical Bulletins 1990-2003 (http://www.ccamlr.org)

139

Figure 7.3. Progressive development and location of IUU fishing for Patagonian Toothfish in the CCAMLR Convention Area and other adjacent areas

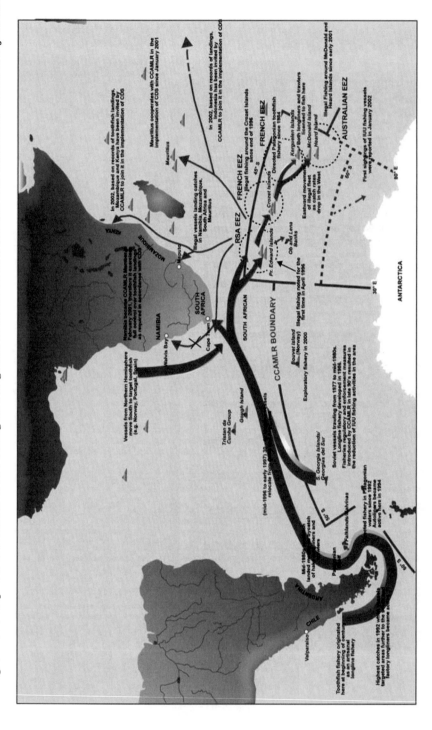

Source: From Sabourenkov and Miller, op. cit n. 38

Figure 7.4. Catches (Tonnes "T") of Toothfish (Dissostichus) in Regulated ("CCAMLR") and Unregulated ("IUU") Fisheries in the CCAMLR Area

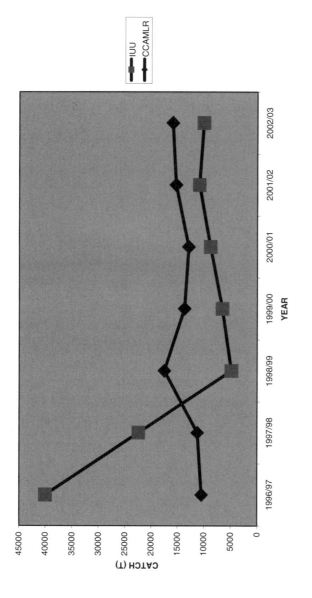

Source: Data from CCAMLR Commission Reports 1997-2003 (http://www.ccamlr.org)

Figure 7.5. Estimated Cumulative financial values (USD million) of "CCAMLR" and "IUU" based Toothfish (Dissostichus spp.) fisheries.

Estimates are based on a landed value of USD 5 000/tonne of H&G product.

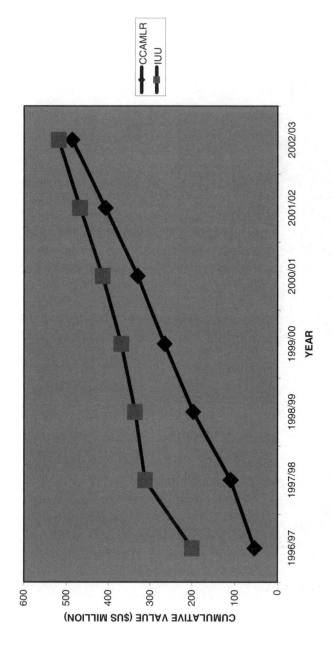

Figure 7.6. IUU Toothfish Catches in CCAMLR Statistical Sub-areas/Divisions

Catches reported by split-year, beginning 1 July one year and ending 30 June the next (*e.g.* 1988/89 split year). Statistical Areas are "48" – Southwest Atlantic Ocean; "58" – Indian Ocean; "88" – Pacific Ocean – Ross Sea (Predominantly D. mawsoni catches).

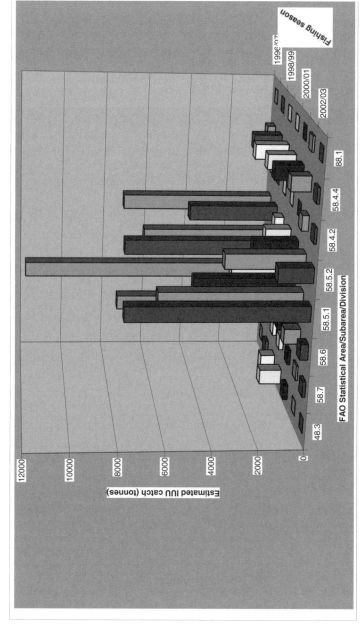

Source: All data from CCAMLR Statistical Bulletins 1990-2003 (http://www.ccamlr.org)

Figure 7.7. CCAMLR Estimated Seabird by-Catch as a consequence of IUU Fishing in the Convention Area

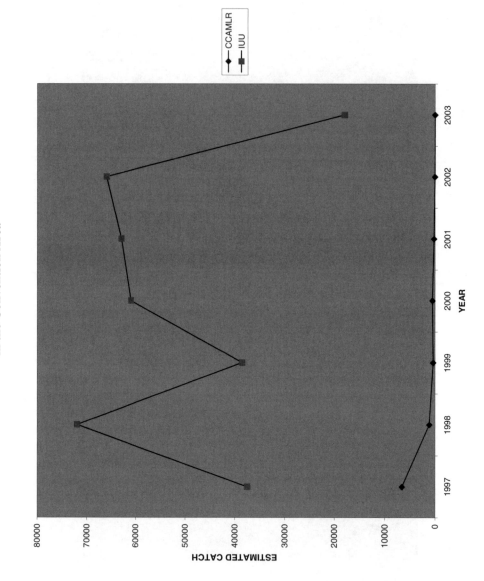

Source: From Miller, Sabourenkov and Ramm op. cit. n. 113).

Figure 7.8. Geographic area of application of the CCAMLR CDS

[Black –CCAMLR Parties; Pale grey – CDS Users; Dark grey – Exporters outside CDS]

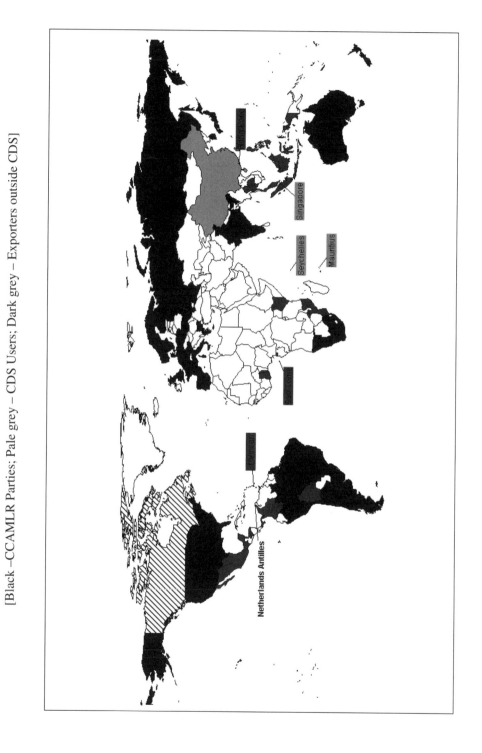

Source: From Miller, Sabourenkov and Ramm (op. cit. n. 113)

CHAPTER 8

GATHERING DATA ON UNREPORTED ACTIVITIES IN INDIAN OCEAN TUNA FISHERIES

Alejandro Anganuzzi, IOTC Secretariat[1]

Background

The Indian Ocean is the basin with the most recent history of industrial exploitation of all the major tuna fishing areas in the world. Although it provided fertile fishing grounds for many of the early residents in the area, it was not until 1952 that longline fleets first entered the eastern Indian Ocean. In what is now a familiar pattern of development for these fisheries worldwide, the first longline vessels enjoyed very high catch rates in the first years of the fishery, yields that quickly turned into more stable catch rates for a number of decades before declining in recent years.

The other major industrial fishery in the region has been the purse-seine fishery, mostly of European origin, which only entered the Indian Ocean as a major player in the early 1980s, and even then was mainly restricted to the western side of the Indian Ocean. Since that time, this fishery has been upgrading its fishing capacity and its production until reaching record levels in very recent years.

The Indian Ocean now ranks second, in terms of productivity, after the much larger central and western Pacific, with perhaps the best economic conditions in terms of access to resources from base ports.

Given these favourable conditions, it was probably only a matter of time before the field of players expanded to incorporate fleets operating at the fringes of the international tuna fishing community, attracted by the large profit margins and a not-yet-developed regulatory framework.

Until the very recent establishment of the Indian Ocean Tuna Commission (IOTC), there was no firm basis for deciding what constituted illegal or unregulated tuna fishing in the high seas of the Indian Ocean. However, it could be argued that unreported fishing activities would undermine any efforts oriented towards achieving long-term sustainability in these fisheries. This document

[1] P.O.Box 1011,Victoria, Seychelles. e-mail: aa@iotc.org. This paper was presented as a background paper at the Workshop.

summarises the efforts to gather information on unreported activities and the development of a regulatory framework along the guidelines of the major instruments of international law.

The early history

The potential of the tuna fisheries was quickly recognised after the arrival of the first purse-seiners in the region in the early 1980s, with catches increasing year after year. The concern to provide rational management of these resources materialised quickly in the establishment of a regional UNDP project, the Indo-Pacific Tuna Management and Development Programme in 1982. From its base in Sri Lanka, the role of IPTP was to develop a centralised data collection point to build the databases that would be necessary to manage these fisheries on a scientific basis.

As IPTP started its work to recover existing data and assist countries in developing new programmes for monitoring their catches, the countries with interests in tuna fisheries in the region began negotiations, with FAO acting as a facilitator, on a new agreement that would institutionalise a regional fisheries body to deal with tuna and tuna-like species in the Indian Ocean, the Indian Ocean Tuna Commission (IOTC).

Negotiations for the establishment of this new Commission, modelled on its sister organisation in the Atlantic, ICCAT, were lengthy, taking most of the following ten years.

During this period, experts working at IPTP painstakingly began to put together a picture of the different fleets operating in the Indian Ocean. Before long, it was clear that some fleets were less than forthcoming in providing information concerning their activities. Efforts then concentrated on establishing the identity, constitution and the *modus operandi* of these fleets. These efforts marked the first period of the fight against non-reported catches in the Indian Ocean.

Identifying the culprits - before a formal definition of IUU fishing

The term "IUU" had not yet been coined or become popular in the forums concerned with the proper management of this incipient fishery. In the absence of a formal management structure to define IUU fishing in the region, the focus was primarily directed on vessels flying various flags that were not reporting data on their activities to IPTP or to any of their responsible governments.

The main objective was to measure the impact of these fleets on the status and productivity of Indian Ocean tuna stocks. IPTP officials compiled reports of activities from port authorities around the Indian Ocean, using data provided by dedicated sampling programmes or, more commonly, based on data from licensing authorities in coastal countries who produced estimates of the number of vessels involved in these activities as well as the catch by species of these fleets.

But the size of the problem, together with a chronic lack of sufficient human resources, means that the picture for those early years is fragmentary at best.

Figure 8.1. Number of Fresh-Tuna Longliners (IUU vs non-IUU) Estimated to be Operating in the Indian Ocean

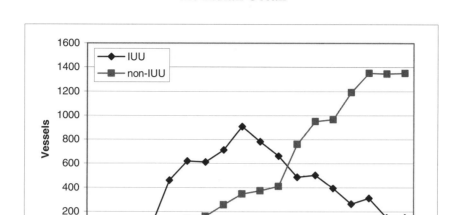

Nevertheless, by the late 1980s it was clear that a large number of deep-freezing longline vessels were operating in the Indian Ocean under various flags of convenience. In some cases, these vessels were originally fishing for southern bluefin tuna, a species that is much more vulnerable to excessive exploitation than its tropical counterparts.

As the catch rates for southern bluefin tuna declined along with the size of the stock, vessels started tapping other species in the Indian Ocean. Soon a profitable operation began with a switch to deeper longlines, a fishing strategy that increased access to the deeper-dwelling stocks of bigeye tuna. Although less appreciated than bluefin tuna as a sashimi species, there was a solid market for bigeye tuna and much higher availability of this species throughout most of the year.

The first non-reporting vessels made their appearance in this fishery. Their area of operations was basically the whole of the Indian Ocean with unloading in Mauritius, Pakistan, Singapore, South Africa and other ports in the region. A significant number of transhipments at sea were also mentioned. Informal reports from Chinese Taipei operators placed the proportion of the catch transhipped at sea at about 50% during the mid-1990s.

Two distinct categories were beginning to emerge from the limited information that was coming to light. The first category was a fleet of large-scale, deep-freezing vessels under various flags of convenience essentially reporting to no government. Important information for assessment purposes – such as catch and fishing effort by area and size of fish caught – was never reported or collected.

An estimated 100 such vessels were operating in these conditions in the area of the south-western Indian Ocean, primarily from Port Louis and Durban (Figure 8.2.). Many of these vessels (but not all) carried licenses for fishing in the EEZs of various countries in the region, and by collecting and comparing license information it was possible to obtain this estimate of the numbers of vessels.

Figure 8.2. Number of deep-freezing tuna longliners (IUU vs non-IUU) estimated to be operating per year in the Indian Ocean

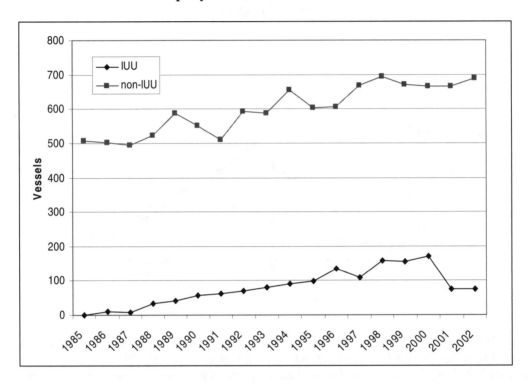

The second component of the longline fleet operating outside any monitoring system was the fleet of small (< 150GRT) fresh-tuna longline vessels (Figure 8.1.). Smaller in size and preserving the fish in ice rather than deep-freezing it, this fleet has a limited range of operation when compared to the larger vessels discussed earlier. Originally this fleet was mobile, coming to the Indian Ocean only at the end of the season for the Pacific bluefin tuna (not to be confused with southern bluefin tuna) in the South China Sea. During its stay in the Indian Ocean, the fleet was based in various ports of the eastern basin, but primarily in the Indonesian ports of Benoa (Bali) and Muara Baru (Jakarta) and, to a lesser extent, in Penang (Malaysia) and Phuket (Thailand).

The total number of vessels was very poorly estimated until recent years, but consistent reports placed the size of the fleet at between 600 to 800 vessels at the time. The major problem with this fleet is that they very rarely reported to the authorities of their original flag and reports to the Indonesian authorities were unreliable or inexistent.

The list of non-reporting vessels was not limited to eastern longline fleets. Purse-seine and longline vessels of then-Soviet origin were operating primarily in the western fishing grounds. Although originally operating under the Soviet flag, by the mid-1980s purse-seiners had moved to various flags of convenience. The estimated number of vessels involved was about 11 purse-seiners (Figure 8.3.). These vessels were rarely seen in Indian Ocean ports and most of their transhipments were carried out at sea.

This was the situation by the mid-1990s, when non-reporting fleets were considered as the main problem in the sound management of these resources, although a formal framework in which to adopt joint actions to ensure sustainability of the tuna fisheries was also lacking.

150

The big breakthrough would come in 1996, when the Indian Ocean Tuna Commission Agreement entered into effect after the accession of the tenth signatory to the Agreement. Now there was a foundation upon which to build a mechanism to rationalise exploitation.

Figure 8.3. Number of purse-seine vessels (IUU vs non-IUU) estimated to be operating in the Indian Ocean

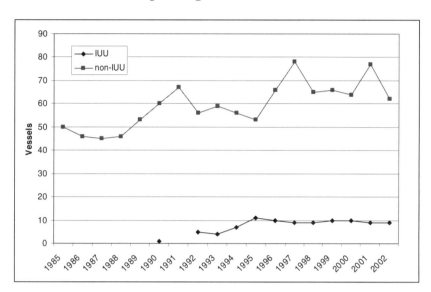

The IOTC years: moving into action

The IOTC took over the responsibility, previously vested in the IPTP, of compiling information on tuna fisheries activity in the Indian Ocean. Its Secretariat, now based in the Seychelles, quickly took steps to find out more about the activities of tuna fleets. In 1999, the Commission approved the establishment of sampling programmes in Thailand and Malaysia, with the co-operation of local authorities, to monitor the activities of small-scale vessels operating in their ports. These programmes provided badly-needed information on average catch rates, essential to estimating the catches of all small-scale non-reporting longliners.

In 1998, the Commission also passed a resolution requesting member countries to provide data on the activities of foreign vessels landing catches in their home ports. This information further improved estimations on the number of vessels in the region.

In 2001, the IOTC, the Overseas Fishery Cooperation Foundation (OFCF) of Japan, Indonesian and Australian officials combined forces to establish sampling programmes in the three main unloading ports of Indonesia, thus closing the information gap in the activities of the fresh-tuna longline fleet.

Figure 8.4. Catches of Tropical Tuna and Billfish for Reporting and Non-reporting Fleets

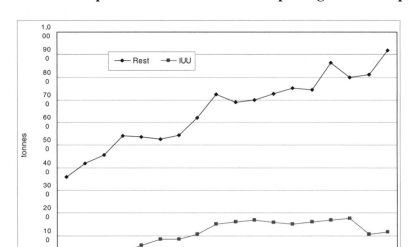

Incorporating this information, the IOTC Secretariat has been able to improve its estimation of IUU fleet catches over the last five years which in turn improves the quality of the data available for the assessment of the main tuna and billfish stocks (Figure 8.4.). This estimation has been facilitated by the fact that, in many cases, the non-reporting fleets operate in a similar way to fleets that are reporting data, reducing the risk of bias.

Concerns about the activities of large-scale longline vessels under flags of convenience prompted the Commission to encourage its Members to take preventive action against vessels suspected of undermining the effectiveness of IOTC management measures.

By 2001 the FAO International Plan of Action on IUU fishing had become a reality, resulting in a concerted plan to take the necessary steps to control non-reporting fleets. In the context of IOTC, this translated into a major initiative taken by the Commission in 2001 with the adoption of an Inspection and Control Scheme at a special session in Yaizu, Japan. The Scheme provided a framework to ensure that IUU fleets, *i.e.,* those whose actions would undermine the conservation measures adopted by IOTC, would be prevented from operating freely in the area.

The Scheme has been implemented in subsequent years through the adoption of various resolutions on an enforcing structure. The major resolutions passed in this respect are listed in Table 8.1.

In addition, the Commission has supported bilateral negotiations between its Members and fishing entities and nations with vessels under flags of convenience, to regularize the situation of those vessels.

Over the last two years, steady progress has been made, and the first results are becoming evident. In the past few months, shipments of fish caught by IUU vessels have been refused entry in the main markets. In Indonesia, an extensive revision of the licensing system that followed a strict policy of re-flagging for fresh-tuna longliners has improved control over the activities of that fleet.

The challenge is now to maintain these encouraging advances in the long term.

The years to come: the new challenges

Agreeing to take a number of concerted actions, as the 22 Members of the IOTC have done in recent years, although an important first step, is not enough. Resolutions are of limited use if they are not accompanied by a strong commitment and efforts to implement them effectively.

This is perhaps the most difficult challenge ahead. Port control measures require that governments are prepared to forfeit sometimes lucrative benefits in favour of fulfilling their international obligations. Control of the markets will work to the extent that markets are concentrated and access to markets is easy to control. The Statistical Document Programme is the main enforcing tool of this basic regulation, working at the level of port control and access to markets.

Programmes such as the Statistical Document Programme, essentially a trade certificate, can serve as a tool for certifying the origin of catches moving across boundaries, if loopholes are eliminated. The effectiveness of such programmes will increase as officials from all countries trading in tuna become more familiar with the mechanics of the programme.

But there is also a need to streamline co-ordination between tuna agencies across the various oceans, to harmonise actions to combat IUU fishing. In the case of highly mobile fleets, stringent measures applied in one ocean encourage the displacement of these fleets to a neighbouring area, unless similar constraints are applied there too.

Table 8.1. Summary of Recent IOTC Resolutions Oriented to Prevent, Eliminate and Deter IUU Fishing

Resolution	Title	Comments
98/04	Concerning Registration and Exchange of Information on Vessels, Including Flag of Convenience Vessels, Fishing for Tropical Tunas in the IOTC Area of Competence	Establishes a record of activities for all vessels fishing for tropical tunas. CPCs should send lists of vessels that have fished for tropical tunas in the previous year.
99/02	Calling for Actions against Fishing Activities by Large-Scale Flag of Convenience Longline Vessels	Urges CPCs to take action (and suggests a number of actions) against longline vessels engaged in IUU fishing activities.
01/01	Control on Fishing activities	CPCs have responsibility for the action of vessels that they authorise, should communicate vessels that are authorised and ensure that they are properly marked, and with documentation in order
01/03	Establishing a Scheme to promote compliance by Non-Contracting Party vessels with resolutions established by IOTC	Establishes actions to be taken by Members in the case of identifying illegal fishing activities in the IOTC Area.
01/05	Mandatory Statistical requirements for CPCs	Establishes the data requirements that constitute proper reporting.
01/06	Concerning the IOTC bigeye tuna statistical document programme	Establishes a trade certification scheme that allows identification of the source of tuna imports.
01/07	Concerning the Support of the IPOA-IUU Plan	Declares IOTC's support for the IPOA-IUU adopted by the Committee on Fisheries of the FAO.
02/01	Relating to the establishment of an IOTC programme of inspection in port	Defines principles that port inspection schemes by CPCs should follow.
02/03	Terms of Reference for the IOTC Compliance Committee	Establishes a Compliance Committee which will review, *inter alia*, information concerning IUU activities by non-Members
02/03	Terms of Reference for the IOTC Compliance Committee	Establishes a Compliance Committee which will review, *inter alia*, information concerning IUU activities by non-Members
02/04	On Establishing a List of Vessels presumed to have carried out Illegal, Unregulated and Unreported Fishing in the IOTC Area	Defines IUU activities in paragraph 1; Establishes a course of action in reviewing information on IUU fishing; defines actions to be taken against those vessels confirmed as IUU.
02/05	Concerning the Establishment of an IOTC Record of Vessels over 24 metres Authorised to Operate in the IOTC Area	Establishes the 'positive list' of vessels authorised by CPCs to fish in the IOTC Area. Defines actions to be taken by CPCs to prevent vessels not in the IOTC Record from fishing in the Area.
02/07	Concerning measures to prevent the laundering of catches by IUU large-scale tuna longline fishing vessels	Defines steps by CPCs to prevent transfer of catches from IUU vessels under the name of authorised vessels.

CHAPTER 9

ESTIMATION OF UNREPORTED CATCHES BY ICCAT

Victor R. Restrepo, ICCAT, Spain [1]

Introduction

The objective of this contribution is to provide a brief overview of the process used currently by the International Commission for the Conservation of Atlantic Tunas (ICCAT) to estimate "unreported" catches. Because the presentation is given at a workshop on Illegal, Unreported and Unregulated fishing, it is useful to emphasise that the scope of the presentation is limited to the first "u" in the acronym (*i.e.*, unreported). ICCAT's scientific body, the Standing Committee on Research and Statistics (SCRS) carries out the estimation of unreported catches referred to in this presentation. It is up to the Commission itself to decide if any particular unreported catch is evidence of IUU fishing or not.

The estimation of unreported catches at ICCAT during the last decade has been closely associated with international trade data. For some species like bigeye tuna (BET), trade data have been reported directly to ICCAT by some Contracting Parties. For bluefin tuna (BFT), which is the basic case study in this presentation, the trade data have been collected through a system known as the BFT Statistical Document Programme (SDP).

The statistical document programmes

The SDP at ICCAT started in 1992 when it was established for frozen bluefin products (the dates in this paragraph refer to the year when the measures[2] were adopted; they generally went into force the following year). In 1993 the bluefin SDP was extended to fresh products, and in 1997 it was amended to also keep track of re-exports. More recently, in 2003, the bluefin SDP was amended again to add information about farmed products and to link the catch information to ICCAT's list of large-scale vessels authorised to fish in the Convention Area (the list is one of the multiple tools used by ICCAT to combat IUU fishing). In 2001, SDPs were also established to track imports and re-exports for bigeye and swordfish. In addition to the above, ICCAT has adopted several other measures related to the validation, interpretation and implementation of SDPs.

---

[1]   ICCAT, Corazón de María 8, 28002 Madrid, Spain, E-mail: **victor.restrepo@iccat.es**

[2]   ICCAT Recommendations and Resolutions can be downloaded from http://www.iccat.es or can be requested from the ICCAT Secretariat.

The ICCAT SDPs collect information on the flag and characteristics of the capture vessel, the area or Ocean of catch and the type and amount of product being traded. They are validated by authorised government officials. Customs officials from Contracting Parties should not authorize the importation of the relevant products (bluefin, bigeye or swordfish) unless they are accompanied by a properly validated statistical document. Twice a year, Contracting Parties should submit summary reports to the ICCAT Secretariat informing about the imports that occurred during the preceding 6-month period.

Case study: Atlantic bluefin tuna

According to the SDP data received by the Secretariat, 50% to 60% of the catch of Atlantic bluefin is traded internationally. Considering that not all importing countries may report back to the Secretariat, the actual proportion of the catch that is traded is probably higher. Most of the international trade in bluefin tuna goes to Japan.

The ICCAT catch database contains a special code called NEI (for "not elsewhere included") which, for the purpose of this contribution, represents unreported catches. NEI codes may be assigned to individual flags by adding a numeric code (*e.g.*, NEI-105); this procedure distinguishes between the unreported catch that is attributed to a country and the catch that is reported by that country.

The calculation of NEI (unreported) bluefin tuna catch follows the formulation:

$$NEI = A - B - C - 0.8\,D$$

where

A = Catch reported to ICCAT

B = Imports to USA

C = Imports to Japan from wild fish

D = Imports to Japan from farming

When the NEI values thus calculated are negative, they are taken as estimates of unreported catch.

A factor of 0.8 is applied to farmed products to allow for a 25% gain in weight from fattening in the farms (1/1.25=0.8). In addition, all product types are converted into round weight (live weight) using the following factors:

Belly meat from wild tuna X 10.28 = round weight

Dressed weight X 1.25 = round weight

Fillets X 1.67 = round weight

Gilled and gutted weight X 1.16 = round weight

Other products X 2.0 = round weight

A conversion factor is not applied to belly meat products from farmed bluefin in order to diminish the possibility of double counting, as bellies are usually shipped separately from other products from the same fish.

The application of the above formula is not fixed over time; it is adapted to current practices. For example, when applied to estimate NEI catches from individual countries, the data are often aggregated among gears or among areas because the biannual SDP summary reports are not very accurate with respect to gear or area specifications. Another example of adaptability is the calculation of "NEI-combined" catches in which data from nine Mediterranean and east Atlantic countries are pooled together in order to reflect current practices of "fishing for farming" and fattening.

The result of the procedure described above to data from 1994 to 2002 suggests that 1% to 5% of Atlantic bluefin catches may go unreported. These estimates are uncertain, however, due to several factors such as: 1) the application of average conversion factors that may be imprecise, 2) the possibility of double-counting through the application of conversion factors to different products from the same fish, 3) the possibility that the SDP for bluefin has not been fully implemented by all importing countries, and 4) the use of highly aggregated data from the biannual reports which does not allow for the validation of details by contrasting individual statistical documents. Despite these uncertainties, the use of SDP data to infer unreported bluefin tuna catches is seen as a very a useful tool.

Other species

The ICCAT statistical document programmes for bigeye and swordfish are at relatively early stages of implementation and have not been used for estimating unreported catches of these species. However, it is likely that the SDP data will be used for this purpose in the near future.

In the past, the SCRS has obtained NEI catch estimates for bigeye tuna based on trade information provided by Japan, following a similar approach to that described above for bluefin. The estimates so obtained suggest that unreported catches were in the order of 5%-10% in the early 1990s, rose to over 20% of the total catch in the late 1990s, and then declined to reach levels of around 5% today. This recent decline in the magnitude of unreported Atlantic bigeye catches is attributed to the effectiveness of various tools used by the Commission to combat IUU fishing, such as positive and negative vessel lists, trade sanctions, etc.

Concluding remarks

ICCAT has used trade data, especially from its statistical document programmes, to estimate unreported catches for bluefin tuna and other species. Although these estimates cannot be exact, due to the multiple assumptions and levels of aggregation that are necessary during computation, they have been very useful in identifying countries that have not properly reported catches to the Commission.

The statistical document programmes at ICCAT do not operate in a vacuum. They are part of a "toolbox" used by the Commission to document IUU fishing activities. This toolbox includes a range of regulations such as vessel lists, transhipment sighting reports and trade sanctions. The interpretation and application of this toolbox has adapted to changes in the fishery and reporting practices, as is evidenced by the many amendments made to the SDPs.

CHAPTER 10

IUU FISHING IN THE NEAFC AREA
HOW BIG IS THE PROBLEM AND WHAT HAVE WE DONE?[160]

Kjartan Hoydal, Secretary, NEAFC

Introduction

Discussions on IUU fishing started in the North-East Atlantic Fisheries Commission (NEAFC) immediately after the FAO had agreed on the IPOA on IUU fishing in February 2001. The first exchanges of view dealt with:

1) Port State Control.

2) The exchange of information on IUU activity between NARFMOs (North Atlantic Regional Fisheries Management Organisations).

3) A fair and equitable treatment of new entrants according to international law.

It was realised at an early stage that it would not be necessary to implement all parts of the IPOA on IUU in the North Atlantic area. In applying the IPOA, the particular situation in the North Atlantic should be kept in mind and form the basis for moving forward. Those elements relevant to the North Atlantic should be selected.

IUU fishing has been on the agenda of the NEAFC Commission ever since, and some aspects have been delegated to NEAFC's Permanent Committee on Enforcement and Control and the Working Group on the Future of NEAFC, which prepares policy proposals to the NEAFC Commission. In the process, NEACF has introduced guidelines for new entrants, discussed lists of IUU vessels and states of flags of convenience and, at the 22nd Annual Meeting in November 2003, adopted the following resolution:

---

[160]   This paper was prepared as a background document for the Workshop.

Resolution

Actions Against Non-Contracting Parties Engaged in Illegal, Unregulated and Unreported (IUU) Fishing in the Regulatory Area

The Commission,

Concerned *that illegal, unregulated and unreported (IUU) fishing compromises the primary objectives of the Convention,*

Aware *that a significant number of vessels registered to non-Contracting Parties engaged in fishing operations in the Regulatory Area in a manner which diminishes the effectiveness of NEAFC management measures,*

Recalling *that the states are required to co-operate in taking appropriate action to deter any fishing activities which are not consistent with the objective of the Convention,*

urges Contracting Parties to take steps towards States identified to have vessels flying their flags being engaged in IUU-fishing in the Regulatory Area by approaching the flag States concerned requesting them to take all appropriate steps to halt the undermining of NEAFC management measures.

The FAO IPOA refers to three separate issues with respect to IUU fishing,

1. § 3.1 Illegal fishing
   3.1.1 conducted by national or foreign vessels in waters under the jurisdiction of a State, without the permission of that State, or in contravention of its laws and regulations;
   3.1.2 conducted by vessels flying the flag of States that are parties to a relevant regional fisheries management organization but operate in contravention of the conservation and management measures adopted by that organization and by which the States are bound, or relevant provisions of the applicable international law; or
   3.1.3 in violation of national laws or international obligations, including those undertaken by co-operating States to a relevant regional fisheries management organization.

2. § 3.2 Unreported fishing
   3.2.1 which have not been reported, or have been misreported, to the relevant national authority, in contravention of national laws and regulations; or
   3.2.2 undertaken in the area of competence of a relevant regional fisheries management organization which have not been reported or have been misreported, in contravention of the reporting procedures of that organization.

3. § 3.3 Unregulated fishing
   3.3.1 in the area of application of a relevant regional fisheries management organization that are conducted by vessels without nationality, or by those flying the flag of a State not party to that organization, or by a fishing entity, in a manner that is not consistent with or contravenes the conservation and management measures of that organization; or
   3.3.2 in areas or for fish stocks in relation to which there are no applicable conservation or management measures and where such fishing activities are conducted in a manner inconsistent with State responsibilities for the conservation of living marine resources under international law.

NEAFC has so far only discussed the IUU activity of non-Contracting Parties. Possible unreported catches, quota overshooting or other activities by Contracting Parties have not been discussed. Some fisheries in the Regulatory area are still not regulated satisfactorily, especially fisheries for deep sea species.

At its meeting in mid-May 2004, the Working Group on the Future of NEAFC will discuss other aspects of the IPOA. *i.e.*, the need for applying IUU measures symmetrically with respect to Contracting and non-Contracting Parties. An overview of NEAFC measures implemented up to now, compared with the measures in the FAO IPOA, is given below.

IPOA §	IPOA Measures	NEAFC Measures
80	States, acting through relevant regional fisheries management organisations, should take action to strengthen and develop innovative ways, in conformity with international law, to prevent, deter, and eliminate IUU fishing. Consideration should be given to include the following measures:	
80.1	Institutional strengthening, as appropriate, of relevant regional fisheries management organizations with a view of enhancing their capacity to prevent, deter and eliminate IUU fishing;	Permanent Secretariat established in 1999
80.2	Development of compliance measures in conformity with international law;	
80.3	Development and implementation of comprehensive arrangements for mandatory reporting;	Scheme 1999
80.4	Establishment of and cooperation in the exchange of information on vessels engaged in or supporting IUU fishing;	Reports from 1999
80.5	Development and maintenance of records of vessels fishing in the area of competence of a relevant regional fisheries management organisation, including both those authorised to fish and those engaged in or supporting IUU fishing;	Yes
80.6	Development of methods of compiling and using trade information to monitor IUU fishing	Not considered
80.7	Development of MCS, including promoting for implementation by its members in their respective jurisdictions, unless otherwise provided for in an international agreement, real time catch and vessel monitoring systems, other new technologies, monitoring of landings, port control, and inspections and regulation of transhipment, as appropriate;	Scheme 1999
80.8	Development within a regional fisheries management organization, where appropriate, of boarding and inspection regimes consistent with international law, recognising the rights and obligations of masters and inspection officers;	Scheme 1999
80.9	Development of observer programmes;	n.a.
80.10	Where appropriate, market-related measures in accordance with the IPOA;	Not considered
80.11	Definition of circumstances in which vessels will be presumed to have engaged in or to have supported IUU fishing;	NCP Scheme 2003
80.12	Development of education and public awareness programmes;	Not considered
80.13	Development of action plans; and	Future WG
80.14	Where agreed by their members, examination of chartering arrangements, if there is concern that these may result in IUU fishing.	Not considered

Every year, the NEAFC Secretariat reports on IUU fishing in the NEAFC Regulatory Area. The latest report is presented below. The main problem in the Regulatory Area is IUU fishing for Oceanic redfish. In 2001, 20% of the catches of redfish in the Regulatory Area were taken by one non-Contracting Party, and this figure rose to 27% in 2002. In addition, a handful of vessels of flags of convenience have been spotted targeting redfish in the Regulatory Area.

IUU fishing in the NEAFC regulatory area: non-contracting parties' activities

The Scheme of Control and Enforcement currently establishes five Regulated Resources in the Regulatory Area (Oceanic redfish, herring, mackerel, blue whiting, Rockall haddock).

Recommended total allowable catches (TACs) for 2002 included co-operation quotas for redfish (1,175 MT) and mackerel (600 MT) for vessels flying the flag of co-operating non-Contracting Parties. For 2003 these co-operation quotas were reduced to 500 MT for redfish and 511 MT for mackerel.

In 2002 and 2003 Estonia authorised two vessels to operate in the Regulatory Area targeting non-Regulated Resources. The declared redfish and mackerel catches were reduced. Vessels are fully complying with the Scheme of Control and Enforcement and Estonian authorities report catches monthly.

In 2002 Japan has authorised one vessel to conduct fisheries in the Irminger Sea and, as in the previous year, the quantities were reduced (9 tonnes of redfish).

In 2002 Latvia has returned to the Regulatory Area with one vessel (formerly German) operating in the Irminger Sea and therefore the likely target is redfish. In 2003 the Latvian vessel has again been observed fishing for redfish. The Secretariat has no information concerning catches.

Six vessels from Lithuania have been observed in the Regulatory Area both in 2002 and 2003. At the 21st Annual Meeting, Lithuania reported catches ten times the allocated "co-operation quota" for 2002 (14,656 MT – these are not final figures).

NEAFC inspectors boarded a Panamanian cargo vessel operating in the Regulatory Area, receiving fish and fish offal (herring, blue whiting) from vessels flying the flag of Contracting Parties.

In 2002 five Belize registered vessels (ex-Russian) were observed targeting redfish in the Regulatory Area. In 2003 three of those vessels were re-flagged in the Dominican Republic. The Secretariat has no information on the catches of these fishing vessels.

Landings in contracting parties' ports by non-contracting parties' vessels

In 2002 a Latvian vessel (DORADO) requested to land catches in a German port. The German authorities refused to authorise landing of redfish based on the fact that the vessel has been observed fishing in the Regulatory Area (point 10 and 11 of the NCP Scheme). Because the vessel also detained onboard catches of redfish allegedly caught in the NAFO Regulatory Area, the German authorities authorised the landing of the NAFO catches.

In 2002 four Lithuanian vessels (RADVILA, ZUNDA, MAIRONIS, NERINGA) also attempted to land redfish in The Netherlands and were only authorised to land redfish allegedly caught in the NAFO Regulatory Area and then proceeded to Lithuania to land the NEAFC catches. It would be interesting to NAFO Contracting Parties to compare these landings with the quotas available for the Baltic States.

Finally, also in 2002, the Danish authorities refused the landing of redfish from a Russian (STARLET 3 - cargo) vessel because according to documents the fish had been caught by Belize fishing vessels (OSTROVETS, OKHOTINO). The same vessel then tried to land such catches in Germany but German authorities refused the landing.

Table 10.1. Observation of NCP fishing vessels in the Regulatory Area

	Observations				NCP Individual Fishing Vessels						
	Total	EU	ISL	NOR	Total	EST*	LTU	BLZ	PAN	LVA	DOM
2001-2002	222	52	157	13	14	1	6	6		1	
2002-2003**	75	46	29		13		5	3	1	1	3

Notes:

* After October 2001 Estonia started automatically transmitting VMS messages to the Secretariat.

** Up to and including April 2003.

Table 10.2. Catches of Regulated Resources (Redfish)

Redfish 2001

	NEAFC RA	EU	FRO	GRL	ISL	NOR	RUS	Total
European Union * **	10,029.0		334.0	4,994.0	1,904.0	595.0		17,856.0
Faroe Islands	3,636.0			4,441.0				8,077.0
Greenland **	498.0			1,350.0	1,460.0			3,308.0
Iceland	2,318.0			14,374.0	25,782.0			42,474.0
Norway	3,853.0			1,217.0				5,070.0
Poland						6.0		6.0
Russian Federation *	24,310.0		54.0	4,050.0				28,414.0
Estonia *	599.0							599.0
Japan *	11.0							11.0
Lithuania	11,389.0							11,389.0
Total	56,643.0	0.0	388.0	30,426.0	29,146.0	601.0	0.0	117,204.0

* - Reported as RED
** - Catches include NAFO 1F

Redfish - 2002

	NEAFC RA	EU	FRO	GRL	ISL	NOR	XJM	XSV	RUS	Total
European Union *	8,479.0	22.0	251.0	7,481.0	998.0	142.0				17,373.0
Faroe Islands **	3,227.7			1,019.8						4,247.5
Greenland **	243.0			1,022.0	3,323.0					4,588.0
Iceland	3,331.0			5,995.0	35,106.0					44,432.0
Norway	1,094.0			4,198.0						5,292.0
Poland	1.4							7.6		9.0
Russian Federation *	23,411.0			2,811.0						26,222.0
Estonia	15.0									
Lithuania***	14,656.0									14,656.0
Total	54,458.1	22.0	251.0	22,526.8	39,427.0	142.0	0.0	7.6	0.0	116,819.5

** - Catches include NAFO 1 F

* - Redfish reported as RED

Table 10.3. Catches of Regulated Resources (Mackerel)

Mackerel 2001

	NEAFC RA	EU	FRO	GRL	ISL	NOR	RUS	Total
European Union		352,683.0	523.0			22,107.0		375,313.0
Faroe Islands		8,829.0	5,888.0			8,967.0		23,684.0
Greenland								0.0
Iceland								0.0
Norway	10.0	386.0	5.0			178,911.0		179,312.0
Poland								0.0
Russian Federation	39,742.0		2,963.0			87.0		42,792.0
Estonia	218.0							218.0
Lithuania	1,949.0							1,949.0
Total	41,919.0	361,898.0	9,379.0	0.0	0.0	210,072.0	0.0	623,268.0

Mackerel -2002

	NEAFC RA	EU	FRO	GRL	ISL	NOR	XJM	XSV	RUS	Total
European Union	78.0	272,412.0	56,799.0			25,739.0				355,028.0
Faroe Islands		5,108.0	5,307.0			9,542.2				19,957.2
Greenland										0.0
Iceland			53.0							53.0
Norway		10,431.0				173,478.0				183,909.0
Poland										0.0
Russian Federation	31,783.0		6,123.0							37,906.0
Estonia										0.0
Total	31,861.0	287,951.0	68,282.0	0.0	0.0	208,759.2	0.0	0.0	0.0	596,853.2

PART III

ECONOMIC AND SOCIAL DRIVERS OF IUU FISHING

The objectives of this session were to identify and discuss the main economic and social reasons behind IUU fishing activities, i.e., the costs and benefits. The session focused on the economic and social drivers of IUU fishing and assessed their relative importance. The discussion also helped identify possible measures that could target individual drivers for an effective and feasible integrated response to the problem of IUU fishing.

CHAPTER 11

ECONOMIC ASPECTS AND DRIVERS OF IUU FISHING: BUILDING A FRAMEWORK

David J. Agnew and Colin T. Barnes of MRAG Ltd., London, England

Executive summary

This report examines the economic and social drivers that influence the development of Illegal, Unregulated and Unreported (IUU) fishing. It does this primarily from the point of view of high seas fishing, especially by vessels flying Flags of Convenience (FOC). These vessels undermine conservation measures agreed by Regional Fisheries Management Organisations, and thus we prefer the term Flags of Non-Compliance with such conservation measures (FONC).

It is difficult to obtain sound information on the historical and existing levels of IUU fishing activity, as a solution applied in one area may simply move the problem to another. In order to determine the effectiveness of measures to combat IUU fishing, it is important to develop good quantitative statistics on the levels of IUU fishing in the entire world's oceans, both those under national jurisdiction and those in high seas waters.

IUU vessels appear to be relatively inexpensive to buy (probably less than USD 1.2 M for a longliner) and running costs are lower since crew wages and conditions are inferior to those on legitimate vessels (with the exception of those pertaining to officers, especially fishing masters). Although some additional costs might accrue with the requirement that these vessels re-supply and tranship at sea, they do not have to pay for licences and expensive safety checks, so on balance they are likely to have lower running costs than legitimate vessels. The additional cost they do have to face, however, is the cost of arrest (forfeit of catch and punitive fines), but against this must be set the probability of being caught. Thus the opportunity cost of engaging in IUU fishing is probably quite low.

The bulk of this report is an analysis of the various incentives to engage in IUU fishing. Our analytical framework is based around the very basic equation,

IUU incentive ~ Profit from IUU fishing = Benefit from IUU fishing – Cost of IUU fishing.

In the analysis, we examine economic and social drivers, including market control, price distortion, effect of the global economy and world fishing opportunities, international regulations, fishing agreements, re-flagging, national fisheries management policy including subsidies and excess capacity and surveillance activities. We also consider the geographical features of IUU fishing areas, the health of other fish stocks, and the financial and operating structure of companies operating IUU vessels.

The analysis points to a number of factors which can create incentives for IUU fishing. More detailed examination of these factors (outside the scope of this report) should make it possible to identify which of them are likely to be most important in creating an economic incentive for vessels to engage in IUU fishing. The ultimate aim of this work should be to eliminate IUU fishing, which would require more detailed analysis to identify how the economics could be manipulated so that the opportunity cost of illegal fishing becomes too high to be sustained. Several solutions are discussed within the analysis. Often, however, the cost of a solution to a particular incentive would also be high for legitimate vessels. We believe that it will be difficult, but not impossible, to find solutions that do not penalise legitimate operators who are following the rules.

Finally, we identify a number of economic and social parameters that are likely to be impacted by IUU fishing. These parameters might be monitored, to complement the quantitative estimates of IUU fishing identified in the second paragraph above, to judge the effectiveness of measures taken to combat IUU fishing.

List of Acronyms

ABC	Australian Broadcasting Commission
ACFM	Advisory Committee on Fisheries Management
CCAMLR	Commission for the Conservation of Antarctic Marine Living Resources/ Convention for the Conservation of Antarctic Marine Living Resources
CCSBT	Commission for the Conservation of Southern Bluefin Tuna
DWFN	Distant Water Fishing Nation
EEZ	Exclusive Economic Zone
EU	European Union
FAO	Food and Agriculture Organization of the United Nations
FOC	Flag of Convenience
FONC	Flag of Non Compliance
ICCAT	International Commission for the Conservation of Atlantic Tunas
ICES	International Council for the Exploration of the Sea
IOTC	Indian Ocean Tuna Commission
IPOA	International Plan of Action
ITLOS	International Tribunal for the Law of the Sea
IUU	Illegal, Unregulated and Unreported fishing
LSTLV	Large-Scale Tuna Longline Vessels
MAGPS	EU Multi Annual Guidance Programme
MCS	Monitoring, Control and Surveillance
NAFO	Northwest Atlantic Fisheries Organisation
NEAFC	North-East Atlantic Fisheries Commission
OECD	Organisation for Economic Co-operation and Development
RFMO	Regional Fisheries Management Organisation
UNCLOS	United Nations Convention on the Law of the Sea
VMS	Vessel Monitoring System

Introduction

This report addresses the OECD project on the economic and social issues and effects of IUU/FOC fishing operations. The project aims to develop a framework for analysing the economic and social effects of IUU/FOC fishing, including:

- review literature on the economic and social effects of IUU/FOC fishing;

- identify key factors affecting incentives for IUU vessels;

- develop an analytical framework to evaluate economic and social effects of IUU/FOC fishing;

- develop a checklist of economic characteristics that should be monitored to understand the key economic features that encourage IUU fishing and to assess its impacts.

We approach this problem by first defining IUU fishing and the scope of the project. IUU fishing covers an extremely broad category of behaviour, and needs some refining in the context of this project. Next, we consider the key economic drivers behind IUU fishing and suggest a framework within which they can be studied and their relative importance evaluated. Finally, we review the economic and social impacts of IUU fishing, and how they might be monitored.

Definitions of IUU fishing

As an activity, illegal, unreported and unregulated (IUU) fishing has been with us ever since fisheries management first started. As an acronym, however, it is much more recent. First used informally during the early 1990s by the Commission for the Conservation of Antarctic Marine Living Resources (CCAMLR)[1] in relation to Southern Ocean fishing, it began life as "IU" (illegal and unreported). Formal use of the term IUU can be found in the report of the Commission's 16th Meeting in 1997 and in a letter to the Food and Agriculture Organization (FAO) that same year, in which the nature and seriousness of these problems were described.[2] IUU fishing is now commonly understood to refer to fishing activities that are inconsistent with or in contravention of the management or conservation measures in force for a particular fishery.

A number of international instruments contain provisions that are relevant to controlling IUU fishing. These include the 1982 United Nations Law of the Sea Convention[3] (the 1982 Agreement), the 1993 FAO Compliance Agreement, the 1995 United Nations Straddling Stocks Agreement[4] (the 1995 Agreement), and the 1995 FAO Code of Conduct for Responsible Fisheries.[5] None of these was set up

---

[1] The Commission established under Article VII of the Convention on the Conservation of Antarctic Marine Living Resources (CCAMLR), 1980. Reprinted in International Legal Materials 19 (1980): 827.

[2] Executive Secretary, CCAMLR to FAO [REF: 4.2.1. (l), 18 December 1997], as cited in G. Lugten, "A review of Measures taken by Regional Marine Fishery Bodies to address contemporary Fishery Issues," FAO Fisheries Circular No. 940. (Rome: FAO, 1999): Footnote 130 at 35. .

[3] United Nations Convention on the Law of the Sea, Montego Bay, 10 December, 1982.

[4] Agreement for the Implementation of the Provisions of the United Nations Convention on the Law of the Sea relating to the Conservation and Management of Straddling Fish Stocks and Highly Migratory Fish Stocks. New York, 4 December, 1995.

[5] See Edeson, W. M. 1966. "The Code of Conduct for Responsible Fisheries: An introduction". Int. J. Mar. Coast. Law 233.

to deal directly with IUU fishing. Concern over the growth of IUU fishing worldwide increased rapidly during the late 1990s. An initiative taken by the FAO Committee on Fisheries in 1999 culminated in the adoption of an IPOA on IUU fishing in March 2001.[6] The IPOA is a voluntary agreement, elaborated within the overall framework of the FAO Code of Conduct for Responsible Fishing.

IUU fishing is defined in paragraph 3 of the IPOA as provided in Box 11.1.

Not all unregulated fishing is necessarily conducted in contravention of applicable international law. This is because many high seas waters and/or fisheries are still unregulated by regional fishery management organisations (RFMOs). Examples of these include the orange roughy/alfonsino fishery in the southern Indian Ocean, and the toothfish fishery on the northern Patagonian shelf edge. While IPOA appears to exempt this aspect of fishing from the definition "IUU", we consider it part of the problem. This is because even in the absence of regulations, states have an obligation under UNCLOS and (after its entry into force in December 2001) the Straddling Stocks Agreement (not to mention the Code of Conduct) to make efforts to ensure such stocks are managed. Thus, while there is no doubt that the orange roughy/alfonsino fishery is currently legitimately unregulated, it certainly should become regulated, and the negotiations for the South-West Indian Ocean Convention address this concern. In fact, it has been argued that there are no areas of high seas fishing that may be considered legitimately unregulated in terms of states' obligations under that Agreement and under Part VII of the 1982 Agreement. However, this appears to be an area of international law about which differences of opinion remain.[7]

---

[6]   See "Implementation of the International Plan of Action to Prevent, Deter and Eliminate Illegal, Unreported and Unregulated Fishing". *FAO Technical Guidelines for Responsible Fisheries*. No. 9, Rome, FAO. 2002, 122 pp. See also Report of the Twenty-Fourth Session of the Committee on Fisheries, Rome, 26 February–2 March, 2001. Document COFI/2001/7, and Kirkwood & Agnew 2002 [G. P. Kirkwood and D. J. Agnew. 2002. "Deterring IUU fishing. Proceedings of the Symposium on International Approaches to Management of Shared Stocks – problems and future directions". Centre for Environment, Fisheries and Aquaculture Science (CEFAS), Lowestoft 10-12 July 2002.].

[7]   See, for example, Freestone, D and Makuch, Z. 1996. "The new International environmental law of fisheries: The 1995 United Nations Straddling Stocks Agreement". *Yearbook of International Environmental Law*. 7: 3-51.

Scope of this report

The objective of this report is to review information on the economic incentives for IUU fishing, and the economic and social impacts of such fishing. In order to do so, it is necessary to define the scope of our review because the areas covered by the FAO definition go beyond the remit of this project.

What we are ultimately interested in is the unauthorised or unrecorded removal of fish from a fisheries ecosystem. Such unauthorised removals damage both fish stock and the ecosystem because they are not accounted for within the fisheries assessment and management system. Indeed, these actions undermine conservation measures promulgated to ensure the rational use of those stocks.

Collateral to this resource damage is economic damage to legitimate, law-abiding fishers. Economic damage may be direct (an IUU vessel may trawl over the gear set by a legitimate vessel) or indirect. Indirect effects are of two kinds. The first is associated with the depletion of the stock that is caused by IUU fishing. Because they are not accounted for, IUU catches usually deplete a resource, leaving less of it for legitimate fishers. The result is the gradual erosion of the fishery's ability to provide a sustainable long-term basis for the use of fisheries and associated environmental resources by fishermen and other stakeholders. Legitimate fishers might therefore suffer a declining allowable catch (as the stock declines) and a declining catch rate. This declining catch rate directly affects the economics of fishing vessels. The second economic effect is bad publicity arising from high levels of IUU fishing, which could make consumers cautious of purchasing even legitimate products from companies engaged in fishing in areas where IUU fishing is widespread, no matter how legitimate their fishing operations may be.

Thus there is a clear cause – the taking of fish beyond what is defined by a management body, or at unsustainable levels – and a clear effect – damage to the ecosystem, which is passed on to legitimate resource users, as well as an economic and social cost. Of course, there are many examples where such damage is being or has been directly caused by management setting quotas that are higher than scientific advice indicates are sustainable (ref: ICES ACFM reports on cod, hake etc. over the past 2 years), but this is not within the scope of our review. Similarly, there are instances where there is no management in high seas waters (the Southern Indian Ocean alfonsino/roughy fishery, for instance). While, as argued above, fishing in these areas is strictly IUU fishing under the definition of 3.3.2, and under the obligations of States Parties to UNCLOS and Straddling Stocks Agreements, it must also be pointed out that there are some states that are not party to these agreements. Hence, this is a problem of international management that is also beyond the scope of this review; it requires action by States with interests in the region, and with obligations under UNCLOS or the Straddling Stocks Agreement.

The remaining IUU fishing problem can be divided into two categories: fishing that takes place *inside* or *outside* areas of national jurisdiction. In the IUU literature, there are two clearly different cases of IUU fishing that take place inside areas of national jurisdiction, *i.e.*, misreporting and poaching (covered by FAO definitions 3.1.1 and 3.2.1). Misreporting is carried out by otherwise legitimate vessels, and while it is likely to be illegal under FAO definition 3.1.1, this will depend on the strength of national laws. It is a well-documented and widespread problem, known to most fisheries management authorities, involving a number of areas such as discarding, high-grading, domestic-use non-reporting, misreporting, and "black fish".[8] Pitcher *et al.* (2002) categorise these catches as unreported discards (which may not be illegal but are not reported by observers), unmandated catches (catches that an agency is not mandated to record) and illegal catches (catches that contravene a regulation: poached fish from closed areas, transhipments at sea, under- or misreported catches including those whose identity is deliberately concealed). Amongst a host of examples from around the world they focus on two, Iceland and Morocco. Using an analytical method they estimate that catches of Icelandic cod may have been underestimated by between 1 and 14% at different times, and haddock by between 1 and 28%. Catches in Moroccan waters may have been underestimated by as much as 50%. Obviously, these levels of IUU fishing in national waters have very serious consequences for domestic fisheries, especially as the level of IUU extractions is not constant from year to year but varies depending on circumstance. For instance, a quota system will inevitably lead to greater incentives for misreporting than management based on effort limitation (Agnew, 2001: sustainability of squid fisheries; Pitcher *et al.* 2002).

This aspect of the IUU problem is very large, and can only be solved by clear management and MCS (Monitoring, Control and Surveillance) action. It is outside the scope of this review.

---

[8]   Valatin, G 2000 "Fisheries management institutions and solutions to the 'black fish' problem". CEMARE Misc. Publ. no. 48, pp. 101-118. *Management institutions and governance systems in European Fisheries*. Univ. of Portsmouth, Portsmouth (UK).

Furthermore, to a large extent the activities covered by definition 3.2.2 are similar to those of 3.2.1 – *i.e.* misreporting, discarding, high-grading, etc. Although these problems are exacerbated by the activities of Illegal/FOC vessels, their solutions are not dissimilar to those applied within waters under national jurisdiction, and will not be covered here.

Flags of Convenience (FOC)

As generally used, the term Flag of Convenience refers to a state that is willing to have a vessel on its national register without undertaking fully its obligations under UNCLOS Article 94 to exert Flag State jurisdiction and control. FOC countries are usually those which have established open registers, accepting vessels from other countries without having a genuine link between the flag state and the vessel or company owning the vessel. Initially, vessels were registered with these countries for reasons more to do with licensing fees, tax evasion, reduced safety requirements etc.[9] While these are all still valid economic reasons for vessels to flag with these countries, an additional incentive is that no effective control is exercised. Under the terminology of the Compliance Agreement the flag state must be able to exert effective control over the vessel, but States referred to as FOC usually fail to do so.[10] It further says that States should ensure that their vessels do not engage in activity that undermines the effectiveness of international conservation measures. Since FOC states are generally not members of RFMOs or other agreements, their flag vessels are not bound by the management regulations enforced by these organisations. Furthermore, while they would normally then be bound *generically* by the provisions of the Compliance or Straddling Stocks agreements, they have usually not signed up to these agreements either. They are therefore effectively beyond the reach of international law.

IUU vessels often fly flags of convenience, or employ re-flagging, as a means of deliberately avoiding fisheries conservation and management measures based on regional arrangements applicable on the high seas. Re-flagging is relatively easy, and IUU vessels may re-flag several times in a fishing season to confuse management and surveillance authorities. One classic example is San Rafael 1, flagged to Belize, which - following an encounter with a fisheries patrol vessel in December 1999 around South Georgia - changed its name to the Sil, then the Anyo Maru 22 and finally the Amur, flagged to Saõ Tome e Principe before sinking around Kerguelen on 9 October 2000.[11] Another is the Camouco, arrested by France in 1999 around the Crozet Islands, and released on bail following a case which was taken to the International Tribunal for the Law of the Sea.[12] After its release the vessel changed its name to Arvisa 1 and subsequently Eternal, only to be arrested again by France on 3 July 2002 for illegal fishing in Kerguelen waters. We mention these cases only to illustrate that IUU vessels often use re-flagging to confuse surveillance, and we do not suggest that any of the above-mentioned flag states should be classified as a Flag of Convenience or Flag of Non Compliance.

---

[9]   European Parliament. Working Document 1 on the role of flags of convenience in the fisheries sector. Committee on Fisheries, 11 April 2001.

[10]   Vukas and Vidas discuss this point in detail, and show how the concept of requiring a genuine link between Flag State and vessel was repeatedly watered down in the negotiations leading up the 1982 UNCLOS agreement. B. Vukas and D. Vidas, "Flags of Convenience and High Seas Fishing: the emergence of a legal framework". In *Governing high seas fisheries: the interplay of global and regional regimes* (Ed. O. S. Stokke) pp 53-91, OUP.

[11]   D.J. Agnew and G.P. Kirkwood 2002. "A statistical method for analysing the extent of IUU fishing in CCAMLR waters: application to Sub-area 48.3". CCAMLR WG-FSA-02/4

[12]   ITLOS press release 35, 7 February 2000. Case of the Camouco, Panama vs. France. For later information on the movements of the Camouco, see ITLOS Transcripts of the Volga case (Russia vs. Australia), statement of Mr Campbell, ITLOS/PV.02/02, 12 December 2002.

However, there are differences between merchant vessel and fishing vessel use of flags of convenience, and between the behaviour of vessels flying flags of convenience in different regions of the world, that have led to the emergence of a new term to describe FOC vessels. For instance, vessels under the Panamanian flag would be regarded as FOC in Antarctic waters, because, as a non-party to CCAMLR, Panama would not be exerting effective control on its vessels in the waters of that RFMO. However, Panamanian-flagged vessels are not FOC vessels in waters administered by ICCAT, as Panama is a member of ICCAT. For these reasons CCAMLR has moved away from the term "Flags of Convenience" and now uses the term "Flags of Non Compliance". In the preamble to Resolution 19, the definition of this term is clear (Box 11.2.):

This clearly indicates the cause of the problem (*i.e.* the lack of State control over vessels which are conducting activities that undermine the effectiveness of conservation measures), but allows States and RFMOs to take action against FONC/FOC States and their vessels only in respect of the violations of specific regional agreements. This maintains consistency with the intent of the FAO compliance agreement and WTO requirements where trade measures are contemplated.

Box 11.2. Preamble to CCAMLR RESOLUTION 19/XXI, Entitled "Flags of Non-Compliance"*

The Commission,

Concerned that some Flag States, particularly certain non-Contracting Parties, do not comply with their obligations regarding jurisdiction and control according to international law in respect of fishing vessels entitled to fly their flag that carry out their activities in the Convention Area, and that as a result these vessels are not under the effective control of such Flag States,

Aware that the lack of effective control facilitates fishing by these vessels in the Convention Area in a manner that undermines the effectiveness of CCAMLR's conservation measures, leading to illegal, unreported and unregulated (IUU) catches of fish and unacceptable levels of incidental mortality of seabirds,

Considering therefore such fishing vessels to be flying Flags of Non-Compliance (FONC) in the context of CCAMLR (FONC vessels),

Noting that the FAO Agreement to Promote Compliance with International Conservation and Management Measures by Fishing Vessels on the High Seas emphasizes that the practice of flagging or re-flagging fishing vessels as a means of avoiding compliance with international conservation and management measures for living marine resources and the failure of the States to fulfil their responsibilities with respect of fishing vessels entitled to fly their flag, are among the factors that seriously undermine the effectiveness of such measures,

* Many of the flags hereby called FONC are commonly referred to as "flags of convenience".

Conclusion

In conclusion, the areas that we will address are FAO definitions 3.1.2, 3.1.3 and 3.3.1. These cover the activities of IUU and FONC vessels in high seas waters covered by RFMOs. However, of necessity we will need to consider that aspect of definition 3.1.1 that relates to piracy by foreign vessels within an EEZ, because the activity of these vessels in high seas waters is intimately linked with their activities in waters under national jurisdiction. Much of the following discussion will therefore focus in the first instance on the drivers for IUU fishing within and outside EEZs, followed by an assessment of its impact.

Review of relevant information

Estimating the extent of IUU/FONC fishing

The problem of IUU fishing has been encountered by most regional fisheries organisations since the 1980s. For instance, in the period between 1985 and 1993 an annual average of 30 – 40 fishing vessels from non-contracting parties were sighted in the regulatory areas of the Northwest Atlantic Fisheries Organisation (NAFO), primarily flagged to Panama and Honduras. Following diplomatic demarches to these countries, some of the vessels were re-flagged to Belize.[13]

NEAFC has also recorded a number of more recent experiences of IUU/FONC fishing. In 2001 non-member Lithuania declared that 14 000 t of redfish had been taken from NEAFC waters. This was taken outside of agreed NAFO quotas of about 100 000 t. Vessels from Sierra Leone have also been sighted in NEAFC waters (Joao, pers. Comm.).

ICCAT has of course experienced the activities of FONC vessels for a number of years. In 1994, a Bluefin Tuna Action Plan was adopted by ICCAT that linked information gathered by the Bluefin Tuna Statistical Document Programme[14] with Contracting Party compliance and non-Contracting Party co-operation with ICCAT's conservation and management regime. After identifying in 1995 that Belize, Honduras, and Panama had vessels that were fishing in a manner which diminished the effectiveness of ICCAT's conservation measures, in 1996 ICCAT prohibited imports by its Members of bluefin tuna products from these three countries (effective from 1997 for Belize and Honduras and 1998 for Panama). This was successful in terms of Panama, which became a contracting party in 1998. Similar sanctions were extended to cover bigeye tuna taken by vessels flagged by Belize, Cambodia, Honduras, Equatorial Guinea and St. Vincent and the Grenadines in 2000. Once again, this move seems to have been effective, and in 2001 ICCAT lifted the import ban on bigeye tuna from St. Vincent and the Grenadines and the bluefin tuna ban from Honduras. ICCAT has estimated that the IUU catch of big eye tuna reached a maximum of 25 000 t in 1998 but has since declined to about 7 200 t (2001). In 1998, the IUU catch was about 25% of the total catch.

The IUU tuna vessels problem is widespread. At the Santiago de Compostella meeting on IUU fishing, Japan presented a paper which suggested that despite various incentives to scrap vessels and move them onto national fleets there are still some 100 IUU large-scale tuna longline vessels (LSTLV) catching an estimated 25 000 t of tuna each year. ICCAT has for some time been concerned about the activities of these vessels, particularly since most of them have crew from ICCAT Contracting Parties and there is considerable evidence of laundering of IUU catch either through links with legitimate vessels or through forging documentation.[15] In response to this concern, at its December 2002 meeting

---

[13]   Reported in Vukas and Vidas, *op. cit.* Citing Joyner and the NAFO annual reports 1994, 1995. Other sighted flags included Cayman Islands, Sierra Leone, St Vincent and the Grenadines, New Zealand, the USA and Venezuela.

[14]   ICCAT resolutions 92-1 and 92-3, implemented in 1993.

[15]   See: Japanese submission at FAO-IPOA meeting; Japanese paper delivered to the Santiago de Compostella meeting. The preambular paragraphs in ICCAT Resolution 01-19 make these concerns very clear: "RECALLING that the Commission makes yearly reviews of various trade and sighting data and based on that information prepares a list of IUU fishing vessels, RECOGNIZING that since IUU fishing vessels change their names and flags frequently to evade the sanction measures against them and that the lists of IUU fishing vessels based on the past trade data are still useful but should not be the sole tool to eliminate the IUU fishing vessels; EXPRESSING GRAVE CONCERN that a significant amount of catches by the IUU fishing vessels are believed to be transferred under the names of duly licensed fishing vessels; BEING AWARE that the majority of crew onboard the IUU tuna longline vessels are residents of the Contracting Parties, Cooperative Non-Contracting Parties, Entities or Fishing Entities; STRESSING THE NEEDS for Chinese Taipei, Japan and Parties concerned to investigate the relation between licensed vessel owners and IUU fishing activities and take necessary actions to prevent licensed vessel owners from being engaged in and associated with IUU fishing activities."

ICCAT enacted a series of resolutions[16] which create both "white" and "black" lists of vessels. Any vessel not on the white list that fishes, tranships or otherwise engages in unregulated fishing is placed, following a series of review procedures, on the blacklist, and there are a number of punitive measures that are activated once a vessel is on this list. IOTC is similarly concerned, but as far as we know has not yet been able to estimate the size of IUU catch in the Indian Ocean.

Since 1992 CCAMLR has experienced large amounts of IUU/FONC fishing, with levels reaching up to 80% of the total catch in some areas of the Indian Ocean.[17] Agnew (2000) for instance, estimates that IUU catches in 1996/97 were restricted to the Indian Ocean and reached 43 000 t. FONC States have been Belize, Panama, Vanuatu, Portugal, Namibia, Vanuatu, Seychelles, Faeroe Islands, South Tomi, St Vincent and the Grenadines, and the Netherlands Antilles. Although many of these states have now acted to stop their vessels fishing in CCAMLR waters, there are also vessels from CCAMLR Members that are engaged in illegal fishing in CCAMLR waters, in particular Russia and Uruguay.[18] Since bringing in a Catch Document Scheme for toothfish, CCAMLR has been able to curtail some of the IUU activity on toothfish, although catches in the Indian Ocean sector are still thought to be very high. The latest estimates from CCAMLR are that IUU-caught toothfish amounted to 11 000 t in 2002, about 45% of the total catch from CCAMLR waters, 99% of this coming from the Indian Ocean.[19] However, examination of trade data by TRAFFIC Oceania suggested that the CCAMLR estimates may have been underestimated in 1999/00, when the Catch Document Scheme came into force.[20] Like ICCAT, CCAMLR brought in two important Conservation Measures regarding lists of vessels engaged in IUU fishing[21] at its October 2002 meeting, although both are "black" lists (CCAMLR chose not to create a "white" list other than its already existing list of vessels licensed by Members to fish in the Convention Area).

Monitoring the effects of IUU fishing

The level of IUU fishing is notoriously difficult to assess. Methods to assess it can be divided roughly into direct and indirect. The direct method relies on statistical methods and actual observations to derive estimates of the level of IUU fishing (*e.g.* Agnew & Kirkwood 2002; Pitcher *et al.* 2002). However, even these methods rely on certain assumptions, such as the value of certain input parameters. The value of these parameters can be treated as uncertain, and in this sense Bayesian approaches may have considerable value. Indirect methods, on the other hand, are based on deductive assumptions. They can be based on occasional sightings of vessels, or on trade data. The use of indirect methods is more widespread.

---

16   02-22, "Recommendation by ICCAT concerning the establishment of an ICCAT record of vessels over 24 meters authorized to operate in the Convention Area", and 02-23, "Recommendation by ICCAT to establish a list of vessels presumed to have carried out illegal, unreported and unregulated fishing activities in the ICCAT Convention Area".

17   Agnew, D J, 2000. "The illegal and unregulated fishery for toothfish in the Southern Ocean, and the CCAMLR Catch Documentation Scheme". Marine Policy 24: 361 – 374.

18   CCAMLR Report, 2002.

19   CCAMLR Scientific Committee Report, 2002, Annex 4, Table 3.2.

20   M. Lack & G. Sant, 2001. "Patagonian toothfish: are conservation and trade measures working?" TRAFFIC Bulletin, Vol. 19, No 1. TRAFFIC Oceania. See Agnew, 2000 op. cit and Green, J. and D.J. Agnew. 2002, ["Catch document schemes to combat Illegal Unregulated and Unreported fishing: CCAMLR's experience with southern ocean toothfish". *Ocean Yearbook 2000*, 16, (in press)] for a discussion of the CCAMLR CDS.

21   CCAMLR Conservation Measures 10-06 (2002) "Scheme to promote compliance by Contracting Party vessels with CCAMLR conservation measures" and 10-07 (2002) "Scheme to promote compliance by non-Contracting Party vessels with CCAMLR conservation measures".

Both methods suffer from the problem that they each require data. As IUU fishing is revealed through the use of one method of assessment, IUU fishers become aware of the danger of allowing such data to be released and therefore move to disguise the data source. One advantage of direct methods is that much of the data are generated by management and surveillance authorities. They are therefore less subject to bias than indirect methods.

It is often thought that the only way to achieve effective control of IUU fishing is through surveillance, and it well known that increasing surveillance leads to increasing avoidance.[22] While it may play a large part, a host of other economic and social considerations also come into play, as shown in the next section. Indeed, some of the economic models currently developed take this into account,[23] and more needs to be done in terms of applying such models for the various IUU situations identified above. In particular see Charles *et al.*

Conclusion

One factor that emerges when examining IUU fishing in a global sense is that it has been widespread over the last 30 years. This time span coincides with the period when international (and national) management regulations were considerably tightened up, being primarily dependent upon closure of the commons as EEZs were declared and codified into international law in the 1982 UNCLOS agreement. However, IUU fishing has not been uniform in its development across the globe. The earliest records appear to come from NAFO, then from ICCAT and finally from CCAMLR. Unfortunately it is not currently possible to really assess the changes in the extent of IUU fishing because much of it is only documented by national or international agencies, and there is no simple global picture. Fighting IUU fishing has been likened to trying to squash a balloon full of air, in that as the problem is solved in one area it pops up in another.

It will never be possible to assess the effectiveness of attempts to eliminate IUU fishing unless there is a global IUU monitoring programme that can show whether the measures taken are having any effect. That global view is currently not available. We would conclude that a necessary precursor to the many current initiatives on IUU fishing would be to monitor IUU fishing (or define methods for its monitoring). As far as we are able to ascertain, although FAO has stated that it will "monitor, to the extent that it is possible, global developments in IUU fishing and report on these developments at UN and FAO fora"[24] this will not necessarily include producing annual statistics on the level of IUU fishing. As we have seen these are difficult to obtain, so considerable effort will have to be exerted to acquire these data.

Analytical framework

In this section the focus is on understanding the economic and social drivers behind IUU fishing activities.

---

[22]   Charles *et al.*, 1999; Milliman, SR 1986 "Optimal fishery management in the presence of illegal activity" J. Envir. Econ. Manage.13, 363-381.

[23]   See Charles, A.T, R.L. Mazany, M. L. Cross, 1999 "The economics of illegal fishing: a behavioural model". Mar. Resource Economics, 14, 95-110; Sutinen, JG; Kuperan, K, 1995, "A socio-economic theory of regulatory compliance in fisheries", Int. Coop. Fish. Aquaculture. Dev: Proc 7th Biennial Conf. of the Int. Inst. Fish. Econ. Trade, National Chinese Taipei Ocean Univ, 1995, vol. 1, pp. 189-203.

[24]   FAO Observers report to CCAMLR, 2002: CCAMLR-XXI/BG/36.

Understanding the economics of IUU vessels

The economics of IUU vessels centre on the vessel operating costs of IUU activities compared to non IUU activities. In addition, factored into IUU fishing activities there will be a **risk factor**, namely the costs of being apprehended, catch confiscation and the potential costs of a fine. In the absence of a competitive fishing environment and with limited regulatory control, IUU fishers will be able to extract a higher level of **economic rent** than transparent non IUU fishing activities. However, the issue of rent extraction as between fishing fleets is one issue. The other issue is the distribution of economic rent as between foreign fishing vessels (IUU and non IUU) and the coastal states in whose waters these vessels are fishing. Several studies show that in many cases local coastal states only receive a fraction of the value of the resource which is taken from their waters.

There appear to be two groups of vessels that are currently engaged in widespread IUU fishing in high seas waters (*i.e.* IUU fishing falling under the scope of this study). The first of these are LSTLV vessels, of which there appear to be about 100, fishing for tuna in ICCAT and IOTC waters. The authors have little knowledge about the economics of these vessels.

Secondly, there are the IUU vessels undertaking longline fishing for toothfish in CCAMLR waters. These vessels may be relatively inexpensive to buy, probably less than GBP 1 million. Information is hard to come by, but there have been a number of cases of contested bonds of arrested IUU vessels brought to the International Tribunal of the Law of the Sea[25] which are relevant. Valuations of vessels in court cases are likely to be lower than the market price, since they are the subject of negotiations on damages. Nevertheless, they do give us some clues. In this regard the Camuoco was originally valued at about USD 3 million (GBP 2 million) by the French authorities which arrested it, but this was contested at the ITLOS court by the applicant (Panama) and it was decided that the value for bond purposes was USD 345 000 (GBP 220 000). Again, in the Monte Confurco case (Seychelles v France) the vessel was originally valued at USD 1.5 million by the respondent (France) and USD 500 000 by the Applicant, and the Court upheld the value of USD 500 000. In the case of the Grand Prince (Belize v France) the respondent (France) valued the vessel at USD 2 million (FF 13 million) and the respondent at USD 360 000, although the court does not seem to have made a judgement between these two figures.

In all these cases there are strong vested interests, for the respondent in having a high valuation (to increase the bail amount) and the applicant having a low valuation (to reduce the amount of bail). The true value of the ship is therefore likely to lie somewhere in between, at an average of about USD 1.2 million. In the more recent Volga case, the value of the vessel was uncontested at about USD 1.1 million (AUD 1.8 million, GBP 720 000). This tends to support a nominal value for an IUU longliner of about USD 1.2 million or GBP 780 000.

Longliners are usually 500-1000 GRT. They are usually staffed with captains from a variety of fishing states, often with Russian engineers.[26] The staff costs for officers will usually be higher than for legitimate vessels, since they are taking certain risks, and will as usual be linked to the value of the catch. Crew costs, however, are much lower, since very cheap labour from Indonesia, China and other developing countries is used, and the crew are paid very poor wages (in the region of

---

[25]   Copies of the court proceedings and judgements in the ITLOS cases can be found on the ITLOS website, http://www.itlos.org/

[26]   The information in this section comes from a variety of confidential sources, but also the Australian Broadcasting Corporations' 4 Corners programme, "The toothfish pirates", broadcast on 30 September 2002 and "The Alphabet Boats: a case study of toothfish poaching in the Southern Ocean", a publication by Austral Fisheries Pty, PO Box 280, Mt Hawthorn, Western Australia 6916.

USD 100/month). As a result, total staff costs are likely to be 25-30% of total catch value. Routine running costs for an IUU vessel will be somewhat similar to those for a legitimate vessel, around GBP 800 000 per year.[27] Vessel operating costs will be lower, however, as in many cases the vessel may not be fully insured and the crew may not be operating under the health and safety and insurance norms that apply to non IUU fishing vessels.

A few years ago most IUU vessels fishing for toothfish were thought to be acting relatively independently, although several would have been owned by a single fishing company. That fishing company would often be operating several legitimate vessels as well as a vessel engaged in IUU fishing. CCAMLR reported that a large number of vessels with a great many flags were engaged in IUU fishing in 1996–1999, and, as exemplified by the case of the San Rafael 1, the activities of these IUU vessels can best be described as opportunistic.

Set against these costs will be the profit from IUU activities. In terms of toothfish these are likely to be between GBP 3 million and GBP 4 million per year (USD 4.5–6 million), based on a fishing year of about 200 days, currently likely catch rates and market prices of toothfish. It can be readily seen that the likely profit far exceeds the costs, even if a vessel was to be arrested and confiscated once a year.

More recently, however, a disturbing development has been the engagement of an organised IUU fleet of vessels with common ownership and control links to two major companies based in the Far East – Pacific Andes and P. T. Sun Hope Investments (Jakarta), although Pacific Andes officially denies this. The Austral Fisheries press release states that "the 'alphabet' boats are, of course, technically operated and controlled by their skippers while being owned by dummy companies in (at various times) the British Virgin Islands, Russia, Belize, Bolivia and elsewhere".[28] We would emphasise that at the moment these are simply allegations from Austral Fisheries.

The development of highly complex company ownership structures has several effects which skew the economic balance sheet for these vessels. Firstly, laundering IUU catch along with legitimately obtained catch (Pacific Andes is a major purchaser of fish caught by legitimate vessels) will allow the price of IUU fish to be higher than would otherwise be the case. There is considerable evidence of fraud in the documentation accompanying toothfish catch documents, as there is in the certificates of registry that are now required by Japan for tuna imports. Secondly, it is not sufficient to simply examine the economics of a single vessel (as we have done above), when a company runs a series of legitimate and IUU vessels, because single vessels can quite easily be sacrificed to the overall benefit of the fishery. There are certainly allegations that the two vessels arrested by the Australian navy in February 2002 (the Volga and the Lena) were the oldest and most dispensable in the IUU fleet fishing around Heard Island. Thus, the actual disincentive of arrest may be much less (for the company) than would be assumed for a fleet. Finally, of course, it is much easier for a fleet and large company operation to afford the administration required for rapid re-flagging, re-configuring and other disguising tactics.

The authors have no direct information on the economic operations of the LSTLVs, but we would assume that their operations are developing the same level of co-ordination as the toothfish vessels, given the increased sophistication of the fraud reported by Japan.

---

[27]   These are figures obtained from discussions with the toothfish industry. A more comprehensive analysis of licensed vessel operation is given in "Évaluation des accords de pêche conclus par la Communauté européenne". Ifremer/Cemare/CEP. Contrat Européen no. 97/S 240-152919, 1999.

[28]   Page 3 of the Austral Fisheries document.

The above summary of the economics of the operations of IUU vessels sets the scene for a discussion of the major economic drivers behind IUU fishing, discussed in the next section.

Economic incentives to engaging in IUU activities

Before embarking on this section, a distinction should be made between economic incentives for companies and vessels on the one hand, and individuals on the other. The drivers for entities and individuals may not be the same, and these differences will be recognised where they occur. However, it may be more useful to present an analytical framework for investigating the economic drivers for IUU activities by reference to generalised categories. Thus the category "world economic outlook" would affect both companies and individuals, as would "disparity between developed and developing world economies".

Our analytical framework is based around the very basic equation,

(1)   IUU incentive ~ Profit from IUU fishing = Benefit from IUU fishing – Cost of IUU fishing

Each of the economic drivers will act differently on this equation. For instance, one might reduce costs, thereby increasing the incentive, while another might increase the value of the catch, thereby achieving the same result.

It is not sufficient simply to analyse the effect on IUU fishing of certain drivers. The objective of undertaking such an analysis is to identify areas where further research would be best directed, and ultimately to find ways in which equation 1 can be tipped into negative profit, thereby reducing any incentive for IUU fishing and assisting its elimination. However, account also needs to be taken of the effect each driver has on legitimate fishers. There is no point in adopting a solution such as a total moratorium on exploiting a particular species if it adversely affects legitimate fishermen more than IUU fishermen. Therefore, our analysis also takes into account how the various drivers affect legitimate fishermen.

At this point, our analysis is simply qualitative. There is very little information on which to make quantitative analyses. However, we think that such information could be acquired, and useful economic models developed, to investigate the relative importance of each of the drivers in influencing the general equation above.

Social drivers

There are a number of social drivers behind IUU fishing. Closer control over the EEZs of coastal states will mean that distant water fishing nations (DWFNs) may have problems in employing fishing crews. In the case of countries such as Chinese Taipei/China and Korea there is therefore an incentive to take the risk of IUU fishing because the relative risks and costs of arraignment may be low. In some of the fishing nations, over-exploitation of their own fishing grounds causes a displacement effect to the EEZs of coastal states and the high seas. Fishing operators may also engage in IUU fishing because of the more limited health and safety controls and other controls over working conditions and workers' rights.

In the case of low-income countries with semi-industrial fishing fleets, IUU fishing may be considered as a relatively low-risk, cost-effective way of maintaining fish supplies for the country. IUU fishing may therefore have a number of social drivers in the case of these countries, including employment, protein supply and food security. Because it evades controls, payment of access rights and social security, IUU fishing is therefore an attractive option.

Markets and trade

Market control/access and the regulatory environment

It is often thought that increasing restrictions on market access will have a deterrent effect on IUU fishing. This may be the case, but we need to understand exactly how such deterrence might take effect. Firstly, it should be noted that trade-based measures have so far only been adopted in respect of tuna and swordfish (ICCAT, CCSBT, and recently IOTC) and toothfish (CCAMLR). The toothfish scheme is fundamentally different from tuna schemes, which are directed on species falling wholly under the control of the RFMO, and are primarily trade documentation schemes, in which documents are issued in respect of products entering trade. Following the success of the ICCAT Bluefin Tuna Statistical Document Programme, a linkage was made in the Bluefin Tuna Action Plan to prohibit imports from non-members whose vessels diminish the effectiveness of ICCAT conservation measures. In 1996, this was extended to allow the prohibition of imports from ICCAT Members who exceed their catch limits.[29] However, a statistical document scheme is not an essential precursor to the imposition of trade measures; sufficient information may already be in existence to provide evidence of the undermining of conservation measures. Thus ICCAT maintains a Swordfish Action Plan, which together with its resolution 96-14 can be used to prohibit imports of swordfish from Members or non-Members. Similarly, ICCAT Resolution 98-18 is aimed at catches of tuna by large-scale longline vessels, and has been used to prohibit the importation of bigeye tuna from one ICCAT Member and four non-Members (FAO, 2002).[30] This is in the (then) absence of specific Statistical Document Schemes for these species.

The toothfish scheme operated by CCAMLR, on the other hand, is a catch certification scheme, with documents being issued at the point of capture or landing. A second major difference is that not all toothfish come under the control of the RFMO, as significant high seas stocks of toothfish fall outside the CCAMLR Area.[31]

There is evidence from the CCAMLR situation that fish certified using a catch document or other trade tracing document may command higher prices than uncertified fish, but that this premium may not be particularly high and therefore it may not act as a sufficient incentive to switch to certified sales (yet). Evidence from CCAMLR suggests that the current premium on fish carrying CCAMLR Catch Documents is only 20-30%. This is encouraging because it was always acknowledged that one of the aims of the scheme would be to create a price differential which would act as an economic disincentive for IUU vessels. Unfortunately, the level of economic penalty associated with IUU catches does not seem to be high enough, on its own, to dissuade IUU fishers.

Another area of potential economic leverage is the cost of fraud. Certainly fraud is taking place, and gaining in sophistication, as evidenced by the Japanese experience with the difficulty of ensuring that tuna from ICCAT IUU-listed vessels is not imported. However, what is important in the balance of equation 1 is the cost of this fraud, which must be increasing. This cost will also include the cost of financing corruption where state officials are involved in either tacitly or actively assisting fraud. Another avenue open to IUU companies would be to disguise their fish through re-packaging and re-labelling. Although there are genetic methods of identifying the species from fish products, these

---

[29] ICCAT Resolution 96-14.

[30] Implementation of the international plan of action to prevent, deter and eliminate illegal, unreported and unregulated fishing. *FAO Technical Guidelines for Responsible Fisheries No 9*. Rome, FAO, 122 pp.

[31] See Agnew, 2000 *op. cit.*, and Kirkwood & Agnew, 2002 *op. cit.* Definitions of the differences between trade and catch documents are given in the Report of the Expert Consultation of Regional Fishery Management Bodies on Harmonisation of Catch Certification, La Jolla, 9-12 January 2002. FAO.

methods are usually expensive and not routinely available for customs authorities. Therefore, attempts to disguise fish products may go unnoticed. On the other hand, such disguising would have to be followed by mixing IUU and legitimate fish for sale within a country to prevent the value of the fish from being considerably reduced.

Increased market control has costs for legitimate vessels as well as for IUU vessels. They have to structure their company activities so as to obtain all the relevant documentation and ensure that their fish are appropriately dealt with by landing and import authorities (including those in states which may not be party to a particular RFMO and/or the scheme operated by that RFMO). For instance, the new ICCAT measures to combat fraud are based on turning the Statistical Document into a Catch Document,[32] and require more rigour in applying the catch document. Such rigour is also required by vessels using the CCAMLR catch document. Vessels may have to carry additional costs associated with verification, such as on-board observers, regular inspections, VMS, etc. For instance, the cost of a VMS unit is about USD 3 000, of an observer is USD 300-500 a day. Finally, there are costs associated with the import action since many instances of IUU fraud involve the use of a false name. Where this name is the same as a legitimate vessel, costly delays in importing products may occur while potential fraud cases are eliminated, once more adding a burden on legitimate vessels.

Species value, price distortion

Obviously, the higher the price of fish, the higher the benefit to both IUU and legitimate fishers will be. In the short term, market forces can be expected to increase the value of fish as volumes decrease due to declining stock sizes and quotas. This will disproportionately advantage IUU fishers at the expense of legitimate fishers, because the latter will be constrained by quotas or limitations on effort, whereas IUU fishers will not. This is a dangerous feedback, because as the resource becomes scarcer, the legitimate quota declines still further, creating greater market pressure for increases in value.

The imbalance that is apparent in this equation is the fact that in such a feedback system market forces are likely to be unconstrained, whereas the deterrent effect of arrests will be severely constrained. We will discuss the relationship between the extent of IUU fishing and the cost of MCS activities later, but it is important to realise here what effect declining stocks have on MCS activities.

The first is financial. Reducing stock sizes leads to reduced revenue to government from a fishery (either in the form of licence sales or tax receipts). This in turn leads to decreased MCS budgets at a time when costs are increasing. Unless additional funds are made available by increasing fines for IUU activity, this can lead to the inability of a management body to adequately police its waters.

The second effect has to do with presence. There is some evidence from CCAMLR that the presence of legitimate vessels can have a deterrent effect on IUU vessels. Legitimate vessels may have observers on board who have a statutory obligation to report all vessel sightings. Legitimate vessels also find their interests coinciding with those of management authorities when it comes to informing

---

[32]   Resolution 02-25 by ICCAT concerning the measures to prevent the laundering of catches by IUU large-scale tuna longline vessels, paragraph 1 of which reads: "Contracting Parties, Cooperating non Contracting Parties, Entities or Fishing Entities (hereinafter referred to as the 'CPCs') should ensure that their duly licensed large-scale tuna longline fishing vessels have a prior authorization of at sea or in port transshipment and obtain the validated Statistical Document, whenever possible, prior to the transshipment of their tuna and tuna-like species subject to the Statistical Document Programs. They should also ensure that transshipments are consistent with the reported catch amount of each vessel in validating the Statistical Document and require the reporting of transshipment."

on poachers.[33] As stocks are depleted, however, the fishing opportunities of legitimate vessels also decrease, with the effect that they cease to be effective as a deterrent.

In the long term, of course, continued IUU fishing will have a negative effect on both IUU and legitimate fishers, in that catch rates will decline and consequently profits will decline.

General market trends and the global economy

While the demand for marine fish products continues to rise steadily, overall supply has been at best static for a number of years and, given the state of the world's marine fish stocks, it is unlikely to increase much above current levels in the near future. Buoyant and increasing fish prices are therefore to be expected. This is an overriding, global driver for IUU fishing because it will clearly lead to increasing benefits (sales). It should, however, have similar effects on legitimate vessels, albeit within the constraints noted above. By contributing to a reduction in the availability of certain species, IUU fishing has a negative impact on food security for coastal states where fish consumption is relatively high.

Like everything else, fishing is heavily influenced by the global economy and by local economy imbalances. Local economy collapses, for instance, are likely to increase the incentive for corruption, decreasing its cost, thus decreasing the cost of this part of the IUU fishing vessel's equation. Large disparities in incomes/economies of developed and developing countries will create a ready and cheap labour pool for IUU fishers (many crew are Indonesian, Chinese or Philippine), once again decreasing their costs. For instance, the illegal trochus fishery in Australian waters in the early 1990s was mostly due to the extreme poverty of Indonesian fishermen,[34] who ran the risk of facing heavy penalties and imprisonment. Illegal fishing in Somali waters is largely due to the ineffective patrolling and enforcement of its EEZ, itself a function of the economic and political situation in the country.[35] A poor economic outlook will also force states to make cuts in surveillance coverage, often an early casualty of worsening economic conditions. Thus, one should look for increasing incentives (support) to control IUU fishing in areas adjacent to states or continents which have severe economic difficulties. The coastal states of West Africa are good examples of where there is a need for such support.

International regulation/management

International regulations

In the normal course of events RFMOs will develop regulations to manage their fisheries. There are also a large number of regulatory issues which are being developed by RFMOs, especially to do with inspection, increased scientific observation, avoidance of by-catch of fish species, avoidance of incidental mortality of birds, avoidance of interactions with marine mammals, etc. These regulations inevitably lead to higher costs for legitimate vessels, and no costs for IUU vessels. Their imposition therefore erodes the profitability of legitimate operations, and increases incentives to engage in IUU fishing.

---

[33] For instance, a licensed Australian trawler spotted a notorious IUU vessel, the Eternal (previously the Arvisa 1, Kambott or Camouco, using several FOC) in French waters around Kerguelen, and after calling the French authorities took up hot pursuit until the Eternal was intercepted by the French naval vessel the Albatross on 3 July 2002, arrested and taken to Réunion. *La Voz de Galicia*, 9 July 2002.

[34] Peachey, G, 1991. "Illegal trochus fishing-what can we do?" Aust. Fish., Canberra, Vol. 50, 8-9.

[35] Hassan, M.G., "Marine resources in Somali waters: opportunities & challenges", *6th Asian Fisheries Forum Book of Abstracts.* p. 93. Asian Fisheries Society.

RFMOs face the difficult question of how to account for IUU fishing. If estimates of IUU catch and reported legitimate catch exceed the total allowable catch, should next year's catch be reduced by that amount to ensure that the fishery is sustainable? It might seem obvious that it should, but this would mean that the cost of IUU fishing was disproportionately higher on legitimate fishers than on the management authorities. IUU fishing is a failure of management, not of the legitimate fishery to behave responsibly. Furthermore, acting in such a way would be somewhat equivalent to acknowledging that IUU fishing was going to be as large next year as it was this. However, the lesson from other areas where total extractions continually exceed the allowable sustainable stock (for instance, most demersal fisheries in Europe) is that such patterns inevitably lead to the collapse of fisheries. This also has economic consequences for legitimate fishers, but it happens in the medium to long term rather than the short term, and is therefore easier to accept.

Externalities

There are a number of externalities that affect IUU and legitimate vessels differently. In addition to having to implement all the above-mentioned international regulations, legitimate vessels must implement general safety and pollution requirements of the IMS/MARPOL, etc. These added costs are not borne by IUU vessels.

The consequences of IUU fishing are discussed in the next section. However, the long-term degradation of resources that result from overfishing, itself a consequence of IUU fishing activity, will lead to fishery closures with consequences on both IUU and legitimate vessels.

Vessel flag transfers

Vessel flag transfers reduce the traceability of vessels and compromise MCS attempts to control IUU fishing, since the legitimacy of hot pursuit ceases if a vessel changes its flag. The costs of re-flagging[36] to various FONC parties is minimal (USD 1 000-5 000, mainly legal costs); it is relatively simple and fast, can often be done at sea, and the benefits are great. Interestingly, however, re-flagging problems seem to have acted against the Grand Prince, in that between the time that she was arrested by the French authorities (12 December 2000) and when the court in La Réunion set the bond of FF 11.4 million, her registration with Belize lapsed. Accordingly, *The Tribunal observed that, in the light of the expiration of the provisional patent of navigation issued by the Marine Registry of Belize or of the de-registration of the Grand Prince, referred to in the note verbale dated 4 January 2001 of the Ministry of Foreign Affairs of Belize, and on the basis of an overall assessment of the material placed before it, the assertion made on behalf of Belize that the Grand Prince was still considered as registered in Belize did not provide sufficient basis for holding that Belize was the flag State of the vessel for the purposes of making an application under article 292 of the Convention.*[37]

For legitimate vessels which need to maintain registration with reputable countries (*i.e.* not FONC parties), transfers of flag are much more costly, and may involve protracted administrative procedures. They are only undertaken when access to a particular fishery is closed to one particular flag.

---

[36]   See for instance www.flagsofconvenience.com.

[37]   ITLOS press release 48.

Fishing agreements

Two types of fisheries agreements are considered here: first, multinational agreements relating to high seas fisheries, elsewhere called RFMOs (Regional Fisheries Management Organisations), and second, agreements between a coastal state or states for access by a third party to fish in their waters.

Membership of fisheries agreements brings benefits to legitimate vessels, as Panama has discovered by becoming a party to ICCAT. However, increased membership brings considerable costs to legitimate vessels of the existing Members of RFMOs, because limited allowable catches have to be divided up between more Members and therefore quota sizes are reduced. This is a very serious problem faced not just by ICCAT, but by all RFMOs, as they attempt to deal with IUU fishing. Various actions by these RFMOs can force the cessation of IUU fishing by certain Non-Contracting Parties, but transferring IUU vessels to legitimate fleets (either by straight transfers or by accession and membership of previously Non-Contracting Parties) increases the capacity of the legitimate fleet. This has direct costs for legitimate fishers. This situation is analogous to that faced by national management authorities; in the end, it is overcapacity which is the largest problem, not necessarily the behaviour of various groups of fishers.

Coastal state fishing agreements may be multilateral, bilateral or private. The largest number of multilateral fishing agreements are those signed by the European Union with African, Caribbean and Pacific (ACP) countries. These agreements support fishing activities in the Atlantic, Indian and Pacific Oceans. Other types of fishing agreements may be bilateral (*e.g.* Chinese Taipei agreements with Mauritius; Japanese agreements with Mauritania) or private, *i.e.*, agreements between fishing companies and third-party states for fishing access. Other options may include joint venture agreements for fishing rights between external fishing companies and local partners (the case of French and Spanish agreements with Namibia and Spanish and Moroccan companies). These types of agreements confer rights of fishing access subject to the provisions of the agreements (fish quotas; types of gear and equipment and vessel size). In the case of EU fishing agreements they may also increase (to a certain extent) the degree of transparency with respect to the number and types of vessels. Agreements in themselves do not necessarily avoid the issues of IUU fishing and even fishing vessels operating under transparent agreements may be operating within one of the constituent elements of IUU fishing (*e.g.* illegal or unregulated).

In addition, in some cases multilateral fishing agreements may bring a displacement effect. For example, the extension of fishing rights to EU vessels within the waters of a number of West African countries may have the effect of pushing IUU fishing into other waters. Furthermore the non-agreement of fishing agreements, for example the non-completion of a fishing agreement between the EU and Morocco in 2001, may have had the effect of promoting IUU fishing within Moroccan waters. A number of studies and consultancy reports have looked at the issue of the economic and social impacts of fishing agreements. These include a study that was carried out for the EU on issues of coherence and complementarity between EU fisheries and development policy with respect to EU fishing agreements and development policy (ADE: 2002).

World fishing opportunities

Although this has already been mentioned in a previous section, it is worth reiterating that the lack of many alternative world fishery resources leads to high opportunity costs for IUU fishing. At the same time, the competition for legitimate fishing opportunities is increasing so the costs associated with those opportunities (such as licensing and other costs such as tonnage payments for certain species which may be defined in fishing agreements *e.g.* the EU fishing agreements with various third-party states) is also increasing.

National fisheries management policy

National management policy

Different countries have very different national management policies. They may adopt input or output controls, have or not have regulations on fishing capacity (by vessel or by GRT/power or by other measures), have heavily detailed or almost non-existent domestic regulations for fishing in inshore and offshore waters. For countries that are very tightly controlled, it is usually also the case that they have very strict regulations concerning the use of their flags by vessels engaged in IUU fishing. Norway, for instance, is particularly strict, being one of the first countries to enact laws denying Norwegian flags and domestic fishing opportunities to vessels with any past involvement with IUU fishing. Others may have very lax laws regarding the use of their flags by vessels engaging in IUU fishing.

Economically, the combination of domestic laws acts as an entry barrier to vessels wishing to engage in IUU fishing. If domestic fishing opportunities are denied to vessels on IUU lists [such as the list created by CCAMLR Conservation Measure 10-07 (2002)] this will be a significant economic cost to those vessels. If that denial is extended to other economic areas, such as the denial of re-flagging opportunities to any vessels associated with a FONC state or the prohibition of landings or exports from FONC states (CCAMLR Resolution 19/XXI) this is a further strengthening of economic cost.

Thus, it is clear that – all things being equal – strong domestic legislation will act to combat IUU fishing in the EEZ of a country and will, additionally, force IUU vessels to seek an alternative flag under which to carry out their activities. This strong management policy would of course extend to such areas as the control of fishing capacity. If fishing capacity is not controlled rigorously by national management policy to be equal to the resources that can be exploited by the national (flag) fleet, there will be an economic incentive for those vessels not making enough money in national fisheries to engage in IUU fishing. Thus, those countries with weak domestic regulations and national management policy are likely to be the source of vessels engaging in IUU fishing. An extreme example of this is, of course, FONC states.

Subsidies

Subsidies benefit legitimate operations because they depress the operating cost curve and change its shape. In effect, the operating costs of a vessel are reduced. This benefit is not available to IUU operators, except when beneficial ownership of an IUU vessel is held by a company receiving subsidies for legitimate vessels. Subsidies also tend to encourage overcapacity by hiding the real economic cost of fishing, and therefore act to exacerbate the situation discussed in the section on the "Health of other stocks" below.

Subsidies may also be given to companies to sell vessels (*e.g.* EU payments for the decommissioning of fishing fleets). If these vessels subsequently become available to the IUU market, the subsidies will act to artificially depress the cost equation for IUU companies, sometimes by as much as 30%. Most of the IUU fleet currently consists of old vessels no longer capable of competing with the modern fleets operating in regulated fisheries. This is especially the case for the LSTLVs transferred off the Chinese Taipei and Japanese flags since 2000. However, there are some signs that new longliners are being purpose-built for the IUU fishery on toothfish. The number of such vessels available is increased and to some extent their purchase costs are further decreased by the continued practice of some countries to provide subsidies for building new and more efficient fishing vessels.

Excess capacity/idle capacity

Excess or idle capacity will, as shown in the section on the "Health of other stocks", lead to lower costs of vessels and crews to IUU vessels. While the EU has paid subsidies under the MAGP schemes in an attempt to decommission vessels, in other cases subsidies paid by the EU (regional development, vessel refitting) have encouraged the transfer of excess/idle capacity from EU waters to the fishing waters of the Eastern Atlantic and the Indian Ocean. In other cases, fishing vessels which are excess to need may be re-flagged, sometimes on numerous occasions, and may end up in IUU fishing activities.

Excess capacity has the potential to be an extremely powerful driver for IUU fishing, because it will act on every scale, from the individual to the vessel to the company. Vessels not offered scrapping incentives will face large costs which can only be mitigated through engaging in IUU fishing. Even when scrapping funds are made available, fishermen are likely to face much reduced employment prospects, through two mechanisms. Firstly, even if they re-train, an experienced fisherman will become an inexperienced other professional. Secondly, fishing communities are likely to face multiple job losses through the multiplier effects of loss of fishing opportunities, so the job market in these areas will be depressed. Vessels engaged in IUU fishing will therefore find that their costs are doubly reduced, firstly by not having to remain idle at the dockside and secondly because the labour market will be very cheap.

There is now considerable and growing concern, especially in the southern hemisphere, that the northern hemisphere's overcapacity problem will increasingly become a very strong driver behind the growth of IUU fishing.

Corruption

Corruption is a significant factor in gaining IUU access to EEZ waters in various parts of the world. The pressure for corruption will also grow when complex or expensive tracing or certification schemes are in place to try to curtail IUU fishing, since the level of sophistication in fraud will increase accordingly. Corruption is a direct cost to IUU vessels, not being relevant to legitimate vessels. In other cases even where countries may have fishing agreements there may be close relationships between the government and business interests in the third-party country and business and local bureaucracies in the countries seeking access to those waters. There is some evidence of these trends in a number of countries which have fishing agreements with the EU. Corruption is a reflection of lack of transparency, absence of good governance and market imperfections. It is in effect a payment for fishing access and rights.

Monitoring, control and surveillance (MCS)

Increased MCS leads to increased costs of IUU fishing. In Charles *et al.*'s model of illegal activity, they found that at low levels of enforcement fishers respond to increases in enforcement by increasing avoidance, but at higher levels of enforcement it becomes uneconomical to continue to do so. Thus the cost of avoidance eventually becomes greater than the benefit from fishing (the greater the time and effort spent avoiding detection, the less time can be spent actually fishing).

Increased MCS may also have an effect on legitimate fishers, but this is usually low, especially where they have VMS on board and so inspection authorities know where they are all the time. In fact, it should be the case that increased surveillance considerably benefits legitimate fishers, since it not only protects the long-term sustainability of their resource but it reduces the supply of their product and any undermining that this might have on product value.

MCS is, unfortunately, of little use in true high seas/RFMO situations, especially with regard to FONC. Although under UNCLOS these vessels and states have an obligation to act in ways which do not undermine conservation measures, there is no right of arrest of such vessels on the high seas by third parties. Arrests and prosecutions can only be brought by the flag state.[38] Thus, in these situations, increased MCS only acts to increase the costs of IUU vessels in so far as an RFMO has an agreement to deny port, landing or transhipment facilities to vessels sighted engaging IUU fishing; or in prohibiting trade in their landings or undertaking other actions in conjunction with IUU lists.

This may not increase the costs of the IUU vessels very much, and it comes at such a high cost to the MCS vessel that it is often not seen as viable to undertake high seas MCS activities. There is also a serious problem with the distribution of MCS costs within RFMOs. Some, such as NEAFC, share costs and inspection duties, but others, such as CCAMLR, have no arrangements for this – in which case costs are borne completely by the MCS vessel.

Fishing activities

Areas of fishing geographical constraints: the juxtaposition of EEZ and high seas

Since there is no third-party power of arrest on the high seas, all such arrests of IUU vessels take place either in EEZ waters or in waters adjacent to an EEZ under hot pursuit rules. The juxtaposition of EEZ and high seas areas is thus a vital economic driver for IUU vessels, and it manifests itself in several ways.

First, let us take the example of a resource that occurs in both an EEZ and in high seas adjacent to that EEZ, but over which no RFMO has authority. Any vessel can then use the high seas area as a refuge, undertaking excursions into the EEZ. Unless a patrol vessel actively engages with an IUU vessel while it is inside the EEZ it cannot undertake hot pursuit and arrest in the adjacent high seas. The risk to the IUU vessel is therefore much lower than if the resource was only available in the EEZ. Some such refuges are notorious in providing a refuge to poachers, for instance the donut hole in the Bering Sea or the waters of the South-West Atlantic which provide a refuge for squid poachers.[39]

Once again the benefit equation is skewed in favour of IUU vessels, because they cannot be arrested once in high seas waters, whereas a national vessel can be. Such an arrest would be dependent upon other evidence a state may have that its flag vessel was engaging in IUU fishing in EEZ waters or on the high seas. This would include any fishing that is contrary to its licence – for instance, continuing to fish in the EEZ after a fishery has been closed – then moving into high seas waters.

On the other hand, other aspects of this issue can economically favour legitimate vessels. For example, because port states can prohibit access to IUU vessels, legitimate vessels can be expected to have lower market access costs than IUU vessels when they fish legitimately in high seas waters adjacent to an EEZ. IUU vessels will generally have to pay higher costs for transhipment, or travel to and from high seas fishing areas, if ports in the immediate vicinity of the fishery are closed to them. By expanding the definition of FOC to FONC it is possible to include vessels that are currently

---

[38]   Freedom of the high seas is enshrined in Article 87 of the 1982 Agreement. Third parties may only arrest vessels through hot pursuit (Article 111) or which are stateless (unflagged).

[39]   See A.J. Barton, D.J. Agnew & L. Purchase 2002. "The Southwest Atlantic: achievements of bilateral management and the case for a multilateral arrangement." *Proceedings of the Symposium on International Approaches to Management of Shared Stocks – problems and future directions.* Centre for Environment, Fisheries and Aquaculture Science (CEFAS), Lowestoft 10-12 July 2002.

flagged to Members of a RFMO. Both CCAMLR and ICCAT have measures that impose costly sanctions on Members' vessels as well as Non-Members' vessels if they appear on the IUU lists.

Quality of MCS

IUU vessels can fish up to EEZ boundaries, and where there is insufficient MCS they may well enter into those parts of the EEZ which are farthest away from the coastline without fear of arrest, and therefore without penalty, and in these circumstances the cost-benefit of IUU fishing in EEZs can be modelled on both macro- and micro-economic levels.[40] It will clearly be strongly influenced by the probability of arrest and the size of the fine in the event of arrest (*i.e.* illegal fishing will occur if the marginal value of the catch, net of the expected marginal fine, exceeds the marginal factor cost – see Charles *et al.* 1999). The strength of MCS activities is of critical importance in deterring IUU activities. However it should also be noted that it is not merely the quality of MCS that is important but also the commitment by national states to the implementation of MCS and the accompanying laws on fisheries and the marine environment. This is not restricted to third-party flag IUU vessels, which may be FONC vessels, but should extend most strongly to licensed vessels. Strengthening national laws and the use of new technologies such as VMS or onboard monitoring of vessel activities will considerably assist MCS authorities, and should increase the cost of doing business as an IUU vessel.

One of the problems facing MCS authorities is the level of penalty that can be applied when an IUU vessel is arrested. In response to large-scale IUU fishing around Kerguelen for toothfish, France has arrested a number of vessels and has fined them with large bonds. In three of these cases, the flag state of the IUU vessel has taken France to the International Tribunal on the Law of the Sea (ITLOS), seeking immediate release of the vessel and considerable reductions in the level of the bond set. In the first case, regarding the Camuoco (Panama *vs.* France), France had set a bond of FF 20 million (USD 3.1 million). Despite drawing attention to the seriousness of IUU fishing around Kerguelen (estimated by France to be in excess of USD 56 million to that date) on 7 February 2000 the Tribunal found that the bond set by France was too high, and reduced it to FF 8 million (USD 1.2 million). The following factors were cited by the Tribunal in reaching its decision that the original bond was unreasonable:[41]

> The Tribunal, in a previous judgment in the 1997 M/V "Saiga" (Prompt Release) case, had determined that: "the criterion of reasonableness encompasses the amount, the nature and the form of the bond or financial security" and that the "overall balance of the amount, form and nature of the bond or financial security must be reasonable".
>
> The Tribunal, in today's Judgment, reiterated that conclusion and elaborated on a number of factors that are relevant in an assessment of the reasonableness of the bond or financial security. The Tribunal considers the following to be of relevance:
>
> The gravity of the alleged offences;
> The penalties imposed or imposable under the laws of the detaining State;
> The value of the detained vessel and of the cargo seized; and
> The amount of the bond imposed by the detaining State and its form.

---

[40] See, for example, P.J.B. Hart, "Controlling Illegal Fishing in closed Areas: The case of mackerel off Norway". Proc. 2nd World Fisheries Congress, Brisbane, 1998; and A. T. Charles, R. L. Mazany & M. L, Cross, "The economics of Illegal Fishing: a behavioural model." *Mar. Resource Economics* 14, 95-10, 1999.

[41] ITLOS press release 35; also see ITLOS press release 42 and 48 in this section.

In a second test case (18 December 2000), the Tribunal again decided that a FF 56.4 million (USD 8.7 million) bond set by France on the Seychelles-flagged Monte Confurco was not reasonable, and reduced it to FF 18 million (USD 2.8 million). However, in the final French case (regarding the Belize registered Grand Prince, 20 April 2001), the Tribunal found "that it had no jurisdiction under article 292 of the Convention to entertain the Application". The Tribunal stated that the "documentary evidence submitted by the Applicant fails to establish that Belize was the flag State of the vessel when the Application was made". France's bond of EUR 1.7 million (USD 1.7 million) was therefore upheld (Belize had asked for it to be reduced to EUR 206 149).[42]

A similar case has recently been brought by the Russian Federation against Australia. This stems from the arrest on 7 February 2002 of the Volga, which was boarded by Australian military personnel from a military helicopter on the high seas in the Southern Ocean for alleged illegal fishing in the Australian fishing zone. The vessel was directed by an Australian warship to proceed to Perth, where it was still detained. The crew of the vessel were repatriated to their respective home countries after a period of detention, with the exception of three officers who remain in Perth under court orders. The catch which had been on board the vessel at the time of boarding was sold by the Australian authorities for the amount of AUD 1 932 579.28. The Australian authorities set the amount of the security for the release of the vessel and the crew in the amount of AUD 4 177 500. The Russian Federation requested the Tribunal to order the Respondent to release the Volga and the officers upon the posting of a bond or security in an amount not exceeding AUD 500 000. What is particularly interesting about this case is that Australia actually made the arrest in high seas waters adjacent to its EEZ around Heard Island.

In making its judgement, the ITLOS tribunal has obviously learned from its previous experiences. It set a bond consisting of the value of the vessel, fuel/lubricants and fishing gear (AUD 1.9 million). Significantly, they did not consider that the proceeds of the sale of fish and bait from the vessel, which is being held in trust by the Australian authorities pending the outcome of domestic proceedings, should form part of the bond. This departs from their previous judgements, and is an important principle because it means that the company must find an *additional* AUD 1.9 million for a bond guarantee. However, they disallowed an application by Australia to include within the bond AUD 1 million for a VMS system on board the vessel. This would have been a "good behaviour" guarantee pending full trial in Australia, because – as was pointed out during the ITLOS hearing – IUU vessels are usually repeat offenders. For instance the Camuoco – which following the January 2000 ITLOS hearing of Panama *vs.* France was released on bail – was arrested on 3 July 2002 by French authorities around Kerguelen Island (again), this time under the name 'Eternal' (previously 'Arvisa 1', previously Camuoco). However, at least one judge disagreed with the court finding, and opined that such a good behaviour mechanism would be appropriate, given the high level of re-offending of such vessels.

It will be clear from the above that the level of bond considered appropriate by the Tribunal is lower than the likely annual profit of an IUU vessel (estimated in section, "Understanding the economics of IUU vessels", as USD 4.5–6 million/year). However, it is also clear that what is most important to ITLOS is the value of the vessel and its cargo, not the overall damage that the vessel can do to the resource. This is an important factor influencing the benefit side of equation 1.

---

[42]   *La Voz de Galicia*, 13 April 2002. Ultimately, the fine was not paid, and France sank the vessel off Réunion in early 2002.

Transhipment/steaming costs

As mentioned in the section, "Market control/access and the regulatory environment" above, closure of ports to IUU vessels increases their cost and can therefore reduce their profits. Several RFMOs are now developing lists of IUU vessels, with the intention of prohibiting a range of benefits being given to those vessels, including port access, flagging, access to licences for legitimate fishing in their EEZ, imports, chartering etc.[43] These should all have the effect of increasing IUU vessel costs of steaming, transhipment, hiring of crew, etc. In principle there will be a radius of action within which it will still be profitable for IUU vessels to fish. Once deterrent measures are reinforced – such as improved implementation of MCS and the control of landings – transhipment and operating costs will eventually increase to the point where IUU fishing becomes considerably less profitable or even unprofitable.

Health of other stocks

If we ignore the (probably remote) possibility that some vessel owners and crew may simply prefer to fish illegally, we could conclude that vessels engage in IUU fishing solely for financial and regulatory reasons. This also immediately implies that vessel owners would prefer to engage their vessels legally in regulated fisheries rather than in IUU fishing, as long as the opportunity to do so exists and legal fishing is sufficiently profitable. However, for a substantial and increasing number of vessels, the conditions of this proviso are not met. As estimated by FAO,[44] nearly 70% of the world's fisheries are either fully exploited, over-exploited, or in various stages of recovery from over-exploitation. Management responses to this have led in many cases to substantially reduced allowable catches, and at last action is also being taken to reduce the over-capacity that exists in most of the world's major fishing fleets. In the absence of heavily subsidised decommissioning schemes, and with ageing vessels being replaced in regulated fleets by (heavily subsidised) newer and more efficient vessels, inevitably owners of vessels unable to maintain past levels of profit will look for other options.

In previous eras, pressures such as these led to vessels looking offshore for new fishing opportunities. For example, the establishment of Exclusive Economic Zones (EEZs) led to many distant water fleets being excluded from fisheries in waters of coastal state jurisdiction, and the response was the development of then-unregulated fisheries on the high seas. This legitimate avenue is now no longer open to many of these vessels, since most of these resources are now regulated by RFMOs and many are also subject to substantial levels of exploitation.

Becoming IUU is thus sometimes the only way that a vessel or company can gain access to very limited resources. This is a very strong benefit that is not shared by legitimate vessels, because in order to remain legitimate they must refrain from IUU activities. These strong benefits (incentives) apply as much to individuals as to companies. Unemployment in the fisheries sector is likely to become considerably worse in the medium and short term in some OECD countries (especially Europe), as the true environmental and economic cost of past poor management policies becomes evident. This unemployment is (and will be) an important driver of IUU fishing, as there is simply no benefit to be gained from being a legitimate fisher as the opportunities for such fishing are not present.

---

[43]   See for example CCAMLR Conservation Measures 10-6 and 10-7, ICCAT measures 02-22 to 02-24.

[44]   FAO. 2000. *The State of World Fisheries and Aquaculture 2000.*

For this reason, there continues to be a strong emphasis on state control over nationals involved in IUU fishing (either as crew or as company beneficial owners) under UNCLOS Article 94.[45]

Company/vessel operations

Vessel economics

The financial operating costs of IUU vessels are likely to be lower than for legitimate vessels. We have attempted to estimate operating costs in the section "Market control/access and the regulatory environment", and IUU vessels will undoubtedly have lower insurance (or no insurance) costs, low compliance costs, low registration/flagging costs – especially if they are FONC flagged – as well as lower crew costs, including social security. In addition IUU vessels will not be paying the taxes and port dues which legitimate fishing vessels may incur nor will they be paying the vessel charges and tonnage charges which may be set in fishing agreements.[46] It may be that in some theatres the purchase of IUU fishing vessels is used as a means of disposing of money from other illegal operations, such as drugs. Indeed, although wildlife crime (including IUU fishing) was until recently thought to be opportunistic rather than organised, there is evidence that it is now much more organised and may have links to other aspects of organised crime such as drug and armament smuggling.[47]

It is possible to assess the cost function for IUU fishing, which would be considerably lower than non-IUU fishing. Capital costs in terms of replacement costs would be reduced as replacement values and depreciation would not be included. It is likely that there would be lower levels of capital investment. Recurrent costs (crew, maintenance) and other maintenance costs are also likely to be lower, and if fuel is obtained through informal channels, this may be cheaper through the avoidance of tax. In addition, IUU vessels will bear the costs of tonnage levies and in many cases where there are private agreements they may well pay lower costs than vessels operating in a transparent fashion under fishing agreements.

Size of company/global companies

We have previously made reference to this factor in reference to the Austral Fisheries publication on "the alphabet boats" and their links to the large multinational company Pacific Andes. Large companies have several advantages over small ones, including:

- The ability to launder IUU catch with legitimate catch.

- Access to worldwide markets, so that they can split consignments and confuse customs authorities.

- Access to bulk processing facilities, with further opportunities for disguising/hiding IUU catch.

---

[45] For example, "IUU fishing and state control over nationals", presented by D. A. Balton at the Santiago de Compostella conference on IUU fishing, November 2002, and the EU Plan of Action for the Eradication of Illegal, Unreported and Unregulated fishing.

[46] Such payments are set for different sizes of vessels and for species in the EU fishing agreements. These do not necessarily mean that the levels of such payments are correct. There may well be under-reporting or misreporting of catches by quantity and species.

[47] *International Environmental Crime: The nature and control of black markets.* Royal Institute of International Affairs, 2002, workshop report. G. Hayman & D. Brack. Sustainable Development Programme, Royal Institute of International Affairs 10 St James's Square, London SW1Y 4LE, UK.

- Complex company ownership structures, which are costly for MCS authorities to trace and easy to change.

- The ability to disguise fleet movements through rapid re-flagging, name changing, and modification of vessels which may thwart legal cases (*e.g.* in the case when two vessels are identical but carry different flags, it is practically impossible to prove unless a vessel is boarded when sighted in a particular area).

- Large fleets can indulge in "sacrifice games" where a fleet of efficient vessels is augmented by one or two slow inefficient vessels which are used as decoys. After their arrest the efficient fleet is practically assured of a period of fishing uninterrupted by a patrol vessel.

- Access to sophisticated communications and early warning systems.

These factors all tend to reduce the costs that an IUU vessel would usually expect to pay.

It should be borne in mind that for legitimate vessels some of the same advantages might apply when they are owned by a large company.

Dual-flag operations

In addition to these factors concerning company size, the make-up of the fleet in a company is particularly important. Companies attempting to operate fleets of both IUU and legitimate vessels can expect to experience lower operating costs (through paying less in licence fees and other access requirements) than companies operating only legitimate vessels. For this reason, a number of companies are suspected of operating this strategy. However, an added risk factor should be taken into account when considering the costs of companies that adopt this strategy, *i.e.*, the increasing propensity of licensing authorities to take into account the overall beneficial ownership of vessels when considering their applications for licences. This trend, which is likely to strengthen, could well redress the balance of the equation and make this a costly rather than a beneficial strategy.

Conclusions

The preceding discussion has identified the various economic drivers behind IUU fishing. These include the factors that are likely to affect the economics of IUU vessels and companies, as well as the factors that are likely to increase the incentive for legitimate vessels to engage in IUU fishing.

In an analytical framework we would anticipate that each of these factors would be examined in detail, through a combination of case studies and models, as appropriate. This should make it possible to identify which of them are likely to be the most important drivers of IUU fishing in various circumstances, for instance for different species, areas, socio-economic classes, high seas and EEZ fisheries. Judgements could then be made about what actions, addressing which drivers, would be most likely to yield results in the fight against IUU fishing.

Economic and social impacts of IUU fishing

The biological and ecological impacts of IUU fishing are well known, and fairly self-evident. Large-scale IUU fishing undermines conservation measures directed at conserving stocks and ensuring the long-term sustainability of fisheries. It is doubly insidious as, because it is extremely difficult to monitor, its effects are also very difficult to predict because reliable estimates of total extractions cannot be used in stock assessment models. Thus, a management authority may not even know that the stock is in danger until it is in a poor state. IUU fishing is, effectively, over-fishing and will ultimately lead to stock collapses, the result being that the resource is of no value to either legitimate or IUU fishermen.

IUU fishing also damages the ecosystem and associated species. As we have pointed out above, IUU fishermen do not respect the various control measures put in place to ensure responsible fishing by legitimate fishers, with the result that they may kill large numbers of other fish as by-catch, with birds, seals and whales as incidental mortality. These deaths also go unreported. There are, for instance, significant problems with by-catch of sharks in tropical tuna fisheries, and with interactions between sharks, orcas and longline fisheries. These are barely reported by legitimate fisheries, let alone IUU fisheries (ref: recent workshop in Apia). Indeed, there are anecdotal reports of IUU vessels shooting orcas in an attempt to protect fish from them.

These biological effects create significant economic and social impacts, which are explored below.

The economic impacts of IUU fishing

The macroeconomic impacts of IUU fishing

The macroeconomic impacts of IUU fishing agreements are those that will affect the level of a national or regional economy. It is fair to say that in terms of loss of economic rent and other revenues to the national economy, the major macroeconomic impacts of IUU fishing will be on low and middle-income countries that have EEZs with important fish resources and whose EEZs lie adjacent to important high sea fishing zones. It is also in the middle- and low-income countries that there are resource constraints in terms of financing and implementing adequate MCS and fisheries law. A number of publications have looked at the economic impacts of the activities of distant water fishing fleets, including IUU activities.[48] One of the problems in assessing the impacts of IUU fishing is that in the absence of adequate MCS, many countries have no idea of the extent of IUU fishing within their EEZs and in adjacent waters. It is only when it is brought to their attention – often by industry groups – that they recognise it and act to curb it.[49]

The development of IUU fishing within a country's EEZ and in adjacent high seas areas may have a number of specific impacts. These are summarised in Table 11.1.

---

[48]   Acheampong, A. (1997). "Coherence between EU Fisheries Agreements and EU Development Cooperation: The Case of West Africa". ECDPM Working Paper No. 52, Maastricht: ECDPM; Brandt, H. (1999). "The EU's Policy on Fisheries Agreements and Development Cooperation. The State of the Coherence Debate. German Development Institute", Report and Working Paper 1/1999; Milazzo, M. (1998). "Subsidies in World Fisheries – A Reexamination". World Bank Technical Paper 406; MRAG (1998). "The Impact of Fisheries Subsidies on Developing Countries". Report to DfID, Contract No. CNTR 98 6509. In association with Cambridge Resource Economics and IIED; Tsamenyi, M. and Mfodwo, K. (undated). "The Fisheries Agreements of the African Atlantic Region, an Analysis of Possibilities for WWF Intervention". Draft; WWF (undated). "The Footprint of Distant Water Fleets on World Fisheries". WWF Endangered Seas Campaign.

[49]   See, for example, "Authorities reassert fight against illegal fishing". Fisheries Information Service, 18 December 2002.

Table 11.1. The Macroeconomic and Social Impacts of IUU Fishing

PARAMETER	INDICATORS	IMPACTS
Contribution of fishing to GDP/GNP	Value added; value of landings	IUU fishing will reduce the contribution of EEZ or high seas fisheries to the national economy and lead to a loss of potential resource rent.
Employment	Employment in the fishing, fish processing and related sectors	IUU fishing will reduce the potential employment that local and locally based fleets may make to employment creation and the potential for employment creation. This is likely to be a major factor only in respect of EEZ IUU fishing.
Export revenues	Annual export earnings	By reducing local landings and not paying access dues, IUU fishing will reduce actual and potential export earnings. This, of course, will have potentially serious implications for surveillance activities, where these are supported wholly or partly by export revenues (or port revenues, see below).
Port revenues	Transhipment fees; port dues; vessel maintenance; bunkering	IUU fishing will reduce the potential for local landings and value added.
Service revenues and taxes from legitimate operations	Licence fees, revenue of companies providing VMS, observer facilities etc., and exchequer revenue from company taxes.	IUU fishing will reduce the resource which in turn will reduce the other revenues that would accrue from companies providing legitimate fishing services. This includes company taxes
Multiplier effects	Multiplier impacts on investment and employment	IUU fishing will reduce the direct and indirect multipliers linked to fishing and fishing associated activities, with the loss of potential activities.
Expenditure on MCS	Annual expenditure on MCS linked to IUU fishing.	The existence of IUU fishing will put budget pressures on MCS/fisheries management.[50]
Destruction of ecosystems	Reduction in catches and biodiversity of coastal areas	Loss of value from coastal areas *e.g.* inshore prawn fishing areas and from mangrove areas that might be damaged by IUU fishing. Reduction in income for coastal fishing communities.
Conflicts with local artisanal fleets	Incidences recorded of conflict between IUU fishing vessels and local fishing fleets.	Reduction in the value of catches for local fishing fleets. Possible increased health and safety risks because of conflicts between artisanal and industrial fleets.
Food security	Availability of fish for local consumption (food and protein balance sheets)	The reduction in fish availability on local markets may reduce protein availability and national food security. This may increase the risk of malnutrition in some communities.

The actual impact of IUU fishing on low- and middle-income countries will depend on a number of different factors. These include:

- The dependence on the fisheries sector for government revenue, export earnings, employment etc.

- The efficacy of MCS and the commitment to the control of IUU fishing.

- The size of a country's EEZ and the importance of high value fish stocks with ready markets.

---

[50] Costs of fisheries management are often high but unquantified. A useful discussion is given in "The cost of fisheries management", W.E. Schrank, R. Arnason & R. Hanneson, Ashgate, Aldershot, UK, 2003.

Countries such as the Seychelles, which are highly dependent on fisheries (notably tuna fisheries) for export earnings, licence fees, transhipment and port duties and which have a large EEZ would therefore suffer more from IUU fishing than, say, Tanzania. While Tanzania has a relatively large coastline and EEZ, fish production from marine resources at present plays a much smaller role in the national economy.

The effects of IUU fishing are often a vicious circle. Lack of resources for surveillance and enforcement at the market place or at sea will enable IUU fishing to develop, which itself will lead to lower revenues from fishing licences or other linked activities, which then feeds back to lower government resources. For instance a large proportion of fish caught in the Russian sector of the Bering Sea is reported to be caught and sold without passing through state-approved channels, which means that little income from fisheries is being harnessed by the government for re-investment in the industry or for enforcement. It also means that billions of US dollars are being lost to IUU operators annually.[51]

The microeconomic impacts of IUU fishing

The macroeconomic impacts of IUU fishing will not simply remain at the levels of the national and regional economies but will filter down to the microeconomic level, *i.e.*, to the level of villages, communities and households. In developing countries these impacts may be significant. In terms of the economic activities of fishing communities and villages, IUU fishing may have negative impacts on the revenues for fisheries, on operating costs as well as on biological stocks. There is evidence that in some cases, *e.g.* off West Africa and Mozambique, the activities of DWFNs and IUU fishing may have a direct impact on the livelihoods of fishing communities by reducing stocks or damaging gear and equipment, as well as posing a threat to the activities of artisanal fishermen, with the risk of collisions and health and safety issues. Furthermore, if IUU fishing damages biological stocks, it can reduce the availability of certain species of fish (*e.g.* small pelagics).

The social impacts of IUU fishing

The social impacts of IUU fishing are inextricably linked to the economic impacts. Where IUU fishing has a negative impact on biological stocks and marine resources, fishing revenues, licence fees, port income and associated value added, it will also have negative social impacts, particularly for middle- and low-income countries where in many cases social support and safety nets are not well established and no alternative to fishing exists. At the national economy level, negative economic impacts of IUU fishing will translate into a number of social impacts, summarised in Table 11.2. below.

It may also be thought that IUU fishing suffers from social feedback akin to the "moral hazard" described in relation to IMF funding.[52] In the case of the IMF, critics argue that the knowledge that IMF financing will be made available in the event of a financial crisis makes the crisis more likely to occur. In the case of IUU fishing, a similar social effect might take place in that the knowledge that IUU fishing is taking place might make fishermen less keen to participate in responsible fisheries. The moral hazard in IUU fisheries relates to the problem of asymmetric information where one party, *e.g.* the coastal state, does not have information on landings, catches and other data relating to fisheries exploitation by IUU fishing vessels.

---

[51]   Vaisman, A., "Trawling in the mist. Industrial fisheries in the Russian part of the Bering Sea" TRAFFIC International, Cambridge (UK), 2001.

[52]   T. Lane and S. Phillips. IMF Financing and Moral Hazard. Finance & Development, June 2001, Volume 38, Number 2. Also at http://www.imf.org/external/pubs/ft/fandd/2001/06/lane.htm.

This might, in extreme cases, lead to a classic race to fish (*i.a.* the *'tragedy of the commons'*)[53] in which a certain level of IUU fishing encourages more IUU fishing in a race to get all the fish before the stocks are depleted. Although this is only likely to occur for species with very low carrying capacities (*e.g.* orange roughy), it could nevertheless be a significant social and economic driver for IUU fishing. Similarly, it may be seen to be an undesirable but unavoidable consequence of publicity about IUU fishing, its effects, and attempts to eliminate it.

Table 11.2. Possible Negative Social Impacts of IUU Fishing at the National Level

PARAMETER	INDICATORS	IMPACTS
Employment	Employment rates in marine fishing communities	IUU fishing may lead to lower employment if it has a negative impact on stocks and the activities of artisanal and local coastal fishing activities. Less opportunity for new generations of fishers to participate in fishing
Household incomes	Gross and net household incomes	Conflicts with local fishing fleets and over-exploitation of certain species may lead to reduction in household incomes and therefore exacerbate poverty. Possible negative impacts on income distribution.
Gender issues	Employment of women in fishing and fish marketing	IUU fishing may have a negative impact on shore fishing by women and on marketing opportunities for women who in many societies have an important role in basic fish processing and marketing.
Nutrition and food security	Availability of fish on local markets at affordable prices.	In some cases, IUU fishing's negative impact on fish stocks and availability may have a detrimental impact on the availability of fish, an important source of protein in some countries.

Summary and conclusions

We have identified above:

- Ways of monitoring the extent of IUU fishing – essential for monitoring the effectiveness of measures taken to eliminate it. It is not clear that any organisation is currently doing this.

- Factors acting as the economic drivers of IUU fishing.

- An analytical approach, involving analysis of these factors through case studies and/or models, showing which of them is most important and pinpointing where, in the economic equation, IUU operations are most vulnerable.

- A system of indicators of the economic and social effects of IUU fishing which could be monitored to assess the damage caused by IUU fishing.

We have not been able to assess, in this brief study, the global extent of IUU fishing, although we have been able to give some recent estimates for various RFMOs and from other case studies. We would suggest that getting a more global picture is essential for future progress in this field.

---

[53]   Hardin, G. 1968. "Tragedy of the Commons", *Science,* 162: 1243-1248.

Neither have we made any attempt to identify which of these factors are most important. It is probable that different factors are important in different situations. For instance, in situations where a whole fleet is idle because of collapsed fish stocks, the cost of re-flagging or penalties on arrest may be a very small part of the economic equation compared to the high personal cost of scrapping.

CHAPTER 12

THE COST OF BEING APPREHENDED FOR FISHING ILLEGALLY: EMPIRICAL EVIDENCE AND POLICY IMPLICATIONS

U.R. Sumaila, J. Alder and H. Keith,[1] Fisheries Centre, University of British Columbia, Canada

Abstract

We first present a conceptual model for the analysis of the costs and benefit aspects of the risk inherent in IUU activity, then proceed to develop and present a map of IUU incidences as reported in the Fisheries Centre's *Sea Around Us project* IUU global database. This map shows that IUU activities are quite widespread geographically. We next present an analysis of the cost and benefit aspects of risks of IUU fishing, which reveals a number of interesting results, including the fact that for the cases analysed as a group even the high probability of being apprehended does not change the current favourable calculation of the potential net benefits of IUU fishing activities. Finally, we discuss three case studies using our conceptual framework, which allowed us to make some valuable deductions.

Introduction

Illegal fishing is conducted by vessels of countries that are parties to a fisheries organisation but which operate in violation of its rules, or operate in a country's waters without permission, or on the high seas without showing a flag or other markings (FAO 2001). Unreported catches are not reported to the relevant authorities by the fishing vessels or flag state, whether or not they are parties of the relevant fisheries organisation. This category includes misreported and underreported catches (FAO 2001). Unregulated fishing is normally conducted by vessels flying the flag of countries that are not parties of or participants in relevant fisheries organisations and therefore consider themselves not bound by their rules (FAO 2001).

Illegal, unregulated, unreported (IUU) fishing occurs not only on the high seas, but also within exclusive economic zones (EEZ) that are not 'properly regulated'. IUU fishing leads to the non-achievement of management goals and sustainability of fisheries (Pitcher *et al.* 2002; Corveler 2002). When stock assessments are performed on fisheries, reported catch and effort data is used. However, the underreporting of illegal catches results in the absence of a significant part of the annual catch that is not included in the assessment (Pauly *et al.* 2002; FAO 2000a). The depletion of many stocks, for example, of Patagonian toothfish (*Dissostichus eleginoides*) has occurred partly because of the

[1] We thank our colleagues, especially Louisa Wood, Robyn Forrest and Jordan Beblow (for the incidence Map), Reg Watson, Tony Pitcher, Daniela Kalikoski and Daniel Pauly for providing us with insights, information and data, Kevin McLoughlin, James Fox and Ilse Keesling for their assistance with the Indonesian case study; and Sachi Wimmer and Denzil Miller for their assistance with the Antarctic case study. We thank the *Sea Around Us project* (SAUP) and the Pew Charitable Trusts for making this work possible by initiating the IUU Global database.

inaccuracy of the catch data. Significant decreases in some fish stocks have become an increasing concern, especially because further restrictions on legal fishing can also exacerbate illegal fishing.

The issue of IUU fishing has therefore been receiving increasing attention among scholars and fisheries managers, as well as governmental, intergovernmental and non-governmental organisations. For instance, the FAO has begun the implementation of an International Plan of Action (IPOA) where all states and regional fisheries organisations are introducing effective and transparent actions to prevent, deter and eliminate IUU fishing and related activities (FAO 2003). A good understanding of the economics of IUU fishing is important in order to design appropriate measures. What are the cost and benefit aspects of risks inherent in IUU activity? This paper explores these questions. It discusses the possible drivers of risk and the costs associated with fraud, avoidance and apprehension in relation to IUU fishing activities. A model is presented and substantiated using case studies to help establish how IUU fishing vessels take such costs and benefits (monetary and social) into account when deciding on whether to engage in IUU fishing or not.

The rest of this paper is organised as follows. The first section conceptualises a model for fishers' decisions on IUU fishing. The literature is briefly reviewed, followed by a presentation of the key drivers of IUU fishing from the point of view of the violator. The formal model is detailed in Appendix 1. The third section, IUU Incidence and Case Studies", presents a global picture of IUU incidence, along with a presentation of three case studies to illustrate the scope and diversity of IUU fishing. In the last section we conclude with a discussion of the points presented.

Conceptualising a model for fishers' decisions on IUU fishing

Since the first formal economic model developed by Becker (1968) on the subject of criminal activity, the economic literature has advanced several reasons to explain why people engage in such activities. Becker (1968) and the papers immediately following him argued that criminals behave essentially like other individuals in that they attempt to maximise utility subject to a budget constraint. The economic argument was very strong in this explanation of illegal activity, embodied in what has come to be known as deterrence models (Kuperan and Sutinen 1999; Charles *et al*. 1999). These models argue that an individual commits a crime if the expected benefits or utility from doing so exceeds the benefits from engaging in legal activity. The models focus on the probability and severity of sanctions as the key determinants of compliance. More recent literature has come to recognise additional motivations, namely, that moral and social considerations play a crucial role in determining whether an individual engages in illegal activity or not (Tyler 1990; Sutinen and Kuperan 1999). With regard to IUU fishing there is evidence to support the hypothesis that moral and social considerations, as well as economics, play a role in the degree of IUU fishing that an individual decides to engage in (Kuperan and Sutinen 1999; Bergh and Davies 2004). However, the case studies discussed later in this paper indicate that moral and social considerations are weak in the case of distant water fleets, which are the predominant operators on the high seas.

Following Becker (1968), Kuperan and Sutinen (1999), Sutinen and Kuperan (1999), and Charles *et al*. (1999), we assume more explicitly that the following direct drivers and motivators play a role in whether or not fishers decide to engage in IUU fishing:

1. Benefits that can be realised by engaging in the illegal activity.

2. The probability that the illegal activity is detected or the "detection likelihood driver". This depends mainly on the level of enforcement or the set of regulations in place.

3. The penalty the fisher faces if caught.

4. The cost to the fisher in engaging in avoidance activities. This depends on the set of regulations in place and the size of the budget allocated by the fisher to this activity.

5. The degree of the fishers' moral and social standing in society and how it is likely to be affected by engaging in IUU fishing.[2]

Benefits from IUU fishing as a driver

For many fishers, the potential benefits of IUU fishing motivate them to engage in the illegal activity. To some extent the higher the economic return in a 'legal' fishery, the lower is the tendency to engage in IUU fishing. In other words, if a fisher is doing well financially, *i.e.*, making a sizeable profit from fishing 'legally', then the probability of cheating is low, whereas if the fisher is losing money and there is the potential to derive benefits from 'illegal' fishing, then the probability of cheating increases. There is also the factor of greed, *i.e.*, the fisher may be making a profit but still engages in IUU fishing because of the desire to increase profits. The following factors are important in determining the potential benefit to the fisher if they cheat:

- Catches – other things being equal, the more catch that can be realised by engaging in IUU fishing the higher the probability that a fisher will engage in IUU fishing.

- Catch per unit effort or the time it takes to catch the fish is also a consideration, since the more time spent searching for fish to and from the fishing grounds, the higher the cost as well as the increase in the probability of getting caught.

- Price – this is related to catch and if prices are too low then in most cases there will not be a financial incentive to cheat. This logic breaks down when food security is a driving factor. However, for the purposes of this study food security is not the focus.

- Cost of fishing, which includes consideration of the cost of labour, capital, fuel, licence and royalty payments, etc.

The expected penalty drivers

Detection likelihood driver: Other things being equal, the higher the probability of getting caught the lower the incentive to cheat, and hence, the higher the risk that the violator will be caught. The major factors that contribute to this driver are, *i)* the effectiveness and efficiency of the enforcement system; *ii)* social acceptance of cheating in society; *iii)* awareness of the regulations; and *iv)* the level of non-governmental or private organisation involvement in detecting infringements.

The avoidance driver: A rational fisher engaging in IUU fishing in a situation where there is some degree of enforcement will take measures (such as engaging in transhipment of catch) to reduce the chances of being detected; this is denoted avoidance activity.

The penalty driver: The severity of the penalty when caught is also an important driver in the decision of a fisher to cheat. Other things being equal, the more severe the penalty the lower the likelihood of cheating; this driver is related to the detection likelihood driver in that if there is no enforcement then the severity of the penalty is meaningless. For example, when a net ban was instituted in Florida, the county with the highest level of NON-compliance was also the county that either dismissed the most cases or imposed the minimal economic penalty to net fishers (Kely 2002).

---

[2] It is worth noting that here we are not dealing with small-scale fisheries, where community cohesiveness allows for social control (see example, Ruddle, 1989).

The types of penalties that are applied include: *i)* the amount of the fine; *ii)* confiscation of the boat; *iii)* confiscation of the catch; *iv)* exclusion from the fishery; and *v)* history of prosecutions/application of the penalty. For example, in Senegal the fines are doubled for foreign fishing vessels that repeatedly operate outside of the fishing access arrangements.[3] In the state of Victoria in Australia, first time offenders are served with a Penalty Infringement Notice (PIN), however, the penalty for repeat offenders can include seizure of the catch and vessel, imprisonment and other penalties (Parliament of Victoria 2000).

Moral and social drivers

Many have observed that the deterrence model alone does not adequately explain why people engage or choose not to engage in illegal activities such as IUU fishing; rather moral and social factors also play a crucial role (Tyler 1990; Sutinen and Kuperan 1999). It has been observed that a given population of fishers, for example, can be classified into *i)* chronic violators, *ii)* moderate violators and *iii)* non-violators (Kuperan and Sutinen 1999). Chronic and non-violators generally make up a small portion of a given population. The former have the tendency to undertake IUU activities no matter what, while non-violators will not engage in IUU fishing under any condition. Moderate violators, on the other hand, will only bypass regulations if the potential economic gain is high enough to cover the potential penalty they may face, given the size of the penalty when caught and the probability of being caught. Secondary influences that may affect the decision of moderate violators to engage in IUU fishing are the legitimacy of the regulation (and fishery management organisation), and the norms of behaviour, including both the general behaviour of the fishers and the moral code of the individual fisher (Tyler 1990; Kuperan and Sutinen 1999). Gauvin (1988) and Bean (1990) have estimated that about 10% of fishers in the Massachusetts lobster and Rhode Island clam fisheries flagrantly violate major regulations. The other 90% of fishers normally comply with regulations. These estimates are not just relevant to these two fisheries: Feldman (1993) presents a number of estimates for other fisheries that are similar to these numbers.

A formal model

From the above conceptual framework, we developed a formal model of the economics of IUU in line with the literature (see Appendix 1). According to this model, the objective of the fisher is the maximisation of the potential gains from engaging in IUU fishing moderated by moral and social considerations. If the fisher engages in IUU activities in a fishery in which there is almost no regulation, then the fisher faces close to zero probability of being caught – implying that the expected penalty the fisher faces is also close to zero. In this situation there will be very little need, if any, to undertake avoidance activities. Moreover, the IUU fisher will choose the level of IUU activity such that the marginal revenue from the activity is greater or equal to the marginal cost of engaging in the activity, which in this study equates to the sum of the marginal cost of fishing and the marginal moral and social cost of engaging in IUU fishing. If the fisher undertakes IUU fishing when there is enforcement, then the fisher will choose the level of IUU fishing such that marginal revenue is equal to or greater than the sum of marginal cost of engaging in IUU fishing, and the potential marginal fine if caught.

IUU incidence and case studies

First, we present a general picture of IUU fishing based on the Sea Around Us project (SAUP: www.seaaroundus.org/) IUU database, and then we present and analyse three case studies using the conceptual framework and model developed in this paper.

---

[3]   See http://www.fao.org/docrep/V9982E/v9982e3n.htm.

The three case studies are selected to give a varied coverage of the different situations under which IUU fishing takes place. The Namibia case study gives us the opportunity to describe the level of IUU fishing in waters that went from virtually zero regulation to a situation with a relatively good level of regulation. The Patagonian toothfish example is presented to illustrate how high market prices can be the key driver for IUU fishing.[4] The Northwest Australia case study is presented to illustrate how fishers will shift to illegal practices if there are more abundant and well managed resources in other national waters despite the risk of detection and apprehension.

General picture of IUU fishing in the world

Figure 12.1. below summarises IUU incidence in the world. This is a map developed from the SAUP database on global IUU fishing at the UBC Fisheries Centre. It contains data on discards and unregulated fishing activities that have been extracted from government fisheries department publications (such as annual reports and media releases) and databases, and data on illegal fishing activities that have been described in the media (e.g. Intrafish, FIS), fisheries management reports and peer-reviewed literature (see Pitcher et al. 2002). The data is spatially referenced by FAO area or sub-areas depending on the level of detail provided. The analyses (Figure 12.1. and Table 12.1.) presented here are based on incidences that are published and are therefore possibly biased to those cases where a large fine is handed down or the offence had a significant impact on the environment or fishers. It is worth noting that both the database and the map are 'living' research products as they are constantly being improved as more data is accumulated (see www.seaaroundus.org for updates).

Figure 12.1. represents the spatial distribution of vessels incriminated in IUU activities. Most of these observed/reported IUU activities are in the EEZ of the country detecting the infringement. Our data indicates that fewer IUU activities are reported in the northern hemisphere. This may be a reflection of the resources expended on monitoring, control and surveillance. Nevertheless, the map does indicate that even with the limited information we currently have, IUU fishing is widespread spatially.

---

[4]   We find this point to be interesting and important to make, even though the current paper focuses on risk issues.

Figure 12.1. Number of Incriminated Vessels for Fishing Illegally Between 1980 and 2003

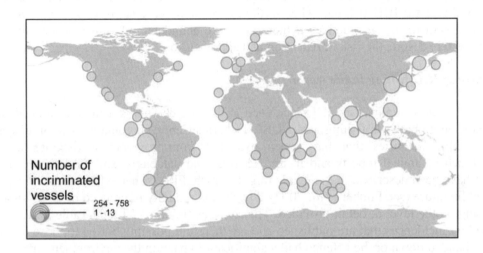

Source: Based on *Sea Around Us* IUU database; www.seaaroundus.org.

Cost and benefit aspects of risks inherent in IUU activity

Table 12.1. is a representation of the model presented in Appendix 1, except that the moral and social components are not included. This is because for the cases presented in the table, these drivers of IUU fishing are at best very weak. We have also implicitly assumed that the cost of any avoidance activity by a given vessel is included in the vessel's variable cost (see below), and the benefit of such action to the vessel is to reduce the effectiveness of monitoring, control and surveillance (MCS) activities (that is, reduce θ) for the vessel. The table lists a number of IUU fishing vessels that have been apprehended while illegally catching fish in different parts of the world. The first entry for instance, is a Spanish vessel apprehended by Australian authorities. The vessel, at the time it was apprehended, contained 116 tonnes of Patagonian toothfish with an estimated market value of USD 630 000. This vessel was fined USD 435 000. The 'Catch' and 'Fines' Columns are completed with actual data. The numbers in italics in the 'Value' Column are calculated using the reported IUU catch and the global price of the fish in question. US prices (computed using data at http://www.st.nmfs.gov/commercial/landings/gc_runc.html) are used as proxies for global fish prices. This is reasonable given that recent studies have demonstrated that prices for many fish species tend to be co-integrated (Asche *et al.* 1999). The variable cost of fishing as a percentage of landed value was calculated using information in Lery *et al.* (1999).

Recall that θ denotes the probability of detection of IUU fishing – it is therefore crucial in the calculation of the cost and benefits of the risk inherent in IUU fishing. The current lack of data does not allow us to say what the value of θ is for the cases in Table 12.1., but it is probably safe to say that many of them will have probabilities of detection that are well below 0.2 or a 1 in 5 chance of being detected. More work to determine prevailing detection probabilities for IUU activities in different fisheries around the world will be very useful in furthering the current analysis. This will also increase the utility of this work to fisheries managers in their effort to tackle the problem of IUU fishing.

Given the data situation, we explore the question of whether the potential benefits of engaging in IUU will be greater than the potential costs when $\theta = 0.2$, given the fines imposed, the value of the catches, and the variable cost of fishing (assuming fixed costs to be sunk). In other words, will the ratio of potential total costs to expected revenue from IUU fishing be greater than or equal to 1? Table 12.1. shows that only four of the 16 cases proved to be uneconomical, with a 1 in 5 chance of being detected. Similar calculations when $\theta = 0.05$ and 0.1 showed that the total potential cost exceeds the expected revenue only for Case 15.

Another interesting question explored is, what fines should have been imposed on each of the cases in Table 12.1. to make the costs aspects of risk at least equal to the benefits aspects for an MCS system when the probability of detection, $\theta = 0.2$. The calculations show that on average, for the cases studied, current penalty levels will have to be increased 24 times to ensure that IUU fishing is uneconomic. The equivalent numbers when $\theta = 0.05$, and 0.1 are 173 and 74, respectively.

Table 12.1. Cost and Benefit Aspects of Risks of IUU Fishing when there is a 1 in 5 Chance of being Apprehended (*i.e.* $\theta = 0.2$)

Cases	Vessel/ Gear	Arresting country	Fishery	Catch (t)	Catch value (USD)	Expected revenue[1] (USD)	Variable cost[2] (USD)	Fine[3] (USD)	Expected Penalty[4] (USD)	Total cost[5] (USD)	Total cost /expected revenue[6]	New Fine[7]
1	Longline	Australia	Patagonian toothfish	116	630 000	504 000	(0.70%) 439 091	435 000	87 000	526 091	1.04	0.75
2	Trawler	Not reported	Cod & haddock	24	1 138	916	(0.66%) 747	22	4	752	0.83	38
3	Boat/dive gear	Australia	Abalone	11 000	75 000	60 000	(0.70%) 52 500	26 250	5 250	57 750	0.96	1.40
4	Longline	Chile	Patagonian toothfish	33	610	488	(0.45%) 273	420	84	357	0.73	2.55
5	Trawler	Not reported	Finfish	Not reported	6 250	5 000	(0.70%) 4 375	2 250	450	4 825	0.97	1.4
6	Trawler	Russia	Cod and haddock	48	1 138	910	(0.66%) 747	22	4	752	0.83	38
7	Trawler	Argentina	Fish including anchoveta	2 685	485 985	388 788	(0.62%) 300 399	24 138	4 828	305 227	0.79	18
8	Pots	Japan	Crab	60	47 820	38 256	(0.62%) 29 648	7 414	1 483	31 131	0.81	5.8
9	Longline	Not reported	Patagonian toothfish	200	2 200 000	1 760 000	(0.70%) 1 533 333	100 000	20 000	1 553 333	0.88	11
10	Bottom trawler	Mexico	Shrimp	5	27 575	22 060	(0.56%) 15 337	5 455	1 091	16 428	0.74	6.2
11	Pots	Russia	King crab meat	0.214	2 456	1 965	(0.66%) 1 621	34	7	1 628	0.83	50
12	Bottom trawler	Russia	Alaska Pollock	6	11 022	8 818	(0.39%) 4 304	1 171	234	4 539	0.51	19,4

Table 12.1. Cost and Benefit Aspects of Risks of IUU Fishing when there is a 1 in 5 Chance of being Apprehended (*i.e.* $\theta = 0.2$) (cont.)

Cases	Vessel/ Gear	Arresting country	Fishery	Catch (t)	Catch value (USD)	Expected revenue[1] (USD)	Variable cost[2] (USD)	Fine[3] (USD)	Expected Penalty[4] (USD)	Total cost[5] (USD)	Total cost /expected revenue[6]	New Fine[7]
13	Gillnet	Russia	Greenland halibut	132	119 328	95 462	(0.59%) 69 833	690	138	69 971	0.73	185
14	Longline	Canada	Sablefish	2.72	12 063.2	9 651	(0.70%) 8 408	15 385	3 077	11 485	1.19	0.40
15	Longline	Mauritius	Patagonian toothfish	200	440 000	352 000	(0.70%) 306 667	2 400 000	480 000	786 667	2.23	0.38
16	Longline	Uruguay	Patagonian toothfish	201	2 122 560	1 689 600	(0.70%) 1 472 000	1 632 000	326 400	1 798 400	1.06	2.60

Notes:

1 Expected revenue = $\theta *0 + (1-\theta)*$catch value. This captures the fact that when apprehended catch from IUU fishing is usually confiscated.
2 Variable costs are the cost of operating the vessel as distinct from the fixed costs of acquiring the vessel.
3 Reported fine imposed, assumed to be the total fine including the confiscation of catch/vessel, flag state's fine, where applicable.
4 The product of the probability of detection (in this example 0.2) and the fine imposed.
5 The sum of variable cost and the expected penalty.
6 The ratio of the potential total cost of IUU to the potential value of engaging in IUU. A value of 1 and above implies engaging in IUU activity is not a profitable proposition.
7 The number of times the reported fines need to be multiplied by in order to make the potential gain equal to the potential cost of engaging in IUU when $\theta = 0.2$. This gives an average multiple of about 24. Similar calculations for $\theta = 0.05$ and 0.1, shows that multiples of 173 and 74 are needed. From the results presented above one can make the following observations:

- Given the current combination of fish price, IUU catch levels, variable fishing cost levels, and the level of fines imposed in vessels caught engaging in IUU fishing, the current fine levels will not serve as a deterrent for two-thirds or more of the cases reported in Table 12.1 when the probability of detection is equal or less than 0.2.
- For most of the cases, the probability of detection must be well above 0.2 for it to serve as a deterrent.
- The reported fines for the cases analyzed will have to be increased many-fold, even for fisheries that are monitored, to ensure that there is a 1 in 5 chance of being detected, for the fines to serve as serious deterrents to IUU fishing.

209

The Namibian EEZ

Background

Namibia has an extensive coastline bordering the highly productive northern Benguela current ecosystem, which is dominated by pelagic fish, mainly sardine, anchovy and horse mackerel. The demersal ecosystem is dominated by valuable stocks of hake. The food web off the Namibian coast is mainly represented by seals as the top predators, hake, squid, snoek, and chub mackerel as the piscivorous species, and horse mackerel, round herring, saury, sardine and anchovy as the main pelagic prey, and lightfish, lanternfish and goby as the main demersal prey (Shelton 1992; Palomares and Pauly 2004).

IUU fishing before independence

Before independence in 1990, the Namibian EEZ suffered illegal, unreported and unregulated fishing because it was virtually a free-for-all fishing zone. There was little or no surveillance of most fishing operations in Namibian waters, hence there was a massive race for the fishery resources of Namibia mainly by distant water fishing fleets (DWFs) beginning in the 1960s (Anon. 1994). Fleets from the former USSR and Spain arrived in 1964; followed by Japan, Bulgaria and Israel in 1965; Belgium and Germany in 1966; France in 1967; Cuba in 1969; Romania and Portugal in 1970; Poland in 1972; Italy in 1974; Iraq in 1979; Chinese Taipei in 1981; and South Korea in 1982 (FAO Yearbooks of Fishery Statistics for hake). Sumaila and Vasconcellos (2000) demonstrate that the impacts of this were huge and negative, resulting in the over-exploitation by distant water fleets with the consequence that the newly independent Namibia inherited an altered ecosystem whose productive potential was severely reduced (Willemse and Pauly, 2004). In addition, the country suffered huge socio-economic losses during this period due to the activities of DWFs.[1]

Fishing activities in Namibian waters were not regulated so reporting of catches was very poor, and also many who would normally not fish there without permission, fished there illegally anyway. This 'free for all' situation implied that all the direct drivers of IUU fishing were skewed in favour of fishers who want to undertake IUU fishing activities – what we call 'the IUU Fisher's Paradise'. The potential of gaining additional revenue from IUU fishing without any risk of being caught is high. Penalties are non-existent, and the violators enjoyed zero cost of engaging in avoidance activities. In terms of our model, the situation in Namibia's EEZ during this period is captured by the optimality condition expressed by equation (3).

The revenue side of this equation was quite high due to the huge quantities of fish caught by distant water fleets in the years prior to independence. The official statistics, which are suspected of being underestimates, shows that 1.4 million tonnes of sardines were caught in 1968. Before these large catches, pre-1968 catches were reported to have been between 100 000 to 600 000 tonnes, most of it taken by distant water fleets. The race for Namibian hake started in 1964 and reached a peak in 1972 when 800 000 tonnes of hake were reported to have been caught. The catches were lower between 1972 and 1980 at about 150 000 tonnes. Catches increased again to around 400 000 tons in 1985, then declined again until 1991 when Namibia took full control of its resources for the first time. Again most of these catches were taken by DWFs. It is reported that up until 1985, 99% of hake catch was landed by DWFs. After 1985, approximately 90% was still landed by DWFs (Anon. 1994;

---

[1] It is probably not possible to discuss DWFs in the legal context before UNCLOS and the establishment of the 200-mile EEZ in 1977, as it cannot be claimed that the fleets were fishing illegally.

Sumaila and Vasconcellos 2000). Horse mackerel was also heavily targeted by DWFs active in Namibia's exclusive economic zone (EEZ) before independence. Annual catches were seldom below 300 000 tonnes, with the peak of 570 000 tonnes landed in 1982, according to the statistics.

It could be argued that the cost side of equation (3) was relatively low compared to the revenue side, implying that the amount of IUU fishing inputs will have to be very high before equation (3) is satisfied. Essentially, under the circumstances prevailing in Namibia's EEZ before independence, and the fact that most of the fishing was by DWFs, it could be argued that moral and social considerations were virtually non-existent. Hence, the only cost that mattered was the fishing cost, which from all indications must have been well below the revenue from IUU fishing. This scenario is in effect the IUU fisher's paradise – zero risk of being caught and penalised, and zero risk of losing moral or social standing in the societies they come from. It should be noted that this result could easily be extended to most high seas IUU fishing situations.

IUU fishing after independence

The new Namibian government that took office in 1990 put fisheries at the centre of its agenda. It made the return of full control (to Namibia) of fishing in its EEZ a primary goal of the government. Just before independence in 1990, more than 100 foreign vessels were fishing illegally in Namibian waters. During 1990 and 1991, eleven Spanish trawlers and one Congolese trawler were arrested for illegal fishing and successfully prosecuted; most of them were forfeited to Namibia by the Namibian courts. It has recently been reported by WWF (1998) that with the announcement of the EEZ regime by the independent government, there was a drop of more than 90% in the number of unlicensed foreign vessels fishing in the area. Namibia achieved this feat by quickly putting in place a fisheries management system with a strong monitoring, control and surveillance component, the primary goal of which was to restrict fishing to only those entitled to do so, and ensure that fishing activities are carried out within legal and administrative guidelines (MFMR 1999). By so doing the government of Namibia quickly moved the IUU fishing environment from an IUU Fisher's Paradise to an IUU fisher's Hell: Suddenly θ and F turned positive, immediately impacting on fishers' risk calculations and decisions on whether or not to engage in IUU fishing. Indeed, the regulators increased θ to close to 1, and F significantly in the beginning to serve as a signal to all IUU fishers that it meant business. To achieve this, Bergh and Davies (2004) report that in 2001 and 2002, 41% and 42% of the fishing industry revenue has been used to pay for monitoring, control and surveillance activities, respectively. More concretely, the annual running cost of the Fisheries Observer Agency (FOA), the organisation responsible for providing observer services to the MFMR, is about NAD 20 million[2] (Per. Comm. Mr. Hafeni Mungungu, CEO of FOA).

The other components of the optimality condition, namely, avoidance, moral and social issues also became elements that carried weight in the risk analysis of a potential IUU fisher. In the first place, because of the now significant value of θ and F, those who planned to engage in IUU fishing would most probably have to engage in avoidance activities too. This increases the total cost to them of engaging in IUU fishing, and therefore has a dampening effect on their incentive for engaging in the illegal activity. Secondly, because DWF fishing was eliminated, restricting fishing to only Namibian-based fishing companies, the moral and social standing considerations became relevant. All of these together resulted in a significant drop in IUU fishing. According to Bergh and Davies (2004), the goal of restricting fishing activity to only those entitled has been fully achieved, while more work is needed with respect to the goal of ensuring that fishing activities are carried out within administrative and legal guidelines, because this goal has only been partially achieved so far.

---

[2]   USD 1 equal to NAD 7.07 (Namibian dollars) in March 2004.

There are many reasons for the success of Namibia in tackling its huge IUU fishing problem after it gained independence. Some of these are specific to the country while others can be generalised to other countries. A key positive factor for Namibian fisheries is the fact that it is a major contributor to the country's national wealth. It is estimated that fisheries contribute over 10% of the country's national income (Lange 2003). This prominence accords the fishing sector high national priority, which allows the Ministry of Fisheries and Marine Resources (MFMR) to get the resources it needs to put in place an effective MCS system. A second point is the fact that Namibia had a number of negative examples from around the world on how not to manage its fisheries because it attained nationhood only recently. This opportunity appears to have been used effectively – to the extent that the Namibian Constitution has sustainability requirements stipulated in it. The legal system was also designed to give the courts the power to deal with illegal fishing activities. The geography of Namibia also played a part. The coast of Namibia is shielded from the population by a strip of harsh desert land resulting in only two major fishing ports along its coast. This meant that coastal fishing communities never really developed along the coast. This had a positive socio-cultural consequence on the management of the resources in that there was no coastal community with long-term claims to fishing rights on the marine resources. Finally, the country took drastic and dramatic initial enforcement of fisheries regulation in its EEZs, which sent a clear signal to potential violators, with a huge positive effect on keeping IUU fishers out of the country's EEZ.

Patagonian toothfish

Background

The Patagonian toothfish is a long-lived, slow growing species. It matures at the age of more than 10 years, lives up to 50 years, can reach lengths of up to 2 metres and weighs up to 130 kg (TRAFFIC 2001). Larger fish normally inhabit greater depths while younger toothfish live in shallower waters (depths ranging from 400 to 3 500 m). It preys on fish, crab, squid and prawns and is preyed upon by sperm whales and elephant seals. Due to its slow growth and late maturity, this species is extremely vulnerable to overfishing. Other Patagonian toothfish market names are Bacalao de profundidad (Chile), Butterfish (Mauritius), Chilean Sea Bass (USA, Canada), Robalo (Spain) and Mero (Japan) (TRAFFIC 2001). It is worth noting that until the late 1980s, the then Soviet Union caught the largest quantities of toothfish (CCAMLR Article XXIV). At present the main catch countries are Chile, Argentina, France, Australia, UK and South Africa (TRAFFIC 2001). Most IUU catch is landed in Mauritius, as the catch documentation scheme has effectively eliminated IUU catch landings in CCAMLR member countries (TRAFFIC 2001). Toothfish catch is exported primarily to Japan and the US, as well as Canada and the EU (TRAFFIC 2001).[3]

The Convention on the Conservation of Antarctic Marine Living Resources (CCAMLR) was established in 1982 with headquarters in Hobart, Australia. Its aim was to deal with the depletion of krill and other fish stocks in the Southern Ocean, in particular the Patagonian toothfish stocks. There are 39 participating countries on the Convention, of which 24 are member countries. CCAMLR governs most of the waters in the Antarctic region. Although there are regulations set by CCAMLR as conservation measures, large quantities of toothfish are still caught illegally in the EEZs of the Sub-Antarctic Island territories and in the Southern Ocean area managed by CCAMLR. Unregulated and unreported catches occur inside and outside of the CCAMLR area (TRAFFIC 2001). Any country within the CCAMLR area governs its own EEZs but operates under regulations (catch limits, gear restrictions) set by CCAMLR.

---

[3]   It should be noted that the IUU trade follows the legal market to the importing countries once it has been landed at a port.

Patagonian toothfish is caught in the Antarctic Southern Ocean, which is divided into three statistical areas defined by FAO and governed by CCAMLR. Area 48 covers the Atlantic Ocean Sector; Area 58 covers the Indian Ocean Sector and Area 88 covers the Pacific Ocean Sector. The Southern part of Area 58 and the southern part of Area 88 are prime target areas for catching Patagonian toothfish. Within the CCAMLR area, toothfish fishing hot spots are located near Prince Edward Islands, South Africa (Sub Area 58.7); Crozet Islands and Kerguelen Islands, France (Sub Area 58.6); and Heard and Macdonald Island, Australia (Sub Area 58.5).

The evolution of the toothfish fishery

The Soviet Union started fishing for toothfish in the mid-1980s after the decline of the icefish fishery (Kock 1991; 1992). The development of the legal toothfish fishery followed the collapse of the Austral hake, *Merluccius australis*, and Golden Kingclip, *Genypterus blacodes*, fisheries in Chilean waters and of some of the Northern fish stocks (TRAFFIC 2001). Until 1997, there were virtually no regulations on the amount of toothfish catch, implying that the relevant optimality condition is that expressed in equation (3), with zero probability of being caught. There were catch limits placed on the longline toothfish fishery in 1990 but these were not actively enforced. The incentive to engage in IUU fishing was consequently high since the probability of being caught was zero even within the EEZs in the CCAMLR area. However, in 1997, it was reported that 80-90% of current total toothfish catch was illegal, constituting 2-3 times the legal catch limits for the fish stock. This information forced all countries with EEZs in the CCAMLR area to establish regulations and limits on the fishery, and begin to manage their waters more effectively. F and θ then assumed positive values within most countries' EEZs. θ is likely to be greater than 0.2 in Australian waters where the amount of patrol vessels is extremely high. F was at first very low as most vessels considered the small fines simply an additional operating cost and the resulting fines issued by courts were very small. However, as will be discussed below, new penalty measures issued by Australia, for example, have rendered an F value that is very high, sometimes 1, when vessels are sunk. Other CCAMLR region countries are following Australia's example.

Management schemes

More enforcement and regulation measures were brought to bear on the fishery in 1998 when all toothfish vessels operating within the CCAMLR area were required to carry a vessel monitoring system (VMS) – a satellite-tracking device to trace the co-ordinates of each vessel. Also, all vessels operating in the CCAMLR were required to mark their gear appropriately to decrease the amount of longlines cut when inspectors approached. More rigorous measures were taken in further attempts to decrease the amount of IUU fishing of Patagonian toothfish. In May 2002, CCAMLR implemented the Catch Documentation Scheme (CDS) for all CCAMLR member countries, in all areas and fisheries with vessels catching toothfish. Before the CDS was implemented, South Africa, Uruguay, Spain and Namibia, all of which are members or acceding states of CCAMLR, accepted IUU toothfish at their ports. After the CDS was implemented, Mauritius remained the only country to accept IUU toothfish, as it is not a member country (TRAFFIC 2001). The CDS tracks the trade of Patagonian toothfish at all CCAMLR members' ports (TRAFFIC 2001). The Catch Documentation Scheme aims to identify the origin of all toothfish landed or imported into countries of contracting parties. It was recommended that all toothfish landings be denied if there was no documentation to show that the toothfish had been caught within the convention area and conforming to the conservation measures issued by CCAMLR. Non-contracting parties can be issued a CDS to be accompanied and verified with all landed toothfish. As these new management schemes have been developed since 1997-2002, they have helped reduce the attractiveness of IUU fishing of Patagonian toothfish.

Benefits drivers

There is a strong economic incentive to engage in IUU activities in the Patagonian toothfish fishery because of the strong demand for the fish and the consequent high market price it commands, and the fact that stocks of the fish have been declining over time (TRAFFIC 2001). Toothfish is considered "white gold" by the commercial longline fleets (ISOFISH 1999). The market price of toothfish has increased from approximately USD 6/kg in 1996 to over USD 11/kg in 2000, an increase of almost 100% in just three years (Statistics Canada 2001), and there are still other reports that toothfish sells for even higher prices.

The variable cost estimates from Table 12.1. for toothfish longline are approximately 70% of the total catch value (Lery *et al.* 1999).[4] By using this percentage even on an annual scale, the net value of illegal catch is still very high. As indicated in the next two sections, the level of detection is very low in this fishery making these profits substantial and attractive to fishermen.

Table 12.2. Estimated Annual Legal and Illegal Catches (Values) of Patagonian Toothfish in the CCAMLR Area

(all values in USD million, except price per kg which is in USD)

Year	Legal catch (t)	Illegal catch (t)	Price per kg	Illegal catch Value	Variable costs [1]	Net Value
1996/97	32 736	68 234				
1997/98	27 868	26 829	6.05	162	113	48
1998/99	37 319	16 636	9.11	151	105	45
1999/00	25 242	8 418	11.19	94	65	28

[1]. Variable costs estimated from Lery *et al.* (1999) used from Table 12.1. for longline vessels catching Patagonian toothfish.

Detection drivers

The development of governance over the Patagonian toothfish fishery has increased significantly since the fishery was first established. This case study can be divided into two time periods: before there were any regulations on the fishery and after the regulations were set to conserve the much depleted stocks. There are certainly numerous organisations and countries working together to stop IUU fishing of toothfish. Although many conservation measures have been implemented, due to the large fishing area and the high level of co-operation needed to combat illegal fishing, the detection of IUU fishing in this fishery is still relatively low, which probably implies that little of such activity is currently captured in the SAUP database.

The likelihood of being caught is fairly low outside the CCAMLR area since surveillance is very costly (TRAFFIC 2001). The Australian government apprehended a vessel at a cost of AUD 1 million and 80 days of pursuit (COLTO 2003). CCAMLR does not carry out any enforcement activities itself, but rather each country within the area is responsible for its own waters. Some countries – such as Australia, South Africa and France – are taking rigorous enforcement actions. For example, Australia has prohibited all toothfish longline fishing in its EEZ and patrol with armed vessels (COLTO 2003). The Catch Documentation Scheme and the Vessel Monitoring System are designed to make it difficult for vessels to land illegal toothfish or fish in illegal areas. The main obstacle to decreasing the amount of toothfish catch is the lack of co-operation from all member countries. This case is more complex

---

[4]   These estimates should be treated with caution as costs may differ between "legal" vessels and IUU/FOC vessels.

than the Namibian case because there are so many countries involved. Non-contracting countries who are invited to CCAMLR meetings and who are aware of the concerns about IUU fishing activities for toothfish are still known to issue Flags of Convenience (FOC), for example, Belize and Panama (TRAFFIC 2001).

Since the implementation of the CDS and VMS as well as port inspections, illegal catches have decreased from about 68 200 tonnes in 1996 to 8 400 tonnes in 2000 (CCAMLR 1998; 1999; 2000). The estimated legal reported catch of toothfish was 51% of the total catch in the CCAMLR area and IUU landings were 49% from 1996-1999. After the CDS was implemented in 2000, IUU landings decreased to 25% of the total catch (CCAMLR 1998; 1999; 2000). The decrease in illegal catch could be due to the increased port inspections and the CDS and VMS projects, but unfortunately are more likely due to the underestimation of the catch due to transhipment activities, underreporting and misreporting, as Japan and the US have not observed a decrease in imported catch (TRAFFIC 2001).

The VMS costs are borne by individual CCAMLR member countries. Each country needs one base station to monitor its own vessels at a cost of approximately USD 30-50 000, paid for by the member country.

On-board instrumentation has a capital cost of approximately USD 20 000, which is very small compared to the high prices received for even just one trip catch [see the "Volga" price below (in Penalty driver section)-AUD 1.9 million for one trip catch. (D. Miller, Executive Secretary CCAMLR. Hobart, Tasmania, Pers. Comm. 2003)]. FAO has reported that the operating costs of the FFA VMS are approximately 0.3% of all operating costs or 0.05% of the total value of production per year per vessel (2003).

Penalty driver

The maximum penalty under Australian jurisdiction when caught with illegal toothfish catch is AUD 550 000[5] along with the confiscation of the entire catch on board (Wimmer, Manager – IUU Fishing Fisheries and Aquaculture Department of Agriculture, Fisheries and Forestry Australian Government, Pers Comm. 2003). More recently a new law has been passed that increased the maximum penalty to AUD 825 000 for vessels longer then 24 metres (COLTO 2003) as well as recovering the cost to pursue the vessel. However, in the court system in Australia, it is very rare that a vessel will actually be fined the maximum penalty. As Australia is the leading enforcement country with regard to IUU toothfish fishing, they have managed to apprehend several known pirate vessels. Some penalties that have been enforced are as follows:

- Confiscation of catch, for example, the "Volga" had 136 tonnes of toothfish seized worth AUD 1.9 million. The International Tribunal for the Law of the Sea (ITLOS) delivered its decision on December 23, 2002, which set a bond of AUD 1.92 million to have the vessel released (equivalent to the assessed value of the boat, fuel and fishing equipment) (Rothwell and Stephens, 2004).

- Fines imposed on captain and crew of vessel, for example, an Uruguay vessel the "Viarsa 1" was fined AUD 20 000 to each crew member (crew of five men); the captain of the "South Tomi" a longliner was issued a fine of AUD 136 000 (the highest fine ever issued by Australia);

---

[5]   AUD 1=USD 0.773 in February 2004.

- Sinking of vessel, for example, the "South Tomi" was the first boat to be sunk; the "Lena" has also been ordered to be sunk.

These more extreme measures enforced by Australia are taking into account that previous fines or penalties were not substantial enough to deter the operators from continuing to fish illegally after paying their penalty. Other countries (*e.g.*, Chile, South Africa, France, etc.) have also increased their penalty fines for the conviction of IUU fishing (TRAFFIC 2001). However, although these seem like severe penalties to deter fishers from IUU activities, it is noted that one of the crewmembers on the "South Tomi" was caught again fishing illegally aboard the "Viarsa 1" two years after his boat was sunk (COLTO 2003).

Avoidance measures

Outside of member countries' EEZs, the risk of being detected and prosecuted is zero as there are no enforcement measures in the high seas. The only reported case (that we are aware of) where apprehension occurred outside a country's EEZ was when Australian patrols pursued an IUU vessel from within Australian waters into the high seas before finally seizing the vessel. In order to decrease the risk of apprehension within the EEZs, the avoidance measures taken by the vessels have been primarily in the loopholes of the management schemes enforced, *i.e.*, CDS and VMS (TRAFFIC 2001). The most frequently used avoidance measures that have worked very effectively are:

- Flags of Convenience: operators can buy a flag from a country with the assurance that the issuing country will turn a blind eye to any of the operator's activities. By flying such a flag, the vessel can move through the high seas without complying with any regulations.

- Transhipping catch and landing it under different species names, trans-fuelling and even changing crews at sea to avoid detection at ports (TRAFFIC 2001). A group of boats (the "Alphabet" boats) organised by one country, put the older, less valuable longliners in the path of patrol vessels so that the newer more valuable boats can continue fishing without being caught. The loss of older boats is considered a worthwhile business risk.

- False co-ordinates under the VMS so that the vessel country cannot identify the exact location of the boat (COLTO 2003).

Moral and social drivers

The toothfish fishery is an international fishery where most vessels are operating outside of their national waters. Since this is the case, the moral obligations or social considerations of cheating and fishing illegally are non-existent. The economic incentives of high prices are so enticing that the threat of being "blacklisted" is not enough to deter illegal fishers. However, there are many non-governmental organisations that labour to detect and publicise vessels catching toothfish illegally. TRAFFIC, a wildlife trade monitoring network, and Greenpeace Oceans-Stop Pirate Fishing are currently working to publicise illegal operators and the names of the companies and vessels involved in IUU fishing of toothfish. The Coalition of Legal Toothfish Operators, COLTO, works with these agencies to identify illegal operators. The coalition is also offering monetary rewards of up to USD 100 000 to anyone with information regarding illegal vessels (COLTO 2003). This may seem like a large amount, but the seriousness of the situation means that COLTO is willing to offer this money in the hope of minimising illegal toothfish catch. This has proven quite successful in obtaining valuable information for the apprehension of illegal vessels. ISOFISH, the International Southern Oceans Fishing Industry Clearing House, was developed as a project in 1997 to report on IUU activity over a 3-year period. This data was distributed to appropriate agencies and governments and resulted in a decrease of IUU catch, and promoted the schemes now used by COLTO and several other NGOs.

These actions are likely to improve the risk of violators losing their moral and social standing, thereby influencing the level of IUU fishing they choose to engage in.

Northwest Australia

The discussion here draws heavily on Wallner and McLoughlin (2000) and Fox *et al.* (2002). In the waters off Northwestern Australia there is a long tradition of fishing by Indonesian fishers. In 1974 a Memorandum of Understanding (MOU) between Indonesia and Australia was signed which included the area of the Australian Fishing Zone (AFZ) in which Indonesian fishers (specifically within the 12-mile territorial limit around Ashmore reefs, Cartier Island, Seringapatam reef, Scott reef and Browse Island – MOU Box) primarily exploit resources using small to medium-sized sailing craft. In 1989 the area accessible to Indonesian fishers (MOU Box) was extended to include the waters between the reefs negotiated in the 1974 MOU. While the early 1990s saw an increase in the number of apprehensions in this box, more recently apprehensions in the box declined (Table 12.3). However, overall in the AFZ off Northwest Australia, apprehensions have increased, with over 138 apprehensions in 2003 up from 111 in 2002 (www.fis.com)

Table 12.3. Vessel Apprehensions in the MOU Box 1988-1999

Year	Number of vessels
1988	1
1989	2
1990	2
1993	2
1994	63
1995	21
1996	6
1997	1
1998	7
1999	2

Source: Fox *et al.*, 2002.

The decline in apprehensions may be due to several factors: increased awareness of the MOU Box and its rules, decreasing fish stocks (and therefore less interest in the area), and enforcement activities acting as a deterrent.

The Australian government undertakes regular aircraft and vessel surveillance patrols in the area. These patrols have a multitude of purposes – including detection of vessel fishing illegally in the AFZ. Between July 1992 and November 1994, 38% of the often motorized and large Indonesian vessels sighted by air surveillance were fishing illegally in the AFZ. Research by Campbell and Wilson (in

Wallner and McLoughlin 2000) identified five Indonesian fisheries in the AFZ, including *i)* shark line and longline fishery, *ii)* sedentary species (trochus/trepang) fishery, and *iii)* demersal finfish fishery, while the remaining two fisheries lacked sufficient detail for further analysis. We will structure the rest of the discussion in this section around these fisheries.

Shark line and longline fishery

This fishery is primarily based outside of the MOU Box and fishers are often detected and directed to the MOU Box or apprehended. Recently this fishery has been focusing more on the MOU Box. Although the fishery has been established for a long time, the recent rise in the price of shark fin from IDR 150 000/kg (USD 60)[6] for quality cuts in the early 1990s to IDR 600 000 (USD 75) for first class fin in 2002 has resulted in a surge in fishing activities. The increased value of shark fin has generated an increase in effort and catches in this fishery, and an increase in illegal vessels (motorized) fishing in the MOU Box as well as areas outside of the MOU Box since 1988. A fishing trip for shark fin catches 5 to 6 kg/vessel worth approximately IDR 3.6 million (USD 432).

Benefits drivers

The fishing effort in this fishery in the early 1990s is estimated at about 5 000 boat-days and shark catch at 800 tonnes, with approximately 200 tonnes taken illegally (Wallner and McLoughlin 2000). The apprehension rate for illegal fishers (primarily motorized vessels) is 25% (this equates to 80 vessel incursions per year) and they spend 3-4 days in the AFZ before being apprehended. It is estimated that boats that are not detected spend approximately 7 days in the AFZ. Wallner and McLoughlin (2000) caution that a number of assumptions have been made in deriving these estimates. Some shark fishers earn IDR 400 000 (USD 100) per year fishing primarily in the AFZ. Indeed fishing in Australian waters is an important source of income for many Indonesian fishers (Fox *et al.* 2002).

Table 12.4. Estimates of Shark Fishing Effort and Catch by Boat Type 1992-1994

Boat Type	No. of boat trips	No. of shark fishing days	Mean fin catch per boat (kg)	Wet fin catch per trip (kg)	Wet shark catch per trip (kg)	Annual shark catch (t)
Sailing	160	3 200	30	130	2 600	416
Motorized (illegal)	80	420	26	113	2 260	158

Source: Based on Wallner and McLoughlin, 2000.

Although the catch per illegal boat is less than for legal boats, the number of days fishing per trip is also much lower, 5.4 days/trip compared to legal boats which is approximately 20 days/trip (avoidance behaviour). Shark fin export prices are as high as USD 120/kg. If a vessel goes undetected the value of the catch is (USD 26 x 120) USD 3 120, which makes the trip quite profitable. Fishers can therefore gain nearly the same economic benefit but in much less time.

Sea cucumber (Trochus /Mollusk (Trepang) fishery

Trepang is the principal target species of this fishery, which is focused on the reefs in the MOU Box. There is a nature reserve surrounding Ashmore Reef, which extends to the 50 m isobar and

---

[6]   1 USD = IDR 8 726.50 (Indonesian Rupiah) in April 2004.

therefore attempts to protect sedentary species. Although a vessel is present 9 months of the year, it is thought that during the other 3 months compliance is low. Over the last few years, effort has increased on reefs and shoals to the north of the MOU Box. Trepang catches are quite variable, ranging from less than 100 kg to 1000 kg/vessel trip (median catch 100 kg) with declining catches over time expressed by many Indonesian fishers (Fox *et al.* 2002). Catches of Trochus are also variable, ranging from less than 10 kg to 1000 kg/vessel trip (median catch 14 kg). Again, most illegal activities in the fishery are undertaken by motorized vessels targeting trepang. The average catch of trepang for an illegal vessel is 157 kg.

Table 12.5. Catch and Effort Estimates by Boat Type for Trepang taken from Reefs within and near the MOU Box 1992-1994

Boat Type	No. of boat trips	No. of trepang fishing days	Mean trepang catch per boat (kg)	Wet trepang catch per trip (kg)	Annual trepang catch (t)
Sail	144	4 320	196	2 156	310
Motorized	100	450	157	408	31

Source: Wallner and McLoughlin 2000.

Benefits drivers

The market price of trepang varies from USD 1.80 to USD 35.10/kg dry weight depending on the species. It is estimated that each 'legal' trip generates approximately USD 1 240 per vessel per trip, while for illegal vessels a trip is worth approximately USD 1 100. Illegal boats spend much less time fishing, approximately 4.5 days per trip compared to sail powered vessels which spend about 30 days per trip. While the catch per boat is less for illegal fishers, the daily catch rate is much higher. Many fishers consider the trip worthwhile if they return with a profit of more than IDR 2.5 million (USD 2 500), less than IDR 2.5 million is considered just a success and less than IDR 1.0 million (USD 1 000) is a significant loss and increasing debt.

Demersal finfish fishery

In this fishery, three types of vessel fish illegally in the AFZ:

- Well equipped Chinese Taipei pair trawlers with Indonesian fishing licenses or under a joint venture with an Indonesian company target red snappers and other demersal fish.

- Highly efficient Indonesian longline vessels or "ice boats" which are well equipped, including hydraulic line haulers. They carry ice so that the product is fresh when it lands in the Singapore market. Although the capacity of these vessels is 20t, most detained vessels had caught 3 to 5 t of fish after one week of fishing. Nine boats were apprehended between November 1992 and November 1994.

- Artisanal fishers from Indonesia who use 'low tech' methods. They are the most numerous group and they undertake the longest trips (average of 35 days/trip).

Benefits drivers

The data on illegal vessels in this fishery is uncertain; however, for this study we assume that most vessels fish for a maximum of 7 days in the AFZ before steaming to Singapore to sell their catches in the fresh fish markets. In this study a price of USD 25/kg for the fish is used based on Erdman and Pet-Sode (1996). Therefore the value of the catch when landed in Singapore is approximately (4 t/trip * 25/kg) USD 100 000.

Legal fishers catch approximately 175 kg per trip; most of this is dried and therefore of much reduced value. Assuming a price of approximately USD 5/kg of dried fish gives approximately USD 1 000. However, rarely is a trip just for fish, other more valuable species such as shark and trepang are included. Nevertheless, the total value is much less for legal fishers compared to illegal. Compared to the artisanal fleet, the illegal fleet of trawler and longline vessels take a relatively small tonnage of the demersal reef fish but in a very short time.

Detection Drivers

Australia has an active air and sea surveillance programme and Wallner and McLoughlin (2000) consider the detection rate to be relatively high (25%). The Indonesians also consider the probability of detection to be high (Fox *et al.* 2002).

Penalty Driver

If the vessel is apprehended, it is escorted to an Australian port and the crew detained until the case is heard in the courts. If the captain and crew are found or plead guilty, the vessel is often confiscated and destroyed, which means further hardship for the captain and crew who are in complex financial/debt arrangements with financiers in Indonesia.

According to Fox *et al.* (2002) a typical shark fishing vessel with its gear including fishing lines, hooks and nets is valued at approximately IDR 18 million (USD 1 800 to 2000). If the boat is a single owner-operator venture then the risk is concentrated in a single vessel and spread among the captain and crew. The owner's ability to generate an income is lost if the boat is confiscated and destroyed. If they are not in debt to finance the purchase of the vessel then their only recourse is to work for another vessel as either captain or crew. Their incomes drop from 30-50% of the profits to 10% or less depending on the number of crew. The debt for the cost of the hooks and other supplies (IDR 5 to 6 million) is spread among the crew. If the owner has borrowed funds to finance the vessel then the loan remains and to repay it they often become a captain or crew for the financier who dictates when and where they fish. An indebted captain is often required to sail more frequently and in riskier weather conditions by the financier to pay off the debt. Access to money lenders is costly: 5% per month compare to the bank rate of 18% per annum (Fox *et al.* 2002).

If the vessel is part of a larger fleet under a single ownership the risk of losing the vessel is spread over the fleet and the risk related to the gear is spread over the captain and crew. The impact of confiscating the vessel is much less for these operations since they can purchase a used replacement vessel at a very low price. Often the profit from two or three trips pays for the cost of the vessel for these large fleets.

Avoidance measures

Illegal vessels use faster boats as well as superior communication and navigation technology than legal sailing craft. Many vessels also use hydraulic lines. Vessel owners also stop off at the last Indonesian port, the island of Rote, to remove the engine from the boat so that they are not apprehended in the MOU Box. Much of the avoidance costs are therefore tied up in the technology. Vessels also avoid staying for long periods in the AFZ, usually spending about 25% of the time that legal vessels spend in the MOU waters. Larger vessels will dash into the AFZ, fish for a short period before dashing back into Indonesian or international waters. Other larger vessels act as mother ships and anchor just outside of the AFZ, while smaller vessels take the risk of fishing illegally for short periods of time, returning with their illegal catches to the mother ship (Wallner and McLoughlin 2000).

Moral and social drivers

For many Indonesians, the decision to fish illegally in the AFZ (using motorized vessels in the MOU Box or fishing outside of the MOU Box regardless of the vessel type) is based on the relatively abundant marine resources found in the AFZ compared to the severely over-exploited marine ecosystems in Indonesia, and the consequent prospect of good catches. Fox *et al.* (2002) also noted that "they made a conscious decision to fish there, just as their elders and ancestors had done". They felt they had no alternative, as resources in other areas were no longer available. Some fishers also said that if Australians did not utilise the resources then they thought it was not wrong to fish it (Fox *et al.* 2002)[7].

Unless they are caught, for many Indonesians a single trip can provide the same economic return as a year of fishing in Indonesian waters. In relative terms, the economic return is small compared to the fisheries listed in Table 1, but for the Indonesian fishers it is high enough to motivate them into action. For example, at Taka Bone Rate in South Sulawesi many fishers who remain in Indonesia have annual per capita incomes of less than USD 300 (Sawyer 1992). However, many fishers from Taka Bone Rate join on as crew on vessels going to Australia to fish and there they earn substantially more from a single trip. Many of the fishers on these illegal vessels are deeply in debt and desperate to reduce their debt or to provide funds needed to meet social and family obligations. Indonesia lacks a social safety net for its economically disadvantaged, and therefore the need to meet family obligations is high among fishers. For some fishers there is an additional social driver due to the long history of Indonesians fishing in the area and therefore a sense of moral right to fish irrespective of vessel restrictions. Fox *et al.* (2002) interviewed Indonesian fishers and many expressed the view that Australia has accommodated traditional fishers through a MOU, but only if they fish using traditional vessels, which are usually sail-powered and therefore less efficient and more time consuming than motorized vessels.

Discussion

The economic gains from IUU fishing are often significant enough to motivate fishers to engage in these activities. In some cases, for example, the high valued Atlantic tuna fishery where high prices have lead to an increased amount of IUU fishing, ICCAT has estimated that Flag of Convenience (FOC) vessels take 10% of all tuna catches by IUU fishing, which is unaccounted for in stock assessments. Another case, of course, is the Patagonian toothfish fishery discussed above that has been fished down quite severely because of IUU fishing, to the extent that it is now endangered. In this case, the incentive is very high as Chilean seabass sells on the illegal market for approximately USD 24 per kilo (BBC 2003). As the demand for fish in the market increases and effort limits are being placed, there are more incentives to fish illegally (FAO 2000a). As the restrictions on legal fishing become greater, with quotas set, gear regulations enforced, and stock sizes managed, there is an increase in the motivation to participate in IUU fishing. More attention therefore needs to be paid to this problem, as otherwise the current mismanagement of the world's fishery resources because of inaccurate stock assessment will only intensify.

---

[7]   See also Butcher (2002) for a similar story from Thailand.

In the case of Indonesians fishing in Australia's AFZ, the monetary stakes are relatively low. The high level of apprehensions and consequential loss of vessels, gear and catch is not a deterrent. Some fishers have had more than 22 vessels confiscated and destroyed (Fox *et al*. 2002), and yet the number of apprehensions in northern waters continues to increase. The risk of increasing their debt to financiers does not limit owner-operators and labourers from fishing illegally, and the owners of large fleets can spread the risk over the entire fleet. The lack of marine resources in their own waters, combined with few alternative income-generating activities and the returns of fishing relative to the alternatives still make IUU fishing a better choice.

It is also important to take into account the fact that there are many ways in which fishers can bypass regulations to engage in illegal fishing. Fishers can easily underreport catches and discard many low-value fish. They can also engage in transhipment at sea which is difficult to detect (Angel *et al*. 1994). There are some cases where vessels report catches of one species for another in order to avoid quota non-compliance (Angel *et al*. 1994). Some IUU fishing occurs in the high seas, which, due to its large area, is very difficult to monitor and survey (Bours *et al*. 2003). Most of the illegal fishing (breaches against national fisheries statutes) is detected in the EEZ of countries, especially where there is an aggressive surveillance and enforcement programme. However, this does not necessarily reflect the total IUU situation for two reasons:

1. Regional fish bodies have passed relatively few fishing regulations to control who has access to the resources on the high seas. The North Atlantic and the waters managed by ICCAT are the exceptions where quotas and joint regional enforcement or national enforcement initiatives encourage compliance among member states. However, if a non-member country fishes in the high seas contrary to the regulations, as seen in non-ICCAT countries fishing for tuna in the Atlantic, mechanisms for penalizing offenders are limited.

2. Similarly, regulations regarding by-catch and other non-target species caught on the high seas are generally not covered in regional fishing regulations or in required trip reporting and therefore not well captured in many databases.

In the face of these big challenges, monitoring, control and surveillance activities are still very limited in scope in many fishing areas. From 1979-1993, the estimated observer and aerial surveillance coverage of the high seas was 5%, which is not enough to catch all illegal practices. What is more, with vessels that have been caught, operators cover the fine as operational expenses, and simply purchase another vessel and start all over again (Agnew 2000). Since the net profits of each vessel usually exceed the price of the vessel, abandoning that vessel once apprehension occurs is not a major problem for most operators (Agnew 2000). Many vessels use fake operating companies to avoid having to pay fines when caught. The true identity of the vessel is never detected and the company name changes many times (ISOFISH 2000). Surveillance and enforcement on the high seas will be very expensive, making monitoring systems difficult to implement on a regular basis, especially in developing countries (Agnew 2000).

A number of lessons can be drawn from the case studies. First, learning from the Namibian experience, the incidence of IUU fishing in an area can be reduced significantly by sending strong signals from time to time to potential violators that swift action will be taken against them. Second, when NGOs and non-governmental agencies take action in an IUU related case, the probability of being apprehended increases and the significance of moral and social considerations for the fishers can be enhanced, as demonstrated by the Patagonian toothfish case study. NGOs make it a primary objective to publicise the operators or companies engaging in IUU activity. Although the social obligations are non-existent if the fishers are outside their national waters, information about their illegal activities being made public in the vessels' country of origin could provide an incentive to decrease IUU fishing. Third, the use of vessel monitoring systems is highly effective in tracking vessels, and for the operators themselves is an inexpensive tool. From the surveillance side, the implementation of VMS reduces the amount of surveillance required and therefore more time can be spent on inspections rather than finding the vessel. From the fishers' perspective, VMS increases the probability of being caught, and if they choose to continue to fish illegally, avoidance measures must be increased. IUU fishing therefore becomes less attractive. Lastly, from the Northwest Australian example we learn that measures to deal with IUU fishing when the violators suffer extreme poverty can be very challenging. Under these circumstances fines and other penalties may not act as a disincentive to IUU fishing.

Finally, we can suggest three ways in which this contribution can be extended to make it even more relevant to policy makers and managers. First, the map presented here needs more data to be fed into it. This means that more effort at building the SAUP IUU database is necessary. Second, the improved database can then be used to improve and extend the model calculations presented in Table 12.1. To further enhance the table, more effort at estimating the value of θ for different fisheries is warranted. Lastly, our observation in the last line of the preceding paragraph on how extreme poverty can pose a problem for current measures at reducing IUU fishing demands that this model needs to be extended to make it flexible in tackling IUU fishing.

APPENDIX 12.A. The Formal Model

In this section, we formalize the discussion above into a model. Following on the earlier discussion, we assume that the decision to engage or not to engage in IUU fishing depends on the potential net benefits (*NB*) from illegal fishing moderated by moral and social considerations. Let *NB* be defined in a broad sense by the following function:

$$NB = f(h(A, e, x), \theta(e, A, R), F, m(e), s(e))$$

$$(1)$$

$$NB_h > 0; \ NB_\theta < 0; \ NB_F < 0; \ NB_m < 0, \text{and } NB_s < 0.$$

Where h is the catch from IUU fishing by a given fisher; e stands for IUU fishing inputs; x is the biomass of fish available; A denotes the level of avoidance activity undertaken by the fisher; the variable R is the set of regulations in place; θ is the probability of detection; F is the penalty a violator faces when caught; m denotes the individual's moral standing, which is assumed to be inversely related to the IUU fishing inputs; and s represents the fishers social standing in society. This variable also depends inversely on the degree of IUU fishing undertaken by the fisher.

To be more specific, equation (1) is rewritten as:

$$NB = [ph(A, e, x) - T(e,A)] - \theta(e, A, R,)F - m(e) - s(e) \qquad (2)$$

Where p is the unit price of fish caught; $h_x>0$, $h_e>0$; $h_A<0$; $T(e,A)$ denotes the total cost of IUU fishing; $\theta_e > 0 \ \theta_A < 0$; $\theta_R > 0$. The first and second terms in equation (2) denote the total revenue and total cost of IUU fishing, respectively; $0 \leq \theta \leq 1$ is the probability of the fisher being caught and convicted if found engaging in IUU fishing. When there is only partially successful regulation and enforcement, the value of θ lies between 0 and 1. F denotes the penalty the violator faces if caught, and to obtain the total expected penalty to be paid by violators, the probability of detection is multiplied by F.

The optimality conditions [no 3.2]

The objective of the fisher is assumed to be the maximisation of the potential gains from engaging in IUU fishing moderated by moral and social considerations, that is, the maximisation of equation (2).

If the fisher chooses not to engage in IUU fishing, then *NB* as described in equation (2) is zero. And that is the end of the story.

If, on the other hand, the fisher chooses to engage in IUU fishing in a situation where there is close to no regulation, then the fisher faces close to zero probability of being caught, that is, $\theta \approx 0$, implying that θF is also close to zero. In this situation there will be little if any need for undertaking avoidance activities, A, hence $T(e,A)$ is reduced to $T(e)$ and $h(A,e,x)$ reduces to $h(e,x)$. The first order condition under no enforcement is therefore simply:

$$ph_e = T_e + m_e + s_e \qquad (3)$$

That is, at the optimum solution, the IUU fisher will choose the level of IUU activity as represented by the decision variable, e, such that the marginal revenue from the activity exactly matches the marginal cost of engaging in the activity, which here means the sum of the marginal cost of fishing and the marginal moral and social cost of engaging in IUU fishing. Equation (3) states that it is not enough for the fisher contemplating whether or not to engage in IUU fishing to seek to make the marginal cost of IUU fishing equal to the marginal revenue – the marginal revenue has to be more than the marginal cost to cover the loss of moral and social standing that the fisher suffers as a result of engaging in IUU fishing. In fact, it is possible that for a given fisher, the loss in moral and social standing is high enough to make engaging in IUU fishing not worth it under all possible marginal revenue scenarios. From equation (3) one can conclude that for non-violators, m_e and s_e are high enough for them to outweigh the marginal revenue from IUU fishing under all possible scenarios.

If the fisher undertakes IUU fishing when there is enforcement, that is, when $\theta > 0$, F>0 and by implication A>0, the optimality conditions become:

$$ph_e = \theta_e F + T_e + m_e + s_e, \qquad (4)$$

and

$$-\theta_A F = T_A - ph_A \qquad (5)$$

Equation (4) says that in the optimum, the fisher will choose the level of IUU fishing such that marginal revenue is equal to the sum of marginal cost of engaging in IUU fishing, and the potential marginal fine if caught. Equation (5) stipulates that the marginal gain to the fisher from engaging in avoidance activity must be equal to the marginal cost of avoidance plus the marginal loss in revenues from catch due to avoidance activity. In other words, the fisher weighs the risk of being caught and penalized ($\theta_e F$), the risk of losing moral (m_e) and social (s_e) standing in society, against the expected gain (ph_e) from engaging in the activity. Note that in the case of equation (3) the risk of being caught and penalized is not present.

REFERENCES

Agnew, D.J. (2000), "The illegal and unregulated fishery for toothfish in the Southern Ocean", and the CCAMLR catch documentation scheme. *Marine Policy* 24: 361-374.

Anderson, L.G. (1994), "An economic analysis of high grading in ITQ fisheries regulation programs", *Marine Resource Economics* 9: 209-226.

Angel, J.R., D.L. Burke, R.N. O'Boyle, F.G. Peacock, M. Sinclair, K.C.T., Zwanenburg, (1994), Report of the Workshop on Scotia-Fundy Groundfish Management from 1977-1993, *Canadian Technical Report of Fisheries and Aquatic Sciences* 1979, 175pp.

Anon. (1994), Namibia Brief. Focus on fisheries and research, Namibia.

Asche F, H. Bremmes, C.R. Wessells, (1999), "Product aggregation, market integration and relationships between prices: An application to world salmon markets", *American Journal of Agricultural Economics* 81: 568-581.

BBC (2003), "Hot Pursuit of toothfish 'pirates'", *BBC News*, 2pp.

Bean, C. (1990), "An economic analysis of compliance and enforcement in the quahaug fishery of Narragansett Bay", unpublished Masters Thesis, University of Rhode Island.

Becker, G.S. (1968), "Crime and punishment: An economic approach", *Journal of Political Economy* 76(2):169-212.

Bergh, P.E., S.L. Davies, (2001), "Monitoring, control and surveillance in: a fishery manager's guidebook". Ed. Cochrane, K.L. p.175-204.

Bergh, P.E., S.L. Davies (2004), "Against all odds: taking control of Namibian fisheries", in Sumaila, U.R., S. I. Steinshamn, M. D. Skogen and D. Boyer (Eds.). *Namibian Fisheries: Ecological, Economic and Social Aspects*. Eburon Deft, Netherlands, *in press*.

Bours, H., M. Gianni, D. Mather (2001), "Pirate Fishing Plundering the Oceans", Greenpeace International Campaign against Pirate Fishing, 28pp.

Burch, K. M. (2002), "Due Process in Micronesia", *Roger Williams University Law Review* 8: 43-92.

Butcher, J. (2002), "Getting into trouble: the diaspora of Thai trawlers, 1975-2002", *International Journal of Maritime History*, 14, 2 (December 2002), 85-121.

CCAMLR (1998), Report on the Seventeenth Meeting of the Commission, Hobart, Australia.

CCAMLR (1999), Report on the Eighteenth Meeting of the Commission, Hobart, Australia.

CCAMLR (2000), Report on the Nineteenth Meeting of the Commission, Hobart, Australia.

CCAMLR Article XXIV, www.ccamlr.org/articles.

Charles, A.T., R.L. Mazany, M.L. Cross (1999), "The economics of illegal fishing: A behavioural model", *Marine Resource Economics* 14: 95-110.

COLTO (2003), Coalition of Legal Toothfish Operators, www.colto.org.

Corveler, T. (2002), "Illegal, unreported and unregulated fishing – Taking action for sustainable fisheries", WWF New Zealand, Wellington.

Erdman , M. V., and L. Pet-Soede 1996, "The Live Reef Food Fish Trade in Eastern Indonesia", *NAGA, The ICLARM Quarterly* 19:4-8.

European Parliament (2001), Working Document 1 "On the role of flags of convenience in the fisheries sector", Committee on Fisheries, 4pp.

European Parliament (2001), Revised Working Document 3 "On the role of flags of convenience in the fisheries sector", Committee on Fisheries, 5pp.

FAO (2003), "International Plan of Action (IPOA) to Prevent, Deter and Eliminate Illegal, Unreported and Unregulated Fishing", Preliminary Draft Appendix D, 31pp.

FAO. (2002), "The Costs of Monitoring, Control and Surveillance of Fisheries in Developing Countries", *FAO Fisheries Circular* No. 976, 47pp.

FAO. (2001), *International Plan of Action to Prevent, Deter and Eliminate IUU Fishing,* 16pp.

FAO (2000a), *A Global Review of IUU Fishing. Australia,* AUS:IUU/2000/6, 53pp.

FAO (2000b), "Illegal, Unreported and Unregulated fishing: Considerations for Developing Countries. Australia", AUS:IUU/2000/18, 10pp.

FAO (2000c), "Monitoring Control Surveillance and Vessel Monitoring System Requirements to Combat IUU Fishing, Australia", AUS: IUU/2000/14. 13pp.

FAO (2000d), "The Role of National Fisheries Administrations and Regional Fishery Bodies in Adopting and Implementing Measures to Combat IUU fishing", Australia, AUS: IUU/2000/10. 22pp.

FAO Yearbooks of Fishery Statistics for Hake, Food and Agricultural Organization of the United Nations, Rome, Italy.

Feldman, P. (1993), *The Psychology of Crime*, Cambridge, Cambridge University Press.

Fox, J.J., G.T. Therik, S. Sen (2002), "A Study of Socio-Economic Issues Facing Traditional Indonesian Fishers Who Access the MOU Box", a report for *Environment Australia,* 63 pp.

Gauvin, J. (1988), "An Economic Estimation of Compliance Behaviour in the Massachusetts Inshore Lobster Fishery", Unpublished Masters Thesis, University of Rhode Island.

Greenpeace (2002), Greenpeace Oceans Campaign: Stop Pirate Fishing Homepage, www.greenpeace.org

ISOFISH (2000), "Toothfish Poachers Changing Vessel Names in an Attempt to Avoid Identification by the CCAMLR Catch Documentation XScheme", *ISOFISH Report* No. 4, 7 pp.

ISOFISH (1999), "The Chilean Fishing Industry: Its Involvement in and Connections to the Illegal, Unreported and Unregulated Exploitation of Patagonian Toothfish in the Southern Ocean", *ISOFISH* Occasional Report No. 2, Hobart, Australia.

Kely, K. (2002), "Integrated Systems Theory and the Florida Net Ban", *The Dialogue* 4(2): 3.

Kock, K.H. (1991), "The State of Exploited Fish Stocks in the Southern Ocean: A Review", *Archiv fur RFischereiwissnscaft* 41(1): 1-66.

Kock, K.H. (1992), *Antarctic Fish and Fisheries,* Cambridge University Press. Cambridge, 359pp.

Kuperan, K., J.G. Sutinen (1998) "Blue Water Crime: Deterrence, Legitimacy and Compliance in Fisheries", *Law and Society Review* 32: 309-338.

Lange, G. (2003), "Fisheries Accounts: Management of a Recovering Fishery", in *Environmental Accounting in Action: Case studies from Southern Africa* (G. Lange, R. Hassan, and K. Hamilton) Edward Elgar Publishers, Cheltenham, in press.

Lery, J.M., J. Prado, U. Tietze (1999), "Economic Viability of Marine Capture Fisheries", Findings of a global study and interregional workshop, FAO Fisheries Technical Paper. No. 377. Rome, FAO. 130pp.

MFMR (1999), "Planning in Action 1999-2000 (Our strategic plan)", MFMR, Windhoek. 30pp.

Milazzo, M.J. (1998), "Subsidies in World Fisheries: A Re-examination", *World Bank Technical Paper*, No. 406, Fisheries Series, Washington.

Ministry of Fisheries and Marine Resources, Government of the Republic of Namibia (1999), *Annual Report,* Windhoek.

Munro, G., U.R. Sumaila (2002), "The Impact of Subsidies Upon Fisheries Management and Sustainability: The Case of the North Atlantic", *Fish and Fisheries* 3: 233-290.

OECD (1997), *Towards Sustainable Fisheries: Economic Aspects of the Management of Living Marine Resources,* Paris.

Palomares, M.L., D. Pauly (2004), "Fish Biodiversity of Namibian Waters: A Review Of Currently Available Information", in Sumaila, U.R., S. I. Steinshamn, M. D. Skogen and D. Boyer (Eds.). *Namibian Fisheries: Ecological, Economic and Social Aspects,* Eburon Deft, Netherlands, *in press.*

Parliament of Victoria (2000), "Inquiry in Fisheries Management: Discussion Paper", September 2000. www.parliament.vic.gov.au/enrc/fisheries/discussion_paper.htm.

Pauly, D., V.Christensen, S. Guenette, T.J. Pitcher, U.R. Sumaila, C.J. Walters, R. Watson, D. Zeller (2002), "Towards Sustainability in World Fisheries", *Nature* 418: 689-695.

Pitcher, T.J., R. Watson, R. Forest, H. Valtysson, S. Guenette (2002), "Estimating Illegal and Unreported Catches from Marine Ecosystems: A Basis for Change", *Fish and Fisheries* 3: 317-339.

Rothwell, D.R. and T. Stephens (2004), "The 'Volga' Case (Russian Federation v. Australia): Prompt Release and the Right and Interests of Flag and Coastal States", in Scheiber, H.N. and K. Mengerink, (eds.) *Multilateralism & International Ocean Resources Law,* Berkeley: Law of the Sea Institute, Earl Warren Legal Institute, University of California, Berkeley.

Ruddle K. (1989), "Solving the Common-Property Dilemma: Village Fisheries Rights in Japanese Coastal Waters", in Berkes F. (ed) *Common Property Resources: Ecology and Community-Based Sustainable Development,* Belhaven Press, London.

Sawyer, D. (1992), "Management, Development and Resource Evaluation of an Indonesian Atoll", Masters of Economic Studies thesis, Dalhousie University, Halifax, Canada.

Shelton, P.A. (1992), "Detecting and Incorporating Multi-Species Effects into Fisheries Management in the North-West and South-East Atlantic", in A. I. L. Payne, K. H. Brink, *Statistics Canada* (2001), *Imports of Patagonian Toothfish* 1999 and 2000, Canada.

Sumaila, U.R., M. Vasconcello (2000), "Simulation of Ecological and Economic Impacts of Distant Water Fleets on Namibian Fisheries", *Ecological Economics* 32: 457-464.

Sutinen, J.G., K. Kuperan (1999), "A Socioeconomic Theory of Regulatory Compliance in Fisheries", *International Journal of Social Economics* 26: 174-193.

TRAFFIC (2001), "Patagonian Toothfish: Are Conservation and Trade Measures Working?" TRAFFIC Bulletin offprint, 19:1 18pp.

Turner, M. (1997), "Quota-Induced Discarding in Heterogeneous Fisheries", *Journal of Environmental Economics and Management* 33: 186-195.

Tyler, T.R. (1990), *Why People Obey the Law,* Yale University Press, New Haven.

Wallner, B., McLoughlin, K. (2000), "Review of Indonesian Fishing in the Australian Fishing Zone", Final report to the *Fisheries Resources Research Fund. Bureau of Rural Sciences*, Canberra, Australia.

Willemse, Pauly, D. (2004), "Reconstruction and Interpretation of Marine Fisheries Catches from Namibian Waters, 1950 to 2000", in U. R. Sumaila, S. I Steinshamn, M. D. Skogen and D. Boyer (Eds), *Namibia's Fisheries: Ecological, economic and social aspects*, Amsterdam: Eburon.

Wood, L., J. Beblow (2003), "Estimation of Global Illegal, Unreported and Unregulated Fish Catch", University of British Columbia, Fisheries Centre, 11pp.

World News: "Coalition Calls For Uruguay to Turn in Suspected Illegal Toothfish Boat", (August, 2003).

World Wildlife Fund (1998), "The Footprints of Distant Water Fleets on World Fisheries", *Endangered Seas Campaign,* WWF International, Godalming, Surrey, 122 pp.

CHAPTER 13

THE SOCIAL DIMENSION OF IUU FISHING

Jon Whitlow, International Transport Workers' Federation

Introduction

We all accept that the concept of sustainable development is built on three integral pillars: environmental, economic and social. However, in all the analyses of IUU fishing little consideration is given to the social dimension. The overwhelming concentration is on the environmental impact and on the economic or trade related areas. If the social dimension is addressed, it is generally only to examine the impact of artisanal fishing and food security. While these aspects are important, the failure to address the social aspect has led to a fixation on short-term piecemeal initiatives and a series of sticking plaster type solutions being put forward. Let's be frank — this approach has not solved the problem and is, in our opinion, unlikely to do so. The concentration on issues related to monitoring, control and surveillance may mitigate the problem, but it will not provide a complete solution, which can, in our opinion, only be achieved through the adoption of a holistic approach, which will require addressing the social dimension.

Another limitation is the refusal to look at the interrelationship between merchant shipping and IUU fishing. The vested interests of certain countries and questions of departmental jurisdiction meant that a valuable opportunity to address the issue at the last session of the United Nations Informal Consultative Process on Oceans and the Law of the Sea was lost. Instead of looking at the central problems of lack of flag State control, vessel registration and the issue of a "genuine link" in their totality, attempts were made to seek a separate approach. This is regrettable as the synergies between IUU fishing and the flag of convenience system in merchant shipping mean that such an approach will severely limit the progress that can be made in the area of IUU fishing. The merchant shipping industry is much more regulated than the fishing industry, where many of the key international instruments are poorly ratified or have yet to enter into force. However, despite the comprehensive set of widely ratified international regulations, there are still many problems in the shipping industry. The OECD Maritime Transport Committee has produced a number of documents that may also be relevant to the work on IUU fishing and we would suggest that many of the conclusions are equally applicable. A 2001 *"Report on the Competitive Advantages Obtained by Some Shipowners as a Result of Non-observance of Applicable International Rules and Standards"* showed that there was a positive economic incentive not to comply with international minimum technical standards. A related 2001 study on the *"The Cost to Users of Substandard Shipping"* found that the various costs associated with non-compliance with international standards are borne by numerous parties within the shipping industry, but not by those who use the services of such ships. A 2003 report on *"Costs Saving from Non-Compliance with International Environmental Regulations in the Maritime Sector"* examines the unfair commercial advantage afforded to sub-standard shipowners who fail to comply with international environmental regulations that apply to their ships.

While these OECD studies may be of indirect relevance in demonstrating the fundamental flaws in the regulatory system in which IUU fishing also operates, the 2003 Report on *"Ownership and Control of Ships"*, which examines mechanisms in both ship registers and corporate instruments that can facilitate the cloaking of beneficial ownership, is of direct relevance, as is the current *"Discussion paper on Ownership and Control of Ships: Options to Improve Transparency"*. These reports apply the disciplines of the Financial Action Task Force and the OECD work on the use of corporate vehicles for illicit purposes and on unfair tax competition to the maritime industry.

The social dimension

There are a wide variety of types of fishing. These range from small-scale artisanal fishers fishing on or near the coasts and returning home each day, to more sophisticated sea-going vessels operating well off the coast, to large factory fleets comprised of a variety of vessels operating for extended periods (including as much as one year) in harsh, distant waters. In distant water fisheries many fishers are employed on vessels registered in countries other than their own and the crew may be of mixed nationality. In order to examine the social dimension, it needs to be understood that many fishers are marginal or casual workers.

The independent International Commission on Shipping (ICONS) 2001 report entitled *"Ships, Slaves and Competition"*, although primarily addressing maritime transport, noted that:

> "Fishing vessels are mostly unregulated and are a particular problem for safety of life, environmental pollution and crew abuse. There was a strong call for more international regulation of fishing vessels, particularly to combat the disregard of safety standards and the abuse of crews." (para 2.18)

> "A major problem is the lack of any widely accepted global conventions on safety and personnel requirements for fishing vessels, as well as the lack of enforcement of ILO instruments on labour conditions." (para 2.20)

> "The Commission also heard of the frequent recruitment of passport holders as fishing vessel crews and of the sub-standard living and working conditions imposed on those recruited under such circumstances." (para 2.21)

The ILO Decent Work Programme focuses on four strategic objectives to:

- promote and realise fundamental principles and rights at work;

- create greater opportunities for women and men to secure decent employment and income;

- enhance the coverage and effectiveness of social protection for all; and

- strengthen tripartism and social dialogue.

It is self-evident that there is a substantial decent work deficit in the fishing industry and that this is related to the social dimension of IUU fishing. During the proceedings at the International Tribunal for the Law of the Sea in the CAMOUCO case, the Agent for the government of France:

> "mentioned the deplorable conditions of crew members on board the ships that had been arrested, with crew members often ill, badly nourished and living in unhygienic conditions close to slavery." (ITLOS/Press 34).

Occupational safety and health

Fishing is among the most dangerous of all professions and the Conclusions of a 1999 ILO *Tripartite Meeting on Safety and Health in the Fishing Industry* led to fishing being formally designated as an exceptionally hazardous industry. The international instruments that address vessel construction and the training of crews (the 1993 Torremolinos Protocol to the Torremolinos International Convention for the Safety of Fishing Vessels and the 1995 International Convention on Standards of Training, Certification and Watchkeeping for Fishing Vessel Personnel (STCW-F)) have not entered into force. As a result, there are no agreed international minimum standards in force for larger vessels (over 24 metres) that would enable a port State to exercise control over a foreign flagged vessel.

The lack of an internationally agreed regime for the enforcement of international minimum standards by port States over large distant water fishing vessels is in itself a problem. However, this is exacerbated in the case of IUU fishing operations, given that many of the vessels are old and badly maintained. The fact that IUU operations can lead to forfeiture of the vessel means that there are sound economic reasons for using old and unsafe vessels. This has considerable implications for those who serve on such vessels, both in terms of the facilities and amenities that are not available on such vessels, and also in terms of the safety of life at sea.

Employment on IUU vessels

Fishers may be employed through licensed or unlicensed recruitment and placement services or through other methods that are not consistent with the requirements of the law of the State of nationality or residence. For many years the Philippines has been requesting assistance to prevent its nationals from being employed on foreign flagged fishing vessels through informal networks which are outside the control of the Philippine Overseas Employment Administration (POEA). In these cases the fishers fly out of the Philippines on tourist visas and join the fishing vessel at a foreign port. Singapore has for many years been the first port of call.

There are many documented instances of the fisher having to pay a fee for the job, being responsible for the costs of joining the vessels and the costs of repatriation, and having the contract of employment changed when joining the vessel. The employment may be for up to two or three years, with few opportunities to leave the vessel, and with the fishers being required to transfer to another vessel while at sea. The employment of many of these fishers is a form of bonded labour.

In other cases the fishers may be migrant workers or political refugees, whose status prevents them from being able to exercise what rights they may otherwise have had.

Examples of unfair contractual terms

In this section we provide a number of examples of what we consider to be grossly unfair contractual terms. While we suspect that the vessels were engaged in IUU operations, that may not be the case. However, they are illustrative of how the social dimension affects IUU fishing operations.

The following clauses were found in a contract for 2 years which paid USD 250 per month, with no guaranteed leave or rest periods, no additional overtime pay and a no strike clause:

> "I understand fully that due to limited water supply, drinking water is supplied by ration. Therefore, sea water is to be used in bathing, washing clothes and tooth brushing."

> "Breakfast, lunch and dinner is provided for free. However, things for personal use is not given free. Snack foods such as bread, biscuit, coffee, milk, sugar, soft drinks, beer, liquor, cigarettes, soap etc. should be shouldered by the fisherman."

> "I also understand that the amount of USD 50 will be deducted by my captain to my salary every month. This will serve as my air ticket deposit in case I was not able to finish my contract but this amount should be refunded the moment I finish my contract."

A fisher was paid USD 255 basic salary per month, with no additional payments for any overtime performed, or any additional leave pay, or a share of the catch. The contract stated that the fisher was employed on board for 24 months with no entitlement to shore leave, nor any guaranteed rest periods per day, and was obliged to perform whatever work and whenever it was so decided by the Master. In addition, the employer had the right to terminate the contract at any moment, for whatever reason, with no compensation payment. The fisher was, however, entitled to free repatriation at the end of the contract or if declared unfit for work due to injury and/or illness by a doctor.

We have come across a contract for up to 3 years, where the fishers are to be paid only when on board for a specific season and were not entitled to leave payment. There was a clause providing that if the fisher obtained other employment, the crew manning agent could claim the salary for breach of contract. The agent reserved the right to withhold the last 2 months' salary and only return the money to the fisher if the fisher showed up for the next season. There was nothing in the contract regarding hours of work, rest, holidays, etc. There were a number of other clauses:

> "Every employee is required to cooperate with the company and owners/operators in their efforts to be "innocent owners"."

> "The Company assures employment for up to three (3) years… If employee accepts employment from a competitor company during said 3 years while company continues to offer employment, the company will claim Employee's pay from that company and the last 2 months (discussed in another clause) will be forfeited."

In another contract the period of employment is 13-15 months, subject to an extension or reduction at the discretion of the fishing master, with the amount of the monthly bonus payable also subject to the fishing master's discretion. There is a clause which provides:

> "The crew must work hard and obey the instructions given by the fishing master or the officers onboard."

Another contract provided that the fisher was entitled to receive a lump sum overtime payment of USD 15 per month. There were no clauses on how many hours the fisher was expected to work, nor any provision concerning rest periods, nor any entitlement to shore leave during the duration of the contract. The amount payable for death or incapacity was left to the discretion of the owner. However, there was a provision which provided that:

> "In case of death of crew, the corpse shall be cremated or shall be disposed of in the place where it occurred."

There was also a provision that if the fisher decided to leave the vessel for whatsoever reason any accumulated salary or fish catch bonuses was forfeited, as was also the case if the fisher began to think about striking to defend his rights.

Examples of abuse of fishers

In this section we provide a number of illustrative examples of gross abuse of fishers. For example, removal of the appendix as a condition of employment for Chinese fishers from Yongchuan County (Sichuan province) employed to work on foreign fishing vessels through a manning agency. In one case the fisher had to pay USD 470 in order to secure a place, then USD 49 for the operation, while wages vary between USD 130 and 180 per month.

There are cases where Philippine fishers had to pay approximately USD 450 each to be hired on 3-year contracts, with no right to enjoy any leave, for USD 200 per month and were expected to work 18 to 22 hours per day.

In some cases the alleged abuses are extreme. A Philippine fisher states:

> "I was chained for thirty days, that is for two periods of fifteen days, in a two square meter storeroom. I was not only chained but also beaten up with a baseball bat."

The reason for this treatment was that the fisher was so tired after working twenty hours per day, with just two hours sleep, that he was no longer able to work. The fisher also comments that:

> "Very often we ache all over. To take a bath or wash our clothes, we use sea water. When we ask for little water to drink, it's more likely to invite more maltreatment."

Another Philippine fisher reports:

> "We often had to sleep with our work clothes and sometimes wet working clothes. .... We were denied medical treatment and medicine... We were only permitted to eat what was left after the *** crew had eaten and were left with half finished cups of coffee to drink and food left over... We were required to massage *** officers and crew on a daily basis after our long hours of work. We were punched, kicked and beaten on the head with closed fists by the *** personnel regularly. The *** crew often grabbed our sensitive parts, applied pressure to the extent that we cry in pain. They also squeezed our necks until we fall to our knees."

Another fisher notes:

> "We were taken by force to work even we were sick. We were denied access to medication and treatment... We were given very little food and water. Most often we drink dirty water, so that some of us constantly suffer from severe stomach ache and diarrhoea. We work 20 to

22 hours daily but were only allowed some two-hours night sleep… We were hit like animals every time we commit errors in our work…"

Share system

Traditionally the income of fishers has often been directly linked to the catch and the revenue derived from the sale of the catch. However, we consider that this system leads to unsafe fishing practices and inefficient utilisation of available fish resources. In the context of IUU fishing operations it facilitates cheating the crew, who may be unaware that they are engaged in IUU fishing operations.

We consider that in the long term such "share" systems should be replaced by fixed wage systems that may, as the result of a collective bargaining agreement, possibly be supplemented by bonus systems. There should also be in place a guaranteed minimum wage system that should, in all instances, provide fishers with an income equivalent to that of comparable shore-based workers. Share-based remuneration systems, where they continue to exist, should be fair and transparent, ensure the best possible prices for the catch and enable fishers to verify the basis on which their income is calculated.

Flags of convenience

It is generally accepted that flags of convenience (FOCs) are integral to the problem of IUU fishing, and that the inability of the FOC system to exercise effective control over vessels which fly its flags is central to the problem. The 2003 G8 Action Plan on the Marine Environment and Tanker Safety stresses the need to address the lack of effective flag State control of fishing vessels, in particular those flying flags of convenience. Fishers live and work on the vessel and as international law establishes that a ship has the nationality of the flag it flies it has important ramifications for the crew, with regard to both civil and criminal jurisdiction and for their ability to exercise their human and trade union rights. Article 94 of the United Nations Convention on the Law of the Sea (UNCLOS) sets out the duties of a flag State and requires that the flag State shall effectively exercise jurisdiction and control in administrative, technical, social and labour matters over ships flying its flag. In doing so the flag State is required to conform to generally accepted international regulations, procedures and practices and to take any steps which may be necessary to secure their observance.

The flag State is fundamental to ensuring that fishers enjoy decent work and are not subject to abuse and exploitation. Fishers do not only need protection from violations of international labour standards: all basic human rights and protection from crimes against the person must also be guaranteed on board vessels, even when they are in international waters. In such cases there can be conflicting claims from different States. UNCLOS clearly places the responsibility with the flag State. However, those concerned with the application of international law view FOCs as being likely to undermine the system. The International Law Commission has expressed its concern, stating that:

> "If the ship flew a flag of convenience, the State of registration would have no interest in exercising diplomatic protection should the crew's national Governments fail to do so." (Report of the 54th Session of the United Nations General Assembly).

In view of the fact that only the flag State is entitled to make an application for the prompt release of the vessel and crew under Article 292 of UNCLOS, this is an area of concern.

Transparency of ownership is also important to fishers as this information may be vital if they try to enforce their rights and recover outstanding entitlements. The OECD Maritime Transport Committee Report on *"Ownership and Control of Ships"* (March 2003) states:

"Open registers [FOCs], which by definition do not have any nationality requirements, are the easiest jurisdictions in which to register vessels that are covered by complex legal and corporate arrangements. The arrangements will almost certainly cover a number of international jurisdictions which would be much more difficult to untangle."

While the OECD report was looking at the issue in terms of maritime security, the conclusions are just as relevant for the use of a vessel for illicit purposes, including IUU fishing. The report notes that a number of FOC registers advertise anonymity as a desirable attribute of their register and states:

"However, in many instances, such as in the case of a known terrorist wishing to remain hidden, the normal procedure would be to use a multi-layered approach, employing a variety of methods, spread over a number of different jurisdictions. Such corporate arrangements are common in the off-shore sector, and any investigators, be they from taxation authorities, law enforcement agencies, security forces or others will find the cloaking processes almost impenetrable. Like peeling an onion, isolating and removing one layer simply reveals another, and another, and because these cloaking devices are relatively cheap and easy to create, those who have a need or a desire to do so can hide themselves very deeply indeed."

The issue of the 'genuine link' is critical because it ought to mean that a shipowner has some form of substantive presence in the flag State in terms of assets and resources that can be subject to fines and penalties in the event of serious breaches of regulatory standards. The United Nations General Assembly Resolution on Sustainable Fisheries (A/RES/58/14):

"Invites the International Maritime Organization and other relevant competent international organizations to study, examine and clarify the role of the "genuine link" in relation to the duty of flag States to exercise effective control over ships flying their flag, including fishing vessels." (para 22).

An identical clause is provided in The United Nations General Assembly Resolution on Oceans and the Law of the Sea (A/RES/58/240). This Resolution also:

"Requests the Secretary-General, in cooperation and consultation with relevant agencies, organizations and programmes of the United Nations system, to prepare and disseminate to States a comprehensive elaboration of the duties and obligations of flag States, including the potential consequences for non-compliance prescribed in the relevant international instruments." (para 29).

Conclusions

It is hoped that this paper has demonstrated the need to address the decent work deficit which exists in the fishing industry and that the social dimension is an integral component of IUU fishing. While the elaboration of a new ILO instrument for the fishing sector is important and merits the support of the OECD, the issue of the social dimension cannot be ignored and needs to be integrated into a holistic approach to the elimination of IUU fishing. In addition there is a pressing need to promote the ratification of the IMO fisheries-specific instruments.

The synergies between FOC operations in the fishing industry and in the maritime sector point to the need for an integrated approach. It is regrettable that the issue of the "genuine link" was referred to the IMO, as it logically belongs to the United Nations Division for Ocean Affairs and the Law of the Sea. It is nevertheless suggested that the OECD and its member economies should support the IMO in elaborating the "genuine link" and that this should later be adopted as an implementing agreement, which would complement UNCLOS and secure the effective implementation by flag States of their obligations under both UNCLOS and applicable international law.

The work of the United Nations in preparing a comprehensive elaboration of the duties and obligations of flag States could provide a useful additional tool to combat IUU fishing and in addressing the social dimension. The OECD and its member economies should support this work and ensure that, once it is adopted, it is given a suitable status, perhaps as an integral annex to a General Assembly Resolution.

The link between concealment of the beneficial ownership and control of IUU fishing vessels and the link to vessel registration clearly demonstrate the need to support the work on improving the transparency of ownership and control, which is currently underway within the OECD Maritime Transport Committee. It is suggested that the OECD could promote the negotiation of an agreement or policy statement, which member states could apply to companies which own or operate vessels established in or operating from their jurisdiction. This could be extended to non-OECD economies by securing commitments from them, as has been done in the case of the FATF and tax havens. It is essential that information is readily available on the ownership of IUU fishing vessels and who buys their catch. In many cases it will be multinational corporations, which are subject to OECD and other applicable instruments, and many of them will have adopted voluntary codes and initiatives with regard to the social dimension.

CHAPTER 14

INCENTIVES FOR INVESTMENT IN IUU FISHING CAPACITY

Aaron Hatcher, Centre for the Economics and Management of Aquatic Resources, University of Portsmouth, England

Summary

Considering investments in IUU fishing as "normal" investment decisions, this paper utilises a simple investment model in order to examine the concern that levels of investment in IUU fishing might be driven by a "spillover" of excess capacity from regulated fisheries. The available evidence suggests that this is rather unlikely. If IUU fishing is relatively profitable, as seems to be the case, most of the investment in IUU capacity will occur whether or not "cheap" capacity is available as a result of the subsidised removal of excess capacity from regulated fisheries. It appears that IUU fishing will only be of marginal profitability if costs are significantly increased or revenues significantly reduced as a result of enforcement efforts to deny vessels access to the fishery and/or to lucrative product markets.

Introduction

This paper presents an economic analysis of the fishery investment decision in the context of investments in IUU fishing. The aim is to understand how incentives to invest in IUU fishing may differ from incentives to invest in legal fishing and, in particular, to consider the importance of the cost of *capacity* in such investment decisions. Given the widely acknowledged existence of excess capacity in many regulated fisheries, and the efforts of policy makers to encourage the removal of this excess capacity, we might question whether there is likely to be an associated "spillover" effect on the supply of investments in IUU fishing. The paper attempts to address this issue. To begin with, however, it is useful to review exactly what is meant by IUU fishing and to briefly consider, in economic terms, why it is a problem. It is also necessary at the outset to clarify what we mean by fishing capacity, and then to narrow the focus of the paper in order to facilitate discussion.

IUU means "Illegal, Unreported and Unregulated". The term was first used by the Commission for the Conservation of Antarctic Marine Living Resources (CCAMLR), but is now widely employed, in particular by the UN/FAO. The definition of IUU fishing set out in Paragraph 3 of the 2001 International Plan of Action (IPOA) on IUU fishing (FAO 2001), adopted within the framework of the FAO Code of Conduct for Responsible Fisheries, is included here in Appendix 14.A. It is apparent from this definition that IUU covers a rather wide variety of undesirable fishing operations and practices, from legitimate operations cheating at the margins (for example, exceeding catch quotas or retaining and landing a proportion of under-sized fish) to entirely illegal operations with no entitlements to take fish in any regulated area. The term "unregulated" also includes vessels fishing in areas subject to international regulation but which are flagged to States not party to the relevant Convention. In addition, the FAO definition covers vessels fishing in areas where no national or international regulations apply, but excludes cases in which the relevant flag State nevertheless fulfils all its obligations under international law.

The reason why IUU fishing is a problem might appear straightforward, although, strictly speaking, the direct impact of IUU fishing is undesirable in economic terms only if it imposes a net social cost. This will certainly be the case if the costs imposed, for example in the form of reduced benefits in the future from exploitation of a depleted fish stock, exceed the current benefits to producers and consumers from IUU fishing (which, it should be appreciated, are very likely to be positive). Almost by definition, however, IUU fishing will impose such costs as a result of stock damage since, as a rule, fishery regulations exist in order to restrict fishing mortality to levels which, if not socially optimal, are at least sustainable. We may then assume that if regulations are not complied with, fishing mortality will be excessive. The main problem with IUU fishing, therefore, whether it takes the form of individual vessels exceeding their legal exploitation limits at the margins or vessels having no legal right to fish at all, is that in most situations fishing mortality is increased to a level which is economically damaging. It is arguable that IUU fishing, according to its broadest definition, is as much of a problem in this regard in many regulated fisheries within EEZs as it is in the high seas fisheries of the Antarctic or the Indian Ocean. IUU fishing may also impose economic costs on society in the form of damage to non-target species, in particular highly "environmentally-valued" species such as seabirds and cetaceans. However, while IUU fishing vessels might be especially guilty of such incidental damage (see, for example, Agnew 2000), the problem is by no means confined to this sector.

The indirect economic impacts of IUU fishing could be at least as serious. The visible presence of vessels fishing illegally may encourage other vessels to violate regulations, since it signals weak enforcement and may undermine the perceived stock-related benefits from regulatory compliance. IUU fishing will also significantly reduce the quality of landings data available for stock assessments and hence severely compromise the ability of managers to set proper exploitation targets.

It should be appreciated that in this paper we are not employing the term "capacity" in a strict economic sense, but rather in the sense in which it is commonly employed by fishery managers and policy makers. In economics, capacity is a short-run measure of unconstrained (and efficiently produced) output from a given (fixed) level of capital stock and a given production technology. There are various alternative precise definitions of capacity but the most straightforward, conceptually, is the short-run potential output which maximises profits, given current input and output prices. Clearly, there can be problems in defining capacity in practice and these will be particularly difficult in the context of the fishery, where there are generally multiple outputs and fluctuating prices, and the unpredictable nature of the resource (not to mention the weather) means that output is stochastic and may often be limited to below potential.[1]

In the arena of fisheries management and policy, "capacity" is generally used to mean both potential output by a fishing vessel (or a fleet of fishing vessels) *and* the amount of physical capital which generates that output. Although this may not be correct, all else being equal, for a given stock of capital, existing technology, current prices, etc., capital and capacity can be considered closely related

---

[1]   For a discussion of the concepts of capacity and capacity utilisation in fisheries see Kirkley and Squires (1999), Kirkley, Morrison Paul and Squires (2002) and Kirkley and Squires (2003).

and therefore interchangeable for most practical purposes. For example, the terms "capacity" and "over-capacity" are often used in relation to desired levels of output such as a TAC to refer to the amount of capital (or fleet size) required to harvest that output at lowest cost. Thus, a situation of over-capacity or excess capacity is one in which the TAC could be harvested efficiently by a smaller fleet, or (perhaps more likely in practice) the existing fleet (efficiently) takes a larger catch than that desired by managers.

Given the very broad definition of IUU fishing adopted by the FAO, we do need to restrict our focus somewhat in this paper. What we are principally interested in here are vessels operating outside of any regulatory regime and, for the most part, outside of EEZs. This includes vessels participating illegally in high seas fisheries which are regulated under international fishery conventions, such as the highly valuable illegal fishery for toothfish in the Southern Ocean (see Agnew 2000). Generally, these vessels are either registered in States not party to the relevant convention or in States which *are* party to the convention but which exercise no meaningful control over vessels flying their flag. Although such vessels are often referred to as "flag of convenience" (FOC) vessels, the CCAMLR uses the term "flag of non-compliance" (FONC) to emphasise that the choice of flag State for IUU fishing vessels is determined, to a great extent, by the lack of regulatory control that will be exercised over them (see Agnew and Barnes 2004). We could also, however, include vessels fishing illegally within the EEZs of States which have few resources available to devote to enforcement. What all these vessels have in common is that they have no right of access to the fishery in question and operate free of any effective regulation. Our focus is therefore on the *absence* of management rather than on the inadequate management of vessels which do enjoy a basic access right. The significance of this is that, in the absence of management, the supply of capacity to the fishery will depend only upon market forces, *i.e.*, the free interplay of demand and supply. A corollary to this assertion is that if we need to be concerned with, say, the supply of capacity to a fishery, then *de facto* we have a situation of management failure (given this, it should be apparent that management failure is not wholly confined to illegal high seas fisheries).

In the next section we set up a simple model for a fishery investment decision and in Section 3 we consider how this might look in the context of investments in IUU fishing. Section 4 then addresses the possible effects of different capacity supply prices on the level of capacity in IUU fishing. A final section presents some concluding comments.

A model for fishery investment

Let us assume that the decision to invest in IUU fishing, as we have more narrowly defined it, is taken as a normal investment decision, *i.e.*, it is based upon the expected net return from the investment over its anticipated life. Thus we assume that the investor neither wants to fish illegally for the sake of it, nor is he deterred to any significant degree by a moral objection against such an activity. In simple terms, given the opportunity to purchase a suitable fishing vessel, the decision to invest in either legal or illegal fishing depends only upon the balance of expected returns against the purchase price. To begin with, we will consider what determines those expected returns in a (legal) fishery and hence what should determine the purchase price of capacity in a perfectly competitive market. In the following section, we can then look at how the investment decision may change in the context of IUU fishing.

The present value (*PV*), evaluated over T years, of the stream of annual profits from an investment in an amount of fishing capacity (or physical capital) K at time $t = 0$ is given by

$$PV = \sum_{t=1}^{T} \rho^t \bar{\pi}_t(K), \quad (1)$$

where a discount factor ρ is defined as

$$\rho \equiv \frac{1}{1+\delta},$$

with δ being the appropriate annual discount rate (assumed constant). The (expected) gross operating profit $\bar{\pi}_t(K)$ in year $t = 1,2,...T$ is given by

$$\bar{\pi}_t(K) \equiv p_t \bar{q}_t(K) - \mathbf{c}_t(K), \quad (2)$$

where $\bar{q}_t(K)$ is the (expected) catch at time t and p_t is the (expected) market price received for that catch.[2] Total operating costs $\mathbf{c}_t(K)$ are assumed to be made up as follows:

$$\mathbf{c}_t(K) \equiv c_t^r(K) + c_t^c(K) + c_t^m(K) + c_t^a(K) + c_t^p(K) \quad (3)$$

where $c_t^r(K)$ are normal running costs (fuel, ice, etc.), $c_t^c(K)$ are crew costs (wages), $c_t^m(K)$ are routine maintenance costs (running repairs to the vessel and its gear, plus the provision and maintenance of safety equipment) and $c_t^a(K)$ are administrative costs (which include costs arising from flag State registration, safety certificates, insurance, etc.). The final category of costs, $c_t^p(K)$, includes the (rental) costs of any fishing permits, such as licences or quota allowances. Of course, in many management regimes marketable permits are not used to allocate fishing rights and the vessel may face a fixed catch or effort limit. In this case $c_t^p(K)$ might be zero but the expected catch $\bar{q}_t(K)$ would be constrained to less than the potential for the vessel.

The *total* expected return (*ER*) from investing in K is given by (1) plus the discounted value of the capacity at the end of the period, which we will denote C_T, so that

$$ER = \sum_{t=1}^{T} \rho^t \bar{\pi}_t(K) + \rho^T C_T. \quad (4)$$

If K were a truly riskless asset, then in a perfectly competitive asset market at equilibrium we would expect the initial cost (capital value) C_0 of capacity K to equal *ER*, its expected return (which would in fact be a *certain* return). If it were greater than this, no-one would invest in the asset, while if it were less than this the demand for the asset by potential investors would push the price up to equal *ER*. Fishing, even when legitimate, is by no means a riskless enterprise, however. If we assume that the discount rate δ applied in the above is equal to the market interest rate r for a safe investment (such as a Government bond), then the investor will expect a higher (average) return from investing an

---

[2] For simplicity, we can think of K as defining the size of a given type of fishing vessel, with the expected (average) annual output (catch) q assumed to be an increasing function of K, i.e., $dq(K)/dK > 0$, so that, on average, a larger vessel will produce a higher annual catch.

amount C_0 in fishing than he would from investing in the safe (riskless) asset. Equivalently, for a given investment in fishing the investor would only be willing to pay an amount *less* than *ER*. One way to model this is to deduct a *risk premium*, R, from the expected total return on the fishery investment so that

$$C_0 = \sum_{t=1}^{T} \rho^t \overline{\pi}_t(K) + \rho^T C_T - R, \quad (5)$$

i.e., the market cost of *K* (here C_0) is less than the cost of a safe investment yielding the same expected return. To be clear, C_0 represents the maximum *willingness to pay* (WTP) of investors for fishing capacity *K*. Investors will not pay more than C_0 for *K*, although they would certainly be prepared to pay less than C_0 if such an offer were made. In a perfect market, however, where there are many potential investors, the equilibrium (market) price for *K* will equal C_0, for the reason previously advanced.

It is apparent from (5) that, all else being equal, higher expected profits $\overline{\pi}_t(K)$ mean that C_0 will be higher. A reduction in C_T, the expected resale value of the vessel (capacity) at time *T*, will lower C_0, as will an increase in the riskiness of the investment and hence an increase in R.[3] Note that in a perfect market for capacity, C_0 will *not* be reduced by a reduction in the investment period (*T*); the vessel can always be sold to a new investor. A transfer of ownership does not affect the value of the investment.

For an individual already participating in the fishery, C_0 represents the *opportunity cost* of remaining in the fishery. Assuming an absence of non-pecuniary motivation, C_0 is the minimum amount that would have to be offered to the individual in order to entice him to disinvest, *i.e.*, to exit the fishery. Note that this *includes* any amount received from the disposal of the vessel: indeed, in a perfect market for fishery investments, as we have observed, this would be the entirety of C_0.[4]

Incentives and disincentives for investments in IUU fishing

Having set out a model for investment in fishing, albeit a greatly simplified one, we can now examine how incentives to invest in IUU fishing might differ from incentives to invest in a legal fishery, considering firstly the expected returns from IUU fishing as compared to returns from legal fishing. There are a number of reasons why revenues and operating costs in an IUU fishery are likely to differ from those in a legal fishery (Agnew and Barnes 2004 review the typical modes of operation of IUU vessels). These can be summarised as follows:

---

[3] The analysis of risk and the behaviour of asset markets is a large topic (see, for example, Varian 1992, Chapter 20, and Hirshleifer and Riley 1992). We can think of R as being related to the extent to which higher or lower returns than *ER* are perceived as likely, *i.e.*, to the *variance* of returns. Note that individual investors may differ in their judgement about the riskiness of the investment and also in their attitudes to risk. Hence C_0 may vary across individuals.

[4] This follows directly from our expression for C_0: at the time of disinvestment future expected returns and (hence) the risk premium are zero so that $C_0 = \rho^0 C_{T=0} = C_0$.

Revenues. In general, IUU vessels target only the most valuable species (such as toothfish, tuna, squid, etc.). Expected revenues are therefore likely to be high, even if access to legitimate markets is made difficult by port controls or some type of catch certification scheme (Agnew 2000). Efforts to deny IUU vessels access to legitimate markets can be circumvented in various ways, however, including the transhipment of catches at sea to vessels which do have access to such markets. In addition, the absence of management means that there are no constraints on catches other than those imposed by the natural environment (the stock, the weather, etc.).

Running costs. We may assume that variable inputs such as fuel, lube, ice and so on can be accessed, one way or another, at prevailing market prices, even if direct access to normal port facilities may be denied by some countries. The *use* of fuel, however, may be relatively high due to increased steaming time to distant fishing grounds in international waters and also to the need to undertake evasion activities such as seeking refuge in international waters when fishing illegally within an EEZ.

Crew costs. IUU vessels, in common with FOC vessels generally, tend to be crewed cheaply, *i.e.*, using labour from countries where labour costs are low and where there may be few alternative employment possibilities. On the other hand, as observed by Agnew and Barnes (2004), the more senior crew, such as the skipper and engineer, typically from developed countries, may demand rather higher remuneration than they would in a legal fishery due to the risks involved in IUU fishing (in effect, a wage "risk premium") and the relatively longer periods spent at sea.

Maintenance costs. Potentially, maintenance costs could be increased due to prolonged operation in international waters. Expenditure on non-essential items such as safety equipment is likely to be lower, given the less stringent registration requirements of FOC States in which IUU vessels are generally registered. FOC registration may also mean that there are no pollutant emissions targets to be complied with. In short, there may be a lesser incentive to maintain the vessel to a high standard, although it would surely be perverse to allow the vessel to become inefficient to the extent that increased harvesting costs exceeded any savings on maintenance costs.

Administrative costs. Also likely to be lower as a result of FOC State registration are various administrative costs such as registration charges, the costs of safety inspections and certification, vessel insurance costs, as well as indirect employment costs such as national insurance contributions. Expenditure on port berthing and landings dues may also be lower.

Management costs may be taken to be zero.

Although we assume that IUU vessels are free of any effective regulation, they are nevertheless subject to attempts at apprehension and sanction. The expected annual costs to the IUU investor arising from such attempts are simply given by the expected annual frequency of successful apprehension and sanction multiplied by the expected level of penalties incurred, including forfeiture of catches and any bonds imposed for the subsequent release of the vessel. Given that successful enforcement events may be relatively infrequent, however, particularly for IUU vessels fishing predominantly in international waters (where States other than the flag State have no right of arrest under international law) the expected cost to the IUU investor may be more appropriately deducted from C_0 as an additional risk premium, rather than included as an annual operating cost. Even without the risk of capture, the risk premium for an IUU investment may be somewhat higher than in a legal fishery if, for example, the vessel is less seaworthy because less has been spent on maintenance (or the vessel was in poor condition already) or the skipper is more prepared to take risks with the weather, etc.

Finally, in our "capacity value" equation (5) we have C_T, the (resale) value of the vessel at the end of the investment timescale. Clearly this will depend upon a number of variables, including the initial value of the vessel and how well it is maintained. If the initial investment is in an old vessel in relatively poor condition, C_T may be disregarded entirely so that C_0 depends almost entirely on the expected profits stream.

For fairly obvious reasons, there are no datasets available which would enable us to make a definitive judgement on whether the value of an investment in IUU fishing is higher or lower than the value of an investment in the same quantity and quality of capacity in a similar legal fishery (*i.e.*, a legal fishery for the same or similar species in a comparable area). On the basis of available evidence, however, (again, see Agnew and Barnes 2004) it does appear to be the case that net operating returns in IUU fishing are, if anything, relatively high, and probably comparable (I suggest) to returns in a profitable legal fishery. This is perhaps not surprising, given that IUU vessels, as we have observed, generally target highly valuable species and almost certainly face lower operating costs in a number of respects. Unless enforcement and deterrence efforts are sufficiently successful as to add a very considerable extra risk premium, it is difficult to see how the value of an IUU investment (*i.e.*, the maximum C_0 or WTP) can be very much lower than that of a similar investment in a legal fishery.

Let us assume, for the sake of argument, that expected net operating profits in an IUU fishery are at least as high as they would be in an alternative legal fishery, and that the investment is evaluated over the same timescale. Assume also that the "normal" risks associated with fishing are similar, but that in the IUU fishery there is an "excess" risk premium R_E which stems from the perceived likelihood of one or more costly enforcement events over the investment timescale. Then we can write an expression for the value (the investor's maximum WTP) for a given amount of capacity K in an IUU fishery, which we will denote C_0^I, as

$$C_0^I = \sum_{t=1}^{T} \rho^t \overline{\pi}_t(K) + \rho^T C_T^I - R - R_E. \quad (6)$$

Further, assume that in the absence of any intervention by the authorities, the present value of the depreciated capacity at time T would be the same whether the capacity is used in legal or illegal fishing. Now we can write

$$C_0^I = \left[\sum_{t=1}^{T} \rho^t \overline{\pi}_t(K) + \rho^T C_T - R \right] - R_E$$

which from (5) is simply

$$C_0^I = C_0^L - R_E, \quad (7)$$

where C_0^L is the value of the same capacity in a legal fishery. Thus the maximum WTP for equivalent capacity in an IUU fishery is given by the WTP for a similar investment in a legal fishery, *less the excess risk premium imposed due to enforcement activities*.

Overcapacity in EEZs and investment in IUU fishing

In many regulated EEZ fisheries, the nature of the past management regime has allowed excess capacity to develop, in the sense referred to in the Introduction. Although, in the short run, the impact of this excess capacity may be an increase in catches above the limits set by managers (given that in most regimes enforcement is considerably less than perfect and the allocation of fishing rights is often highly inflexible), sooner or later we would expect profitability in the fishery to decrease as the stock is depleted (see, for example, Munro and Clark 2003). The problem of excess capacity is exacerbated to the extent that fishing capacity (capital) is *non-malleable* (see Clark, Clarke and Munro 1979).[5] If existing capacity has a low resale value, the opportunity cost of remaining in the fishery is reduced and voluntary disinvestment is less likely to take place in the short run, even if current operating profits are low. This means that government intervention is almost certainly required if a significant immediate reduction in capacity is to be achieved (see Appendix 14.B). As noted earlier, it is often suggested that the subsidised removal of excess capacity in this way from regulated fisheries is responsible for a "spillover" of cheap capacity into IUU fisheries and that this may be a significant driver for IUU fishing (*e.g.*, Bray 2000, p.12, Agnew and Barnes 2004, p.20). This might take the form of redundant vessels being sold to IUU investors at "bargain basement" prices, or once-legal operators moving their vessels into IUU fisheries (in economic terms the effect is the same).[6]

Let us examine this suggestion. Recall that in a perfect market for fishing capacity, the equilibrium cost of capacity will equate the opportunity cost of capacity for incumbents (those who have already invested in the fishery) with the cost of the same capacity to new investors. It follows that if the opportunity cost for incumbents in a regulated fishery is low, this must necessarily be linked to *low* expected returns in alternative uses for that capacity, *i.e.*, use in other fisheries (otherwise, any excess capacity could obviously be sold outwith the fishery at a higher price). This may well be the case for alternative *legal* fisheries, which in general may be taken as operating at full capacity (and "new" fisheries, such as those for previously unexploited species, are relatively few and may require new capacity of a quite different technical specification). However, the existence of investment opportunities in IUU fishing, if profitable, would tend to support rather high vessel resale values. If this does not happen, and if IUU fishing is potentially profitable as we have suggested, it could be because the "supply" of potential investors in IUU fishing is greatly exceeded by the supply of secondhand capacity at any given price.[7] Thus IUU investors represent a "thin" market for capacity and collectively take the price of capacity as given (*i.e.*, the demand for capacity from IUU investors has little or no effect on the resale price of capacity, which is determined exogenously). Another (though not exclusive) explanation could be that there exist barriers to trade in vessels between legal fisheries and IUU investors which result in significant transaction costs.[8]

---

[5]   Capital is non-malleable if it has few (or no) alternative uses and hence a very low resale value (possibly only the scrap value). The result is that capital is treated as a *sunk cost* and the opportunity cost of remaining in the fishery is therefore significantly lowered, comprising little more than the present value of expected future operating profits.

[6]   For a simple explanation of the operation of a spillover effect in fisheries see Munro and Clark (2003).

[7]   It may be that relatively few investors are willing to engage in illegal fishing because of normative beliefs against illegal activity or a high degree of risk aversion.

[8]   The alternative explanation would be that demand from potential IUU investors *does* determine the resale price of capacity, but that either expected returns in IUU fishing are inherently low, or the risk from enforcement activities

Figure 14.1. The Demand for Capacity in IUU Fishing at Different Supply Prices

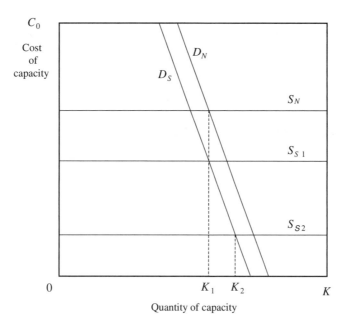

Consider the situation depicted in Figure 14.1. Here, the demand for an amount of capacity K in IUU fishing at a cost C_0 is indicated by D_N in the case of new capacity and D_S in the case of secondhand capacity. The WTP for secondhand capacity is assumed lower than for new capacity (simply because it is older and less efficient), but in both cases demand over the relevant price range is relatively *price inelastic*, reflecting a generally high WTP for investments in IUU fishing so that low capacity prices are not necessary to attract most of the potential investment. Equivalently, most of the demand for IUU capacity would be satisfied at relatively high capacity prices. Suppose that, to begin with, the supply price for new capacity (*i.e.*, new vessels ordered directly from boatyards) is given by S_N while the price of secondhand capacity (in some unspecified market) is given by S_{S1}. For simplicity, it is assumed that whether new or secondhand capacity is purchased the resulting level of capacity in the IUU fishery is the same at K_1.

Now let a supply of "cheap" capacity S_{S2} become available as a result of the exit or subsidised removal of excess capacity from a regulated fishery. Although the price of secondhand capacity is now considerably reduced, the total level of capacity in the IUU fishery only increases by a relatively small amount, to K_2. If IUU fishing is highly profitable, as depicted in Figure 14.1., it would be hard to argue that the main driver for the level of IUU capacity is the availability of cheap capacity spilling over from a regulated fishery. Rather, the main effect of the cheap capacity is to deliver a "windfall"

(and hence R_E) is sufficiently high that the WTP for investment in IUU fishing is significantly reduced. We have suggested, however, that this appears not to be the case.

247

gain to the majority of investors in IUU fishing who would have invested in any case, but at a higher cost.

Now suppose that IUU operations are very marginal, *i.e.* expected returns are sufficiently low that the availability of cheap capacity *is* the main driver for investments in IUU fishing. The situation would now look more like that depicted in Figure 14.2. Little or no new capacity is invested in IUU fishing, while given a "normal" supply of secondhand capacity S_{S1}, only a relatively low level of capacity K_1 enters the IUU fishery. The availability of cheap capacity at S_{S2} now makes a significant difference, increasing the level of IUU fishing capacity from K_1 to K_2.

Figure 14.2. The Demand for IUU Capacity when IUU Operations are Marginal

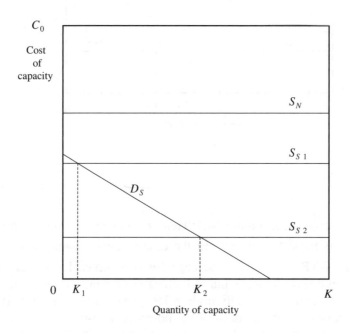

Clearly, the alternative scenarios depicted in Figures 14.1. and 14.2. are hypothetical, but they serve to illustrate the following proposition. If IUU fishing *is* relatively profitable, then the use of secondhand capacity disposed of cheaply from regulated fisheries is largely opportunistic on the part of IUU investors and cutting off this supply of capacity would merely divert much of the demand to more costly secondhand capacity or even to new capacity.[9] Only if IUU fishing operates at a very low level of profitability would we expect the main driver for IUU investments to be the availability of cheap capacity and should we therefore be particularly concerned about the disposal of excess capacity from regulated fisheries.

---

[9] According to Agnew and Barnes (2004, p.20) there is evidence that new vessels are being built for the illegal longline fishery for toothfish.

Earlier in this paper it was suggested that it is in the absence of effective management that, in policy terms, we need to be concerned about the market supply of production factors employed in fishing. The implication here is, of course, that IUU fishing is problematic because current enforcement capabilities are inadequate to deter IUU activities. Further, if IUU fishing is relatively profitable, then to make it less profitable requires that enforcement activities are sufficient to impose very significant additional expected costs. In equation (7) we set out the simple rule

$$C_0^I = C_0^L - R_E,$$

which implies that, in order to make the value of an IUU investment significantly lower than the value of an equivalent investment in a legal fishery, we need to raise the value of R_E considerably. Given $C_0^I << C_0^L$, we might then be in a position (as depicted in Figure 2) where the spillover of cheap capacity from regulated fisheries could be an important driver for IUU fishing.[10]

Conclusion

Available evidence suggests that IUU fishing is relatively profitable rather than being of only very marginal profitability. If this is the case, then any spillover of cheap capacity from capacity reduction programmes in regulated fisheries will certainly deliver benefits to IUU operations, but it is unlikely to be the main driver for the level of capacity invested in IUU fishing. However difficult it may be to achieve in practice, the conclusion is that expected returns from IUU fishing must be reduced to the point where investment in capacity for use in IUU fisheries is no longer perceived as profitable. This could be achieved, for example, by a greatly enhanced probability of costly sanctions for engaging in IUU activities, although alternative approaches such as denying IUU vessels access to output markets would, if successful, also reduce the profitability of IUU fishing very considerably. Either approach requires a great deal of enforcement effort, which is costly to society. There is an inevitable trade-off to be made between increasing the social cost of enforcement and reducing the social cost of IUU fishing. As proposed above, it is only if expected returns in IUU fishing are very low that we should be concerned about the spillover of cheap capacity from regulated fisheries and hence the need to prevent resale of decommissioned capacity at low prices.[11]

---

[10] As an interesting but extreme case, suppose that there is a very high expectation of vessel confiscation on an annual basis (so that $R_E = \rho^T C_T$ where $T = 1$ and hence $C_0^I = \rho\bar{\pi}_1(K) - R$). Now capacity is treated as an annual operating cost and clearly we must have $S_S << C_0^L$ if IUU fishing is to remain viable. A not dissimilar argument might apply if the IUU fishery is expected to be very short-lived, so that the entire value of the investment has to be recouped in just a few years' operating profits. This would not be the case, however, where the vessel could subsequently be transferred to a different fishery.

[11] Unfortunately, intervening in markets can often have undesirable as well as desired consequences. The availability of cheap capacity to legitimate and well-managed fisheries in less developed countries, for example, would be considered a good, but would be cut off by any policy to deny such gains to IUU vessels.

APPENDIX 14.A.

Extract from the *International Plan of Action to Prevent, Deter and Eliminate Illegal, Unreported and Unregulated Fishing (FAO 2001)*.

3. In this document

3.1 Illegal fishing refers to activities

3.1.1 conducted by national or foreign vessels in waters under the jurisdiction of a State, without the permission of that State, or in contravention of its laws and regulations;

3.1.2 conducted by vessels flying the flag of States that are parties to a relevant regional fisheries management organization but operate in contravention of the conservation and management measures adopted by that organization and by which the States are bound, or relevant provisions of the applicable international law; or

3.1.3 in violation of national laws or international obligations, including those undertaken by cooperating States to a relevant regional fisheries management organization.

3.2 Unreported fishing refers to fishing activities

3.2.1 which have not been reported, or have been misreported, to the relevant national authority, in contravention of national laws and regulations; or

3.2.2 undertaken in the area of competence of a relevant regional fisheries management organization which have not been reported or have been misreported, in contravention of the reporting procedures of that organization.

3.3 Unregulated fishing refers to fishing activities

3.3.1 in the area of application of a relevant regional fisheries management organization that are conducted by vessels without nationality, or by those flying the flag of a State not party to that organization, or by a fishing entity, in a manner that is not consistent with or contravenes the conservation and management measures of that organization; or

3.3.2 in areas or for fish stocks in relation to which there are no applicable conservation or management measures and where such fishing activities are conducted in a manner inconsistent with State responsibilities for the conservation of living marine resources under international law.

3.4 Notwithstanding paragraph 3.3, certain unregulated fishing may take place in a manner which is not in violation of applicable international law, and may not require the application of measures envisaged under the International Plan of Action (IPOA).

APPENDIX 14.B.

Capacity Adjustment and "Buyback" Schemes

Given the policy decision to reduce the level of capacity in a fishery through intervention, fishing vessels could then simply be decommissioned without compensation. However, natural justice and political realities generally dictate some form of buyback scheme, usually on a voluntary basis. Under most such schemes, fishermen are invited to bid for funds in return for relinquishing the right to fish with their existing vessel. Bids are then selected according to some chosen "value for money" criterion, but this can be problematic. For example, the opportunity cost of remaining in the fishery is lower for the more unprofitable vessels and hence these are more likely to take advantage of a voluntary decommissioning scheme. Since higher levels of fishing mortality are likely to be exerted by the more profitable vessels, however, this poses a problem for managers seeking to reduce overall levels of fishing mortality while at the same time, presumably, wishing to see overall fleet profitability increase rather than decrease (see, for example, Walden, Kirkley and Kitts 2003). More generally, buyback schemes have been criticised for being costly, being relatively ineffective in practice and rarely dealing with the underlying causes of over-capacity (*e.g.,* Hatcher 1999, Holland, Gudmundsson and Gates 1999).

Buyback schemes vary in their rules for the disposal of redundant capacity. Under the EU's Common Fisheries Policy, for example, a series of "Multi-annual Guidance Programmes" (MAGPs) have, for the last twenty years or so, provided for national buyback schemes within a framework of Community rules and funding (see Hatcher 2000). Community rules have allowed vessels for which fishing rights have been relinquished to be disposed of either by scrapping, permanent transfer to a third country or permanent reassignment to non-fishing use, although Member States could determine more restrictive terms of disposal if they wished. In the UK, for instance, decommissioning rules have always required scrapping (see Pascoe, Tingley and Mardle 2002). Recently, however, Community rules have been changed to remove the possibility of transfer to a third country, with effect from January 2005.[12] According to the European Commission's *Explanatory Memorandum* for the proposed amendment to the relevant Regulation, the existing rules "only result in a transfer of Community over-capacity to third countries and do not correspond to a reasonable use of European tax-payers' money".[13]

---

[12] Council Regulation (EC) No 2369/2002 of 20 December 2002 amending Regulation (EC) No 2792/1999 laying down the detailed rules and arrangements regarding structural assistance in the fisheries sector. *Official Journal of the European Communities*, L358, 31.12.2002, p.49-56.

[13] Commission of the European Communities, COM(2002) 187 final, Brussels, 28.5.2002, p.3.

REFERENCES

Agnew, D. J. (2000), "The Illegal and Unregulated Fishery for Toothfish in the Southern Ocean, and the CCAMLR Catch Documentation Scheme", *Marine Policy*, 24: 361-374.

Agnew, D. J. and C.T. Barnes (2004), *Economic Aspects and Drivers of IUU Fishing: Building a Framework*, OECD document AGR/FI/IUU(2004)2, Paris: Organisation for Economic Co-operation and Development.

Bray, K. (2000), "Illegal, Unreported and Unregulated Fishing", paper presented to the *International Conference on Fisheries Monitoring, Control and Surveillance,* Brussels, 24-27 October 2000.

Clark, C. W., F.H. Clarke and G.R. Munro (1979), "The Optimal Management of Renewable Resources: Problems of Irreversible Investment", *Econometrica*, 47: 25-47.

FAO (2001), *International Plan of Action to Prevent, Deter and Eliminate Illegal, Unreported and Unregulated Fishing,* Rome: FAO.

FAO (2002), *Implementation of the International Plan of Action to Prevent, Deter and Eliminate Illegal, Unreported and Unregulated Fishing,* FAO Technical Guidelines for Responsible Fisheries, No. 9. Rome: FAO.

Gréboval, D., ed. (1999), *Managing Fishing Capacity*. FAO Fisheries Technical Paper, No. 386. Rome: FAO.

Hatcher, A. (1999), Summary of the Workshop on Overcapacity, Overcapitalisation and Subsidies in European Fisheries. In: Hatcher, A. and K. Robinson, eds., *Overcapacity, Overcapitalisation and Subsidies in European Fisheries,* Proceedings of the First Concerted Action Workshop on Economics and the Common Fisheries Policy, Portsmouth, UK, 28-30 October 1998. CEMARE Miscellaneous Publication No. 44. Portsmouth, UK: University of Portsmouth, p.1-6.

Hatcher, A. (2000), "Subsidies for European Fishing Fleets: the European Community's Structural Policy for Fisheries 1971-1999", *Marine Policy*, 24: 129-140.

Hirshleifer, J. and J.G. Riley (1992), *The Analytics of Uncertainty and Information,* Cambridge: Cambridge University Press.

Holland, D., E. Gudmundsson and J. Gates (1999), "Do fishing Vessel Buyback Programs Work: A Survey of the Evidence", *Marine Policy*, 23: 47-69.

Kirkley, J., C. J. Morrison Paul and D. Squires (2002) "Capacity and Capacity Utilization in Common-Pool Resource Industries", *Environmental and Resource Economics*, 22: 71-97.

Kirkley, J. and D. Squires (1999), "Measuring Capacity and Capacity Utilization In Fisheries", in Gréboval, D., ed., *Managing Fishing Capacity*. FAO Fisheries Technical Paper, No. 386. Rome: FAO, p.75-199.

Kirkley, J. and D. Squires (2003), "Capacity and Capacity Utilization in Fishing Industries", in: Pascoe, S., Gréboval, D., eds., *Measuring Capacity in Fisheries*. FAO Fisheries Technical Paper, No. 445. Rome: FAO, p.35-56.

Munro, G. R. and C. W. Clark (2003), "Fishing Capacity and Resource Management Objectives", in Pascoe, S. and D. Gréboval, eds., *Measuring Capacity in Fisheries,* FAO Fisheries Technical Paper, No. 445. Rome: FAO, p.13-34.

Pascoe, S. and D. Gréboval (2003), *Measuring Capacity in Fisheries,* FAO Fisheries Technical Paper, No. 445. Rome: FAO.

Pascoe, S., D. Tingley and S. Mardle (2002), *Appraisal of Alternative Policy Instruments to Regulate Fishing Capacity,* Report to the UK Department for the Environment, Food and Rural Affairs. CEMARE Report No. 59. Portsmouth, UK: University of Portsmouth.

Varian, H. R. (1992) *Microeconomic Analysis*, 3rd Edition, New York: Norton.

Walden, J. B., J. E. Kirkley and A. W. Kitts (2003), "A Limited Economic Assessment of the Northeast Groundfish Fishery Buyout Programme", *Land Economics*, 79: 426-439.

CHAPTER 15

EFFORTS TO ELIMINATE IUU LARGE-SCALE TUNA LONGLINE VESSELS

Katsuma Hanafusa[1] and Nobuyuki Yagi[2]
Fisheries Agency of Japan

Introduction

Longline fishing is the main method employed by tuna fisheries to produce frozen tuna for sashimi and sushi. It accounts for approximately 60% of Japan's total tuna catch, followed by purse-seine fishery that produces around 20% of Japanese tuna. Larger-sized tuna receive higher per-unit market price for sashimi and sushi, and most of them are captured by longline fishing vessels. Purse-seiners tend to catch smaller-sized tuna and their harvests are mostly used for canned tuna production.

Japan had the largest number of tuna longline vessels in the world, but their number is continually decreasing. In 2000, the total number of pelagic longliners (over 120 gross tons) was 529 vessels, indicating a 32% decline from 773 vessels in 1985, mostly as a result of the national fleet reduction programme implemented following the decision by the FAO. Chinese Taipei, however, has substantially increased the number of its longline vessels, followed by China. The number of flag-of-convenience (FOC) tuna long-line vessels is also considered to have increased during the 1990s.

Large-scale tuna longline vessels (LSTLVs) are highly mobile; they operate in the high seas and EEZs of foreign countries, changing oceans and rarely returning to the flag state, except for Japanese LSTLVs, and their catch is delivered directly from fishing grounds to the Japanese market by carrier vessels. For this reason, control and monitoring by the flag state of their fishing operations, in particular the catch amount, is extremely difficult without co-operation from the marketing country.

Various measures to eliminate IUU LSTLVs have been developed and implemented internationally, including trade-related measures. At the same time, a series of direct consultation meetings on the termination of IUU fishing activities have been held with IUU owners, flag governments and Japan.

---

[1]   International Affairs Division, Fisheries Agency of Japan.

[2]   Fisheries Processing Industries and Marketing Division, Fisheries Agency of Japan.

Preliminary estimation of the number of FOC LSTLVs

Since there are no official statistics which directly show the total number of tuna FOC vessels, an attempt has been made in this paper to estimate the number of FOC LSTLVs. Although it is not clear when FOC LSTLVs started, our reports indicate that FOC LSTLVs were first spotted in the Mediterranean Sea in the 1980s. It is believed that the number of FOC LSTLVs in the 1980s was smaller than the figures for the 1990s.

The estimated number of FOC LSTLVs in 1985 was 77, and this had increased to 232 in 2000. The proportion of FOCs in the total LSTLVs is estimated to be around 20% at its peak. Annual changes in the estimated number of LSTLVs are shown in Table 15.1.

Review of market related measures

Measures related to bilateral consultations

Measures to eliminate FOC IUU LSTLVs fishing were first taken by ICCAT in the early to mid-1990s. These measures were designed to focus on flag states but their effectiveness was limited since the vessels changed flags very quickly. Although the flags of FOC LSTLVs vary, almost all owners and operators were Chinese Taipei.

This section describes the history of the evolution of Chinese Taipei FOC/IUU LSTLVs and bilateral consultations between the Japanese side (government and tuna industry) and Chinese Taipei IUU owners, government and other flag governments that have accepted the Chinese Taipei LSTLVs with a history of IUU fishing, in an effort to seek direct solutions. These consultations were held in parallel with the multilateral efforts of tuna RFMOs.

Chinese Taipei - History of Emerging Chinese Taipei FOC LSTLVs

Most of Chinese Taipei's tuna longline fishing vessels were traditionally near-shore fishing vessels landing fresh fish. In the 1980s the number of LSTLVs was around 100 vessels. However, in the late 1980s, as the cost competitiveness of Korea weakened and the Chinese Taipei economy underwent rapid growth, the number of Chinese Taipei LSTLVs producing frozen tuna for sashimi increased drastically, exceeding 300 vessels by the early 1990s. These Chinese Taipei fishing vessels mainly harvested yellow fin and bigeye tunas in the tropical zone of the Indian Ocean.

In a bid to improve this situation, in 1993 the Federation of Japan Tuna Fisheries Co-operative Association and the Chinese Taipei Deep Sea Boat Owners and Exporters Association agreed to limit the annual amount of landing of Chinese Taipei frozen tuna in Japanese markets and, at the same time, agreed to adopt the Export Certification system, under the witness of the fisheries authorities of Japan and Chinese Taipei. Under this system, Chinese Taipei-produced frozen tuna were required to attach an export certificate for each loading, with quantities specified, issued by the Chinese Taipei Deep Sea Tuna Boat Owners and Exporters Association. As Chinese Taipei LSTLVs have no market other than Japan, their catch was fully monitored through the issuance of export certificates. Thus, Chinese Taipei vessel owners were in no way able to misreport their catch. At the same time, this system led to a spectacular improvement in the catch control capability of the Chinese Taipei authorities. However, this export certification system only covered Chinese Taipei-flagged vessels and did not cover other flags, while there was no limitation on the export of LSTLVs flags to foreign countries. Consequently, there was an upsurge in the number of flag-of-convenience (FOC) fishing vessels owned by Chinese Taipei.

Table 15.1. Yearly Changes in the Number of Large-scale Tuna Longline Vessels in the World

Year	Japan	Korea	China	Indonesia	Chinese Taipei	Philippines	FOC	Total	% of FOC
1985	773	156	-	18	(75+)	Unknown	77	1 099+	-
1986	771	167	-	19	(81+)	Unknown	93	1 131+	-
1987	770	189	-	37	(103+)	Unknown	103	1 202+	-
1988	759	199	-	43	(107+)	Unknown	128	1 236+	-
1989	764	196	-	60	(143+)	Unknown	132	1 295+	-
1990	758	203	-	62	(196+)	Unknown	171	1 390+	-
1991	743	194	-	31	497	Unknown	195	1 660	13%
1992	724	185	-	34	522	Unknown	177	1 542	12%
1993	722	174	3	36	681	Unknown	152	1 768	11%
1994	701	184	3	29	693	Unknown	183	1 793	13%
1995	703	201	3	35	699	Unknown	203	1 844	14%
1996	674	200	3	29	705	Unknown	190	1 801	13%
1997	661	202	3	24	714	Unknown	213	1 817	15%
1998	663	209	11	23	601	Unknown	238	1 745	16%
1999	528	202	27	21	600	(21)	248	1 626	18%
2000	529	197	52	28	597	(15)	232	1 635	17%
2001	529	198	98	70(56)	602	(14)	250	1 691	15%
2002	525	192	100	93(79)	612	24(10)	100	(1 557)	6%
2003	517	178	105	14	610	14	30	(1 475)	2%

Notes:
- FOC (1985-2000) and Indonesia (1985-2000) were estimated from import amounts of tuna for sashimi in Japan.
- The numbers for the Philippines and Indonesia in brackets are the number of vessels listed in ICCAT IUU lists.
- FOC (2001-2003) is estimated from numbers of vessels that participated in the Japan-Chinese Taipei Action Plans.
- Total figures after 2002 do not represent the world total.

Source: World Tuna Longline Fishery Conference, August 2003

Even before this arrangement, the Chinese Taipei frequently operated their vessels using the flag of a third country. For example, coastal countries in the Indian Ocean, such as Bangladesh, did not allow Chinese Taipei-flagged vessels to operate in their waters for diplomatic reasons. In some cases, Chinese Taipei fishers obtained FOC vessel registrations from countries such as Panama and Honduras to enable their vessels to operate in these coastal state waters. It was widely known among Chinese Taipei vessel-owners that such operations were profitable to them as they could evade taxation by the Chinese Taipei authorities.

After 1993, through the acquisition of FOC fishing vessels, Chinese Taipei LSTLV operations were practiced rampantly and on a larger scale because Chinese Taipei-flagged longline vessels were subject to the upper limits of the landing amount at Japanese ports. This move was accelerated by the export of secondhand LSTLVs from Japan. During the "bubble economy" era in the late 1980s-early 1990s Japanese fishers built new LSTLVs, and their old vessels were exported as cargo vessels to various countries; these were later obtained by Chinese Taipei fishers who turned these secondhand Japanese LSTLVs into FOC LSTLVs for IUU fishing. Chinese Taipei FOC LSTLVs operations were free from any catch limitations as well as from any tax obligation. Not only newly emerging vessel owners but also traditional vessel owners, who already had many duly-authorised LSTLVs, came to possess many FOC LSTLVs.

Around 1995, a poor harvest occurred in the Indian Ocean, probably due to overfishing, and the Chinese Taipei fishing fleet moved to the Atlantic. As a result, their bigeye catch in the Atlantic increased sharply and, in 1997, ICCAT took measures to restrict the annual Chinese Taipei catch of bigeye tuna to 16 500 MT. Under such conditions, Chinese Taipei vessel owners, who still wanted to increase their production of bigeye tuna in the Atlantic, apparently stepped up FOC operations in the Atlantic.

During the same period, tuna fishing grounds throughout the world experienced poor harvests and concern was loudly expressed about the deterioration of resources. The excessive number of fishing vessels, *i.e.* catch effort, was perceived as an alarming problem. In 1998, the FAO developed an international plan of action (IPOA) on excessive fishing capacity. Notably, the FAO decided in the IPOA that 20-30% reductions in the number of LSTLVs were necessary. Following this decision, Japan scrapped 20% of its LSTLVs between 1998 and 1999. In Japan, there was a growing demand for concerted vessel reduction by Chinese Taipei and the Republic of Korea – both having LSTLVs. At the same time, criticism intensified against Chinese Taipei FOC fishing vessels operating outside the framework of the international management regime.

Consultations on the elimination of FOC/IUU LSTLVs between Japan and Chinese Taipei

Consultations between Japan and Chinese Taipei began in March 1998 and more than 20 consultations were held during a three-year period before the Organization for the Promotion of Responsible Tuna Fisheries (OPRT) was established late in 2000 and specific measures were implemented.

Basic Agreement and Action Plans

A Basic Agreement was reached in the autumn of 1998 between the fisheries authorities of Japan and Chinese Taipei. Consultations continued and the Action Plans were developed in February 1999 in order to implement the Basic Agreement. The major elements of the Action Plans were:

1) Chinese Taipei would aim to reduce the number of their LSTLVs by 10%, from 600.

2) Japan would seek to scrap FOC fishing vessels originating from Japanese secondhand fishing vessels.

3) Chinese Taipei would consider calling back relatively new FOC vessels constructed in Chinese Taipei, to the Chinese Taipei registry.

At that time, it was estimated that there existed 120-130 Chinese Taipei FOC fishing vessels originating from Japanese secondhand vessels and 60-70 newly constructed vessels in Chinese Taipei. However, even after this Basic Agreement, Chinese Taipei FOC vessel owners did not fully recognise this problem, and even continued to construct new FOC LSTLVs.

Negotiations to implement the Action Plans

In February 2001, at long last the Agreement to implement the Action Plans was reached between OPRT and the Chinese Taipei FOC vessel owners. The following programmes were agreed:

1) *Scrapping programme*: Japan would purchase and scrap the FOC fishing vessels originating from Japan under a three-year programme, 2001-2003.

2) *Re-registration programme*: Chinese Taipei owners of FOC fishing vessels constructed in Chinese Taipei would purchase Chinese Taipei fishing licenses and re-register them under the Chinese Taipei registry under a five-year programme, 2001-2005.

A major change during this period was the establishment of the Kaohsiung Foreign Registered Fishing Vessel Association (KFRFVA) – an organisation set up by the owners of FOC/IUU LSTLVs to protect their interests. Through the establishment of this organisation, the real owners of FOC fishing vessels made their appearance, creating a situation where substantial talks became possible. It also became evident that the KFRFVA included many members of other Chinese Taipei tuna organisations, composed of duly licensed LSTLV owners.

KFRFVA joined the Chinese Taipei side in the Japan-Chinese Taipei consultations that took place several times in 1999, during which negotiations were held regarding the purchase price and purchase methods of FOC vessels, as well as a way to incorporate vessels into Chinese Taipei registration.

Scrapping programme

Negotiations over the purchase price and the method of procurement of funds required for purchase were particularly difficult. Conflicting interests between Japanese and Chinese Taipei duly licensed longline fishers and FOC fishers also required repeated negotiations to reach a compromise. In the final stage, it was agreed that the Japanese Government would provide the initial funds required to purchase vessels for scrapping, and that funds would be reimbursed over a long period from contributions made by Japanese and Chinese Taipei longline fishers who would continue fishing operations.

KFRFVA called on its members to participate in the scrapping programme, and owners of a total of 62 FOC LSTLVs – or half of the estimated FOC vessels originating from Japanese secondhand LSTLVs – supported this proposal. The scrapping programme began in 2001, with an initial target of 62 vessels. In fact, scrapping contracts were made for 44 vessels in that year and, by the end of 2003, a total of 43 LSTLVs were disposed of (of which 39 were scrapped and 4 sank accidentally).

Establishment of OPRT

In taking account of the long-term nature of the reimbursement of the fund as well as ensuring that efforts were not just limited to the short-term goal of scrapping fishing vessels, a consensus was formed among fishing industries and fisheries authorities in both Japan and Chinese Taipei for the establishment of a framework to *i)* cope with tuna conservation and management on a long-term basis, *ii)* contribute to the sustainable development of both Japanese and Chinese Taipei longline fisheries, and more broadly, *iii)* contribute to the conservation of tuna at large that extensively migrate throughout the world's oceans. The outcome of this consensus was the establishment of the Organization for the Promotion of Responsible Tuna Fisheries (OPRT). The significance of this organization will be discussed in the section "Measures by private initiatives" below.

Re-registration programme

An additional complicated issue was how to bring FOC vessels back under Chinese Taipei registration. Although many FOC owners were also the owners of duly licensed LSTLVs, the magnitude of their involvement in the FOC operations varied substantially. Chinese Taipei duly licensed LSTLV owners feared that their interest would be impaired substantially by the return of FOC vessels to Chinese Taipei. For example, the per-vessel share of Chinese Taipei ICCAT bigeye quota (16 500 MT) would be reduced or the value of the licence would be lessened. They demanded stringent conditions for bringing the vessels' registry to Chinese Taipei. The owners of FOC vessels, on the other hand, naturally wanted to have their vessels returned to Chinese Taipei at a minimum cost. Such adjustments of interest within Chinese Taipei took more than two years of negotiations. In 2001, the legal system for returning the FOC vessels to Chinese Taipei registration was established. The initial target for Chinese Taipei registration was 67 vessels (of which 2 sank accidentally). As of the end of March 2004, a total of 48 FOC LSTLVs had returned to Chinese Taipei.

LSTLVs shifted to other countries

In parallel with Japan-Chinese Taipei consultations, ICCAT adopted a series of measures against IUU LSTLVs. In 1998, ICCAT adopted the IUU Action Plan that enabled ICCAT to apply trade measures against countries that continued to allow IUU fishing operations to take place. In 1999, ICCAT adopted a resolution to require Contracting Parties to urge its nationals not to associate with IUU activities, including the non-purchase of IUU-caught tuna. In 2000, based on the 1998 Action Plan, ICCAT adopted trade sanctions on bigeye tuna against Belize, Cambodia, Equatorial Guinea, Honduras and St. Vincent & Grenadine.

Measures decided by ICCAT at each annual meeting usually take more than three years before actually being enforced and during this period the owners of FOC fishing vessels looked for other recipient countries and changed their registration. Consequently, the expected effectiveness of IUU counter measures was difficult to enact in a timely manner.

Philippines

The Philippines was targeted as the first destination of such evasion. From 1998, more than 40 FOC LSTLVs were abruptly transferred to the Philippines, by being chartered by Philippine companies (registration was transferred to the Philippines). Consultations were held between the Japanese and the Philippine governments from 1999, and the number of charter LSTLVs decreased to 16 before the 2000 ICCAT annual meeting. Before the end of 2001, all the charter contracts were terminated. At present, only 14 LSTLVs are owned and operated by Philippine companies.

China

The next destination of FOC fishing vessels was China. The number of Chinese LSTLVs was estimated at around 10 in the late 1990s, increasing to over 40 in the first half of 2000, before rising to 60 before the end of the same year. By the end of 2001, it had reached almost 100. Japan and China held a series of consultations over the rapid expansion of the Chinese fishing fleet.

In the case of China, a unique characteristic was that China supplied crew to FOC LSTLVs. Chinese Taipei LSTLVs have employed Chinese seamen for many years because of rising wages and crew shortages in Chinese Taipei. It was said that most of the crew, except for the fishing master, were Chinese. Therefore, unlike other flag-of-convenience countries, China not only made its vessel registration available to Chinese Taipei FOC LSTLVs, but also took them over and easily incorporated them into their own fisheries. Owing to close relations between Chinese LSTLVs and Chinese Taipei vessel owners, Chinese LSTLVs have sometimes operated jointly with Chinese Taipei LSTLVs and FOC LSTLVs, and have acquired operational know-how without much difficulty. However, as a result of Japan-China consultations, China has declared that it would terminate relations between its LSTLVs and IUU fishermen.

Indonesia

The third destination of FOC LSTLVs was Indonesia. Since 2000, a large number of LSTLVs suddenly appeared under Indonesian registration, exceeding 60 vessels in 2001. This prompted Japan to hold several tuna consultations with Indonesia. As a result, only 13 LSTLVs were identified as being actually owned and operated by genuine Indonesian companies. Indonesia de-registered the rest of the LSTLVs.

Seychelles and Vanuatu

Despite extensive consultations as described above, in 2002 there still remained around 100 FOC-LSTLVs. Seventy per cent of those were new, leaving no room for re-registration to Chinese Taipei. Further consultations continued, leading to a new programme to expeditiously dispose of these LSTLVs in accordance with the 2001 ICCAT resolution concerning "More Effective Measures to Prevent Deter and Eliminate IUU Fishing by Tuna Longline Vessels". Japan talked with Vanuatu and Seychelles, the major flag states of the remaining Chinese Taipei FOC/IUU LSTLVs, and reached an agreement with them to bring these LSTLVs under strict control. A total of 69 FOC-LSTLVs committed themselves to comply with the following co-operative management schemes:

Figure 15.1. Transition of the Numbers of IUU Large-scale Tuna Longline Vessels (IUU-LSTLVs)

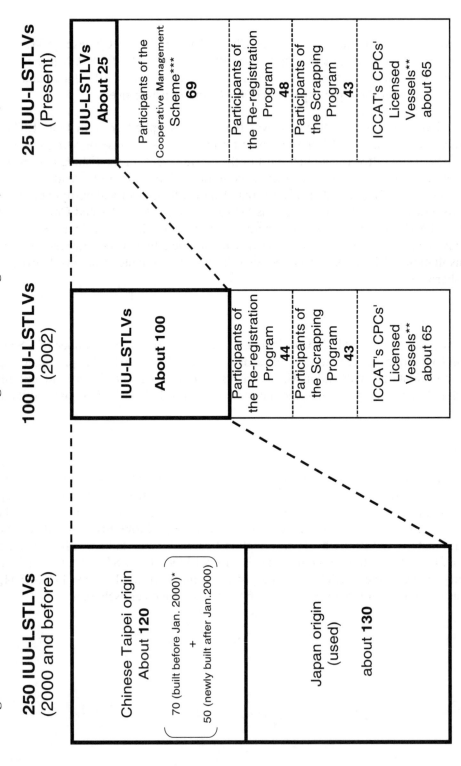

250 IUU-LSTLVs
(2000 and before)

Chinese Taipei origin
About **120**

70 (built before Jan. 2000)*
+
50 (newly built after Jan.2000)

Japan origin
(used)

about **130**

100 IUU-LSTLVs
(2002)

IUU-LSTLVs
About 100

Participants of
the Re-registration
Program
44

Participants of
the Scrapping
Program
43

ICCAT's CPCs'
Licensed
Vessels**
about 65

25 IUU-LSTLVs
(Present)

IUU-LSTLVs
About 25

Participants of the
Cooperative Management
Scheme***
69

Participants of
the Re-registration
Program
48

Participants of
the Scrapping
Program
43

ICCAT's CPCs'
Licensed
Vessels**
about 65

Notes
* Japan-Chinese Taipei Joint Action Program to Eliminate IUU Vesels concluded in January 2000.
** CPCs means "Contracting Parties" and "Cooperative non-Contracting Parties, Entities or Fishing Entites".
*** Cooperative Management Schemes between Japan and Seychelles/Vanuatu were agreed in July 2003

262

i)   Arrangements for the legalisation of FOC-LSTLVs were established between the fishing authorities of the two flag states (Vanuatu and the Seychelles) and Japan, and vessels participating in the scheme must be subject to strict joint monitoring and control measures.

ii)   All of the participating LSTLV owners must obtain Japan's fishing licences for LSTLVs and freeze those licences so as to reinforce and complement the co-operative management scheme mentioned in point i) above, as well as to prevent an increase of overall fishing capacity.

iii) Those LSTLVs are authorised to fish only in an area where, and for species for which, their fishing operations will not pose a problem, in light of regulatory measures and resolutions adopted by the relevant RFMOs. Specifically, 21 Seychelles-flagged LSTLVs may catch yellowfin and bigeye tuna in the Indian Ocean only, whereas 48 Vanuatu flag LSTLVs may fish for albacore in the Pacific Ocean (within which 4 Vanuatu-flagged LSTLVs are also exceptionally allowed to target yellowfin and bigeye tuna in the Pacific).

Despite the above efforts, approximately 25 old FOC-LSTLVs are believed to remain (Figure 15.1.). But many of them may have stopped fishing because of their age or have been transformed into other types of vessels such as squid jigging vessels and transhipping vessels. Thus, it can be presumed that the number of remaining FOC-LSTLVs is, in fact, very small at present.

Measures related to RFMOs

Introduction of Statistical Document Programmes

i) International Commission for the Conservation of Atlantic Tuna (ICCAT)

In the early 1990s, Japan compiled and analysed its trade statistics and the estimated number of FOC LSTLVs operating in the Atlantic and presented the data to the ICCAT. Japan found that its trade statistics did not contain important information required for fisheries management, such as area of catch, vessel name and flag country. This finding led to the adoption of the ICCAT Bluefin tuna Statistical Document Programme.

Initially, ICCAT did not take prompt action to combat the FOC/IUU problem, with no effective measures being implemented in the late 1980s or early 1990s. However, ICCAT measures were accelerated as moves to regulate international trade in bluefin tuna emerged under the Convention for the International Trade in Endangered Species of Wild Fauna and Flora (CITES) from 1991 to 1992 – the year CITES held its Conference in Kyoto, Japan. Sweden presented a proposal to list the western Atlantic bluefin tuna stock in CITES, Appendix I, and eastern Atlantic bluefin tuna stock in Appendix II, on the grounds that the deterioration of bluefin tuna could not be prevented, due to inadequate management by ICCAT. Although the proposal was withdrawn as a result of consultations among the countries concerned, CITES urged ICCAT to reduce quotas and take effective counter-measures *vis-à-vis* non-member states.

ICCAT then took measures against non-member states. First, it introduced the Bluefin Tuna Statistical Document Programme (BTSD Programme) in 1992. This system is designed to collect the information needed for fisheries management through international trade. Flag states are required to validate the area of catch and amount of bluefin tuna for export. The system was designed to identify flag states that accepted FOC/IUU LSTLVs.

Second, in order to restrict the export of bluefin tuna by non-Contracting Parties – which had seriously diminished the effectiveness of ICCAT conservation and management measures by accepting FOC/IUU fishing vessels – the ICCAT Bluefin Tuna Action Plan (BTAP) was then adopted in 1994. Under BTAP, the ICCAT identifies non-Contracting Parties whose vessels have been fishing for Atlantic bluefin tuna in a manner which diminishes the effectiveness of the ICCAT bluefin conservation and management measures, and requests such countries to rectify their fishing activities so as not to diminish the effectiveness of the measures. If those identified non-Contracting parties do not rectify their fishing activities in the following year, ICCAT recommends Contracting Parties to take non-discriminatory trade restrictive measures which virtually prohibit the imports of bluefin tuna from those non-Contracting Parties. According to the 1994 BTAP, ICCAT adopted measures to prohibit the import of Atlantic bluefin tuna from Panama, Honduras and Belize. In 1998, this plan was reinforced to cover all tuna and tuna-like species, under which both Contracting Parties and non-Contracting Parties whose LSTLVS have been fishing tuna and tuna-like species in a manner which diminishes the effectiveness of the ICCAT conservation and management measures are identified and treated in the same manner as under the BTAP.

Third, in order to supplement this measure by controlling the re-export of bluefin tuna to the Contracting Party *via* a third country, the Bluefin Tuna Re-export Certificate was developed in 1997. Finally, in the same way, similar Statistical Document Programmes for export and re-export of bigeye tuna (with the exemption of catches caught by purse seiners and pole and line vessels destined principally for canneries in the Convention area) and swordfish were adopted in 2001.

Since bigeye tuna is the most important species for LSTLVs in terms of financial gain, the expansion of the Statistical Document Programme to bigeye tuna has had a substantial impact on IUU LSTLVs.

ii) Indian Ocean Tuna Commission (IOTC)

In 2001, in order to cope with the problems of IUU fishing by large-scale tuna fishing vessels in the Indian Ocean, a similar Statistical Document Programme for export and re-export of bigeye tuna was adopted.

iii) Inter-American Tropical Tuna Commission (IATTC)

In 2003, in order to address the problem of IUU fishing in the Convention area, a similar Statistical Document Programme for export and re-export of bigeye tuna was adopted.

iv) Commission for the Conservation of Southern Bluefin Tuna (CCSBT)

In 1999, in order to have a better estimation of southern bluefin tuna caught by both Contracting Parties and non-Contracting Parties, and to properly control fishing activities by vessels of non-Contracting Parties, a Statistical Document Programme for export and re-export was adopted.

v) Western and Central Pacific Fisheries Convention (WCPFC)

In April 2004, the WCPFC Preparatory Conference will discuss a similar Statistical Document Programme.

Table 15.2: Sample of IUU Large-Scale Tuna Longline Vessel List

	Flag country	Name of Vessel	Owner's Name	Owner's Address	Expected Area of Catch
89	E. Guinea	Chen Chieh No. 726	Chen Chin Cheng Fishery Co. Ltd. S.A.	E. Guinea	Atlantic
90	E. Guinea	Chen Chieh No. 736	Chen Chin Cheng Fishery Co. Ltd. S.A.	E. Guinea	Atlantic
91	E. Guinea	Chen Chieh No. 8			Indian
92	E. Guinea	Chi Man	Chi Man Fishery S.A.		Atlantic
93	E. Guinea	Chia Ying No. 6	Pesquera Happy Sun S.A.	E. Guinea	Atlantic and Indian
94	E. Guinea	Columbus	Pesquera Columbus S.A.	E. Guinea	Pacific
95	E. Guinea	Dong Yih No. 688	Dong Yih Fishery S.A.	E. Guinea	Indian
96	E. Guinea	Ever Rich	Lin Ching Isang	E. Guinea	Pacific
97	E. Guinea	Exito	Pesquera Exito S.A.	E. Guinea	Indian
98	E. Guinea	Fortuna No. 1	Naviera Fortuna S. de R.I.		Atlantic and Indian
99	E. Guinea	Hai Ming No. 1	Hai Ming Fishery S.A.	E. Guinea	Pacific and Indian
100	E. Guinea	Hai Zean No. 11	Hai Zean Fishery S. de R.I.	E. Guinea	Atlantic
101	E. Guinea	Hai Zean No. 3	Hai Zean Fishery S. de R.I.	E. Guinea	Atlantic
102	E. Guinea	Hai Zean No. 31	Pesquera Hung Lin S.A.	E. Guinea	Atlantic
103	E. Guinea	Hsiang Jang No. 11	Atlantic Fishery S.A.	E. Guinea	Atlantic
104	E. Guinea	Hsiang Jang No. 111	Kwo Jeng Productos Marinos S.A.	E. Guinea	Atlantic
105	E. Guinea	Hsiang Jang No. 112	Kwo Jeng Productos Marinos S.A.	E. Guinea	Atlantic
107	E. Guinea	Hsiang Jang No. 66	Atlantic Fishery S.A.	E. Guinea	Atlantic
108	E. Guinea	Hsin Hua No. 103	Pesquera Hsin Hua Fishery Co. Ltd.	E. Guinea	Indian
109	E. Guinea	Hung Yu No. 212	Pesquera Columbus S.A.	E. Guinea	Indian
110	E. Guinea	Hung Yu No. 606	Hung Yu Fishery Co. Ltd.	Korea	Indian
111	E. Guinea	Hwa Mao No. 202	Hwa Mao Fishery Co. S.A.	E. Guinea	Indian
112	E. Guinea	I Man Hung No. 166	Chun Far Fishery S.A.	E. Guinea	Atlantic
113	E. Guinea	Jin Cheng Horng	Navierage Ko Yuan Fishery S.A.	E. Guinea	Atlantic and Indian
114	E. Guinea	Jiyn Horng No. 116	Jiyn Horng Ocean Enterprise Co. Ltd.	Honduras	Indian
115	E. Guinea	Jiyn Horng No. 116	Jiyn Yeong Fishery S.A.	E. Guinea	Indian
116	E. Guinea	Kae S.A.	Chin Ching Fishery Co. Ltd.	E. Guinea	Atlantic
117	E. Guinea	Kae Shyuan	Chin Man Fishery Co. Ltd.	E. Guinea	Atlantic
118	E. Guinea	Kuang Horng	Chuen Song Fishery S. de R.L.	E. Guinea	Atlantic and Indian

Table 15.2: Sample of IUU Large-Scale Tuna Longline Vessel List (cont.)

	Flag country	Name of Vessel	Owner's Name	Owner's Address	Expected Area of Catch
119	E. Guinea	Lung Soon No. 212	Exito Fishery S.A.	E. Guinea	Pacific and Indian
120	E. Guinea	Lung Soon No. 282	Exito Fishery S.A.	E. Guinea	Pacific
121	E. Guinea	Lung Soon No. 662	Exito Fishery S.A.	E. Guinea	Indian
122	E. Guinea	Pesquera No. 68	Choyu Fishery S.A.	E. Guinea	Atlantic
123	E. Guinea	Shang Shun No. 622	Exito Fishery S.A.	E. Guinea	Pacific
124	E. Guinea	Shin Kai No. 6	Shin Kai Fishery S.A.	E. Guinea	Pacific
125	E. Guinea	Shing yang	Chien Chong Hsin	E. Guinea	Atlantic
126	E. Guinea	Shung Ying	Chen Chong Hsin	E. Guinea	Atlantic
127	E. Guinea	Sun Rise No. 313	Singarope Corp.	E. Guinea	Atlantic and Indian
128	E. Guinea	Viking No. 1	Viking Fishery S.A.	E. Guinea	Atlantic and Pacific
129	E. Guinea	Wei Ching	Wei Ching Ocean Enterprise S.A.	E. Guinea	Atlantic and Indian

Source: ICCAT Report, 1998-99, (II).

Figure 15.2. Tunas Imported to Japan Against the Non-Purchase Guidance

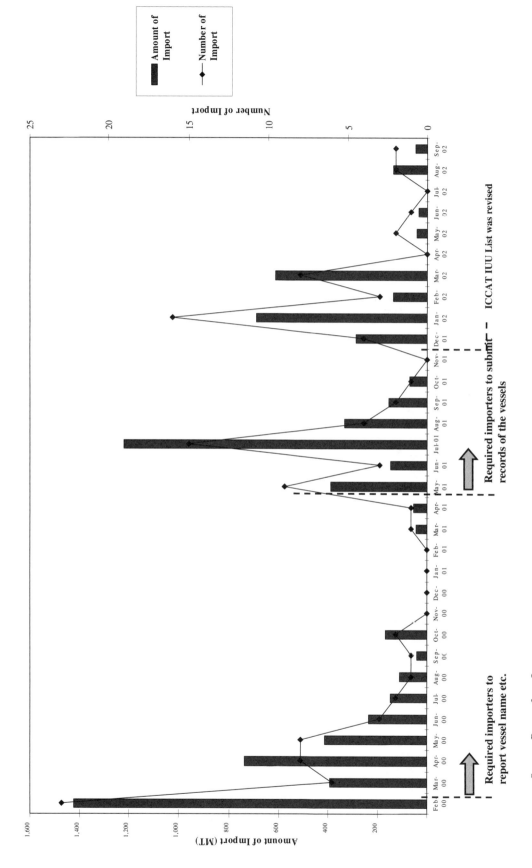

Source: Report from Importers.

267

Figure 15.3. One Implication of Tuna Laundering

Amount of Japanese Bigeye Tuna Imported from Chinese Taipei and Number of Vessels exported over 350 MT Bigeye per Year

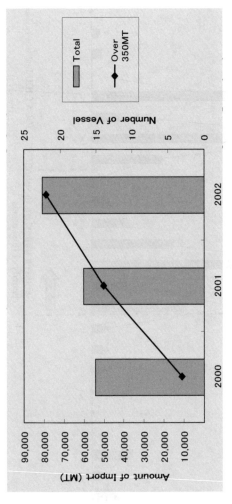

Source: Report from Importers.

Mandatory data requirement of history of LSTLVs

Although the ICCAT adopted trade sanction measures against FOC/IUU flag states, vessels simply changed flags to evade sanctions. Japan compiled a list of LSTLVs believed to be engaged in IUU fishing activities and distributed it to the ICCAT and the IOTC (Table 15.2).

In 1999, a resolution was adopted for the ICCAT Contracting Parties to urge their general public, importers, transporters and other business people concerned to refrain from purchasing, trading and transhipping tuna caught by such IUU vessels.

Soon after this resolution was adopted, IUU LSTLVs started changing their names. Japan required importers to submit information on previous vessel names and owner names to detect the relation between IUU owners and current owners. Since this mandatory requirement was introduced, the import of IUU LSTLV caught tuna disappeared (Figure 15.1.). Instead, imports from Chinese Taipei and China increased substantially. In particular, many of these duly authorised LSTLVs doubled their annual catches despite very poor fishing conditions (Figure 15.2.). This phenomenon strongly suggested a possible at sea transfer of tuna from IUU vessels to duly authored vessels, so-called "tuna laundering" and also pointed to limitations of the effectiveness of measures based on negative listings of vessels. Japan requested that China and Chinese Taipei investigate these incidents. As a result, the unusual record of catch per vessel disappeared. This phenomenon acted as a trigger to establish a positive listing scheme for vessels.

Adoption of Positive Listing scheme for Fishing Vessels

i) ICCAT

Taking into account the high mobility of LSTLVs, in 2000 a resolution was adopted to urge Contracting Parties to submit a list of large-scale fishing vessels (LSFVs larger than 24 metres in overall length), licensed to fish tuna and tuna-like species in the Convention area.

In order to identify tuna and tuna-like species caught by duly authorised fishing vessels and to prevent those caught by IUU fishing vessels from entering the international market, in 2002 ICCAT agreed to establish a list of duly authorised LSFVs, *i.e.* a Positive List. In addition, in order to avoid any adverse effects on tuna resources in other oceans as a result of the establishment of the ICCAT Positive List and the subsequent transfer of vessels to other oceans, requests were made to other RFMOs to establish similar records in a timely manner.

ii) IOTC and IATTC

In 2002, taking into account the ICCAT decision on the establishment of a Positive List, and the consequent shift of LSFVs from the Atlantic, and responding to the request by ICCAT to establish similar records of duly authorised LSFVs, both IOTC and IATTC agreed to establish Positive Lists.

iii) Western and Central Pacific Fisheries Convention (WCPFC)

In 2002, the third WCPFC Preparatory Conference, concerned with the potential redeployment of IUU fishing vessels from other regions, adopted a resolution urging all States and entities concerned to promote co-operation in exchanging information on IUU fishing activities. The WCPFC Preparatory Conference was scheduled to discuss a positive listing scheme in April 2004.

iv) Implementation of the positive listing scheme

In November 2003, Japan implemented a new trade monitoring and controlling system, based on the ICCAT, IOTC and IATTC Positive Listing Schemes on a global scale. Only tuna products caught by the LSTLVs listed in the Positive Lists are allowed to enter the Japanese market. All other members of these RFMOs have a legal obligation to implement the same measure.

However, about 25 LSTLVs still remain. In addition, several hundred FOC/IUU tuna longline vessels, just under 24 metres, *i.e.* 23.9 metres and less, are actively in operation. They shifted target species from bigeye to albacore or shark. The major market for albacore is the USA and for sharks Latin American countries. The FOC/IUU tuna longline vessel owners continue to operate, while Japan cannot detect their activities through its market. They may still continue to practice tuna laundering and the use of forged documents for export to the Japanese market, and it becomes harder to detect such illegal activities.

Measures by private initiatives

The Organization for the Promotion of Responsible Tuna Fisheries (OPRT) was established in December 2000. Its members come not only from fisheries, traders and consumer's organisations in Japan, but also from tuna longline fisheries industry organisations in China, Ecuador, Indonesia, the Philippines and Chinese Taipei. It covers more than 95% of duly licensed LSTLVs in the world, given the fact that Japan is the only country with a sashimi market. The OPRT's objective is to contribute actively, through the Japanese market, to the promotion of conservation and sustainable utilisation of tuna resources throughout the world.

An important role of the OPRT is to compile information on tuna landed in Japan by LSTLVs from member flags and provide such information to flag state authorities as well as to relevant international organisations. The OPRT feeds back landing information to any flag state that seriously wishes to implement fisheries management. The OPRT is also working on the development of a list of LSTLVs of countries complying with resource management.

Emergence of another problem

The success of the foregoing efforts can probably be attributed to the simple nature of the market for LSTLV-caught tuna; Japan is the sole outstanding market of these catches. It was relatively easy to monitor tuna caught by LSTLVs (whether legal or illegal) and to take effective measures against tuna caught by IUU fishing. In addition, it was quite fortunate that Japan was able to find out who actually conducted FOC fishing operations and directly consult with them to settle the matter. The global application of a positive listing scheme played a decisive role in achieving this progress. The highly mobile nature of this type of fishery required such a transboundary global measure, as unanimously advised at the 2003 FAO COFI meeting. Ironically, however, another type of tuna fishery – purse seine fishery – dramatically increased its capacity, and tuna longline vessels of slightly less than 24 metres LOA also increased sharply. This typically occurs in the Western and Central Pacific where no management measures have been implemented so far.

Purse seine fishery

Unfortunately, Chinese Taipei residents are again involved in these two types of capacity expansion. The vessels on the Forum Fisheries Agency (FFA) Regional Vessel Register were reviewed in respect of major fishing members (those who have more than ten purse seiners) during the period of the WCPFC Preparatory Conferences. Figure 15.4. shows the result. The Chinese Taipei FOC vessels were identified based on the Register, as well as from owner names, addresses and other information collected from Japanese trade data. Twenty-eight (28) large purse seiners were identified as currently existing Chinese Taipei FOC vessels. Available information suggests that other vessels also exist, but it is not conclusive enough to identify them as Chinese Taipei FOC vessels. The Chinese Taipei FOC vessels in Figure 15.1. should, therefore, be considered as minimum estimates. Furthermore, most of these purse seiners are large; seven of them are over 2 200 GRT class, each of them catching more than 10 000 MT of tuna annually (more than 40 times the annual catch of a longline vessel). In short, it is surprisingly evident that the Chinese Taipei fishing industry increased its purse seine fishing capacity dramatically by using FOC, whereas all other major fishing fleets were restrained to a stable level of fishing capacity or even reduced their capacity. It was reported that the construction of large purse seiners was still under way in Kaohsiung, Chinese Taipei.

Figure 15.4. Number of Purse Seiners of Major Fishing Members registered in the FFA Regional Register

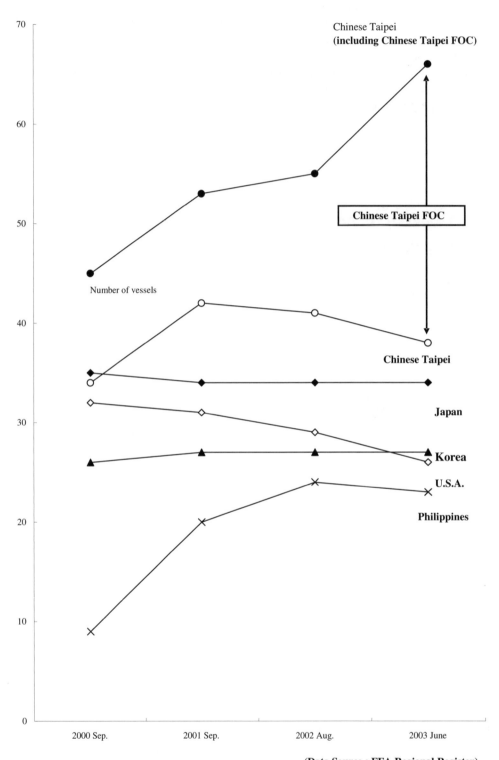

(Data Source : FFA Regional Register)

Only seven Chinese Taipei companies own all twenty-eight FOC purse seiners. All of these companies are located in Chinese Taipei. One of them used to have Chinese Taipei licensed purse seine vessels, while all the other companies currently own Chinese Taipei-licensed fishing vessels, either purse seiners or large tuna longliners. Three of those seven companies were or are engaged in the fishing vessel construction business. It seems obvious that all of the seven Chinese Taipei companies intentionally circumvented the government licensing control by use of FOC so as to continue their excessive fishing for tuna in the WCPFC Convention Area.

As shown in Table 15.3., the Chinese Taipei fishing industry has continued its construction of FOC purse seiners since 1999, when the members of the Multilateral High Level Conference, including Chinese Taipei, adopted a resolution to stop the increase of capacity in the western central Pacific. Particularly after the October 2002 WCPFC Preparatory Conference meeting in Manila, where the resolution was adopted again to restrain the capacity expansion, construction was accelerated further.

Table 15.3. Increase of Chinese Taipei FOC Purse Seiners

Sep-2001			Aug-2002			Jun-2003			Feb-2004		
Name	Flag	GRT	Name	Flag	GRT	Name	Flag	GRT	Name	Flag	GRT
KOO'S 101	MH	996	KOO'S 101	MH	996	KOO'S 101	MH	996	KOO'S 101	MH	996
KOO'S 102	MH	996	KOO'S 102	MH	996	KOO'S 102	MH	996	KOO'S 102	MH	996
KOO'S 103	MH	1198	KOO'S 103	MH	1198	KOO'S 103	MH	1198	KOO'S 103	MH	1198
KOO'S 106	MH	1096	KOO'S 106	MH	1096	KOO'S 106	MH	1096	KOO'S 106	MH	1096
KOO'S 107	MH	1096	KOO'S 107	MH	1096	KOO'S 107	MH	1096	KOO'S 107	MH	1096
EASTERN MARINE	VU	1099	EASTERN MARINE	VU	1099	EASTERN MARINE	VU	1099	EASTERN MARINE	VU	1099
FAIR CRYSTAL 707	VU	1060	FAIR CRYSTAL 707	VU	1060	FAIR PIONEER 707 ※	VU	1060	FAIR PIONEER 707	VU	1060
FAIR WINNER 707	VU	1060	FAIR WINNER 707	VU	1060	FAIR WINNER 707	VU	1060	FAIR WINNER 707	VU	1060
FONG SEONG 666	VU	2234	FONG SEONG 666	VU	2234	FONG SEONG 666	VU	2234	FONG SEONG 666	VU	2234
FONG SEONG 696	VU	2234	FONG SEONG 696	VU	2234	FONG SEONG 696	VU	2234	FONG SEONG 696	VU	2234
ORIENTAL MARINE	VU	1099	ORIENTAL MARINE	VU	1099	ORIENTAL MARINE	VU	1099	ORIENTAL MARINE	VU	1099
Total	11 vessels		FAIR VICTORY 707	VU	1280	FAIR VICTORY 707	VU	1280	FAIR VICTORY 707	VU	1280
			HF 88	VU	1284	HF 88	VU	1284	HF 88	VU	1284
			HSIANG FA 8	VU	1150	HSIANG FA 8	VU	1150	HSIANG FA 8	VU	1150
			Total	14 vessels		KOO'S 108	MH	1099	KOO'S 108	MH	1099
						FAIR CHAMPION 707	VU	1280	FAIR CHAMPION 707	VU	1280
						FONG SEONG 168	VU	2380	FONG SEONG 168	VU	2380
						FONG SEONG 196	VU	2386	FONG SEONG 196	VU	2386
						FONG SEONG 818	VU	1152	FONG SEONG 818	VU	1152
						HSIANG HAO 8	VU	2200	HSIANG HAO 8	VU	2200
						HSIANG SHENG 6	VU	1150	HSIANG SHENG 6	VU	1150
						SHUN FA 8	VU	1150	SHUN FA 8	VU	1150
						TUNA CATCHER	VU	1099	TUNA CATCHER	VU	1099
						Total	23 vessels		EASTERN STAR	VU	2386

Table 15.3. Increase of Chinese Taipei FOC Purse Seiners (cont.)

Sep-2001			Aug-2002			Jun-2003			Feb-2004		
Name	Flag	GRT	Name	Flag	GRT	Name	Flag	GRT	Name	Flag	GRT
									FONG SEONG 668	VU	2386
									TUNA QUEEN	VU	1099
						※ FAIR CRYSTAL 707 was re-named FAIR PIONEER 707.			YUNG DA FA 168	VU	1152
									YUNG DA FA 668	VU	1152
									Total	28 vessels	

Longline fishery

The same review was carried out for longline fishing vessels of major fishing members (those who have over fifty longline vessels) on the FFA Regional Vessel Register. The result is shown in Figure 15.5. The Chinese Taipei fishing industry experienced increases in both FOC and Chinese Taipei-licensed longline vessels operating in the WCPFC Convention area, whereas the numbers of Japanese and Korean longliners remained relatively stable. There was an increase in the number of Chinese longliners, but most of these are relatively small, with low productivity. Their catch of tuna did not increase significantly.

**Figure 15.5. Number of Longliners of Major Fishing Members registered
in the FFA Regional Register**

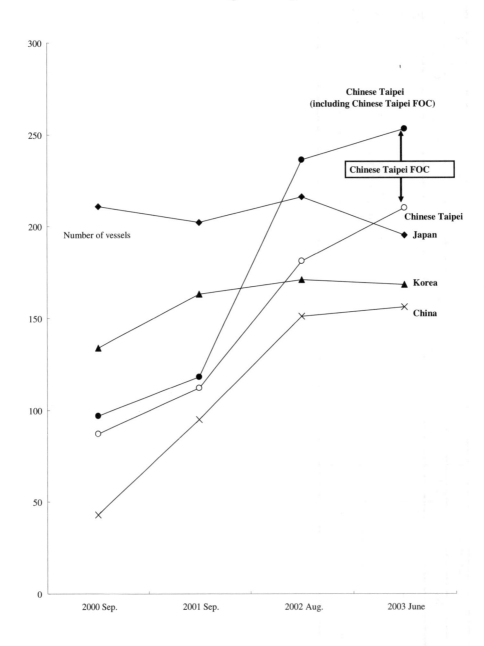

It is now necessary to look not only at transboundary but also at trans-fishery measures to counteract IUU tuna fishing and the over-capacity of all fisheries for tuna resources. Otherwise, the control over one fishery, including the elimination of IUU fishing, could cause the immediate explosive expansion of other fisheries targeting the same tuna resources. At the same time, possibilities for the sound development of fisheries of developing states must be secured under overall capacity control. In the past, developed states' fishing industries took advantage of the developing states' right for fishery development so as to evade capacity control measures. In order to avoid any increase in overall capacity, such fishery development of developing states should be realised through the appropriate transfer of fishing capacity from developed states.

Conclusions and recommendations

Large-scale tuna longline vessels (LSTLVs) produce mainly frozen tuna, and land them at Japanese ports for sashimi use. Since the Japanese market offers the highest price for sashimi tunas, almost all LSTLV products come to Japan.

After Chinese Taipei started monitoring its LSTLV catch through the Japanese market in the early 1990s, many Chinese Taipei vessel owners used flag-of-convenience (FOC) LSTLVs to circumvent regulations. The global problem of IUU tuna longline fishing was caused solely by Chinese Taipei residents. The International Commission for the Conservation of Atlantic Tunas (ICCAT) took a series of measures to eliminate the FOC/IUU LSTLVs during the mid to late 1990s. However, since they focused on flag states and not on people who actually conducted IUU business, their effectiveness was easily undermined by flag hopping.

Japan analysed its trade and other data to identify the real operators of the IUU LSTLVs and started consultations with them. As a result of intense consultations between Japan and Chinese Taipei, including IUU vessel owners, as well as the efforts of the regional fisheries management organisations (RFMOs) to establish a positive vessel listing scheme, the number of IUU LSTLVs has been substantially reduced.

However, Chinese Taipei fishermen switched from large-scale longline fishing (over 24 metres) to small-scale longline (less than 23.9 metres) as well as large-scale purse seine fishing, and continued to catch tuna in an area where no management measures have been introduced, *i.e.* western and central Pacific. The flags of these vessels are developing states. In the past, developed states' fishing industries took advantage of the developing states' right for fishery development to evade capacity control measures. So long as developed countries continue to build new vessels and developing countries continue to accept these vessels under their registry, over-capacity problems will continue and will expand. The IUU tuna fishing problem is part of the tuna over-capacity problem.

Based on past experience, we can conclude that:

1) Measures focused on flag states, including trade measures, have had limited effect.

2) Trade tracking and its resulting accumulation of information by market countries is an enormous task but it provides the most important fundamentals for the creation of effective measures to combat IUU fishing.

3) Direct consultations with IUU vessel owners played an important role in solving the problem.

4) Measures based on positive listings are effective, but tuna laundering and the use of forged documents may still continue.

5) FOC/IUU fishing is part of over-capacity.

6) All FOC flag states are developing states.

7) Even after the elimination of IUU fishing, so long as developing states accept unlimited registration of foreign fishing vessels, the over-capacity problem will continue.

IUU measures should be specific to each fishery and based on trade and other data for the identification of real operators. The following global action is urgently required to solve the IUU and over-capacity problem:

1) The FAO should establish a global record of tuna fishing vessels, compiling existing records of tuna fishing vessels of relevant RFMOs, and RFMOs should co-operate with the FAO to establish such a record.

2) Developed states, parties and fishing entities should stop building new tuna fishing vessels except for those replacing existing licensed vessels with equivalent fishing capacity, whatever flag is used.

3) The FAO should request RFMOs to establish, as a matter of priority, a system to transfer fishing capacity from developed states, parties and fishing entities to developing states smoothly.

4) A nation, party or fishing entity whose residents have caused the rapid expansion of fishing capacity in recent years should cut at least that expanded portion of fishing capacity.

5) RFMOs should develop market-oriented measures for purse seine caught tuna. Countries importing purse seine caught tuna should play a vital role.

REFERENCES

Annual Reports of the International Commission for the Conservation of Atlantic Tunas (ICCAT).

Japanese Delegation, April 2004. Information Paper on Expansion of Fishing Capacity. Sixth Preparatory Conference of the Western Central Pacific Fisheries Convention (WCPFC).

Miyahara, M. (2002), "International Management of Tuna Resources and Flag-of-convenience (FOC) Fishing Vessels" (in Japanese), *Journal of International Fisheries* 5(1) pp19-26, Japan International Fisheries Research Society.

CHAPTER 16

ILO SUBMISSION TO THE WORKSHOP ON IUU FISHING ACTIVITIES[1]

Brandt Wagner, Maritime Specialist, ILO

Summary

This document provides information on work underway by the International Labour Organization[2] to prepare a comprehensive standard (a Convention supplemented by a Recommendation) on work in the fishing sector. This work may be relevant to the issue of IUU fishing.

Introduction

At its 283rd Session (March 2002) the Governing Body of the ILO decided to place on the agenda of the 92nd Session of the International Labour Conference an item concerning a comprehensive standard (a Convention supplemented by a Recommendation) on work in the fishing sector. This standard will revise seven ILO standards (five Conventions and two Recommendations) adopted in 1920, 1959 and 1966 that are specifically aimed at persons working on board fishing vessels (henceforth "fishers"). These five standards concern the issues of: minimum age, medical examination, articles of agreement, competency certificates, crew accommodation, hours of work and vocational training. The standard may also address other issues, such as occupational safety & health and social security. The aim is to ensure "decent work" for fishers, within the context of the ILO's primary goal of promoting opportunities for women and men to obtain decent and productive work, in conditions of freedom, equity, security and human dignity.

The rationale for this revision is to reflect changes in the sector which have occurred over the last 40 years; to achieve more widespread ratification; to reach, where possible, a greater proportion of the world's fishers, particularly those working on smaller vessels; and to address other fishing operations, employment arrangements, methods of remuneration and other aspects. This revision will complement

---

[1]   This paper was submitted as a background document to the Workshop.

[2]   The International Labour Organization is the UN specialised agency which seeks the promotion of social justice and internationally recognised human and labour rights. It was founded in 1919 and is the only surviving major creation of the Treaty of Versailles which brought the League of Nations into being and it became the first specialised agency of the UN in 1946.

the parallel work being done by the ILO to consolidate its standards for seafarers (on vessels engaged in commercial maritime transport) into a comprehensive new standard.

In accordance with the Standing Orders of the Conference, the Office prepared a *preliminary report* intended to serve as a basis for the first discussion of the item on the fishing sector standard by the Conference in 2004. The report gives an overview of the fishing sector and analyses the relevant legislation and practice concerning labour conditions in the sector in various ILO member states. The report and attached questionnaire were communicated to the governments of member states of the ILO, which were invited to send their replies so as to reach the International Labour Office by 1 August 2003. The report, entitled *Conditions of work in the fishing sector: A comprehensive standard (a Convention supplemented by a Recommendation) on work in the fishing sector*, Report V(1), International Labour Conference, 92nd Session, Geneva, 2004, is a available at:

http://www.ilo.org/public/english/standards/relm/ilc/ilc92/pdf/rep-v-1.pdf

It is available in English, French, Spanish, German, Russian, Arabic and Chinese.

On the basis of these replies to the above-mentioned questionnaire, the Office prepared a *second report*. Replies were received from over 80 ILO member states. In accordance with the Standing Orders of the Conference, governments were requested to consult the most representative organisations of employers and workers before finalising their replies to the questionnaire, to give reasons for their replies and to indicate which organisations have been consulted. Governments were also reminded of the importance of ensuring that all relevant departments were involved in the present consultative process, including the departments responsible for social and labour affairs, fisheries, maritime safety, health and the environment. The report also took into account the report of the Tripartite Meeting of Experts on Labour Standards for the Fishing Sector, which had been held in Geneva from 2 to 4 September 2003 in order to discuss issues to be covered in the fishing standard. It provides proposed conclusions with a view to a Convention and a Recommendation. The report, entitled *Conditions of work in the fishing sector: A comprehensive standard (a Convention supplemented by a Recommendation) on work in the fishing sector: The Constituents' Views*, Report V(2), International Labour Conference, 92nd Session, Geneva, 2004, is a available at:

http://www.ilo.org/public/english/standards/relm/ilc/ilc92/pdf/rep-v-2.pdf

It is available in English, French, Spanish, German, Russian, Arabic and Chinese.

Next steps

At the 92nd Session of the Conference (Geneva, 1-17 June 2004) a Committee on Work in the Fishing Sector will be established to consider this agenda item. The report of this Committee will be submitted to the plenary of the Conference, which is expected to adopt conclusions concerning a Convention and a Recommendation for the work in the fishing sector. Immediately afterwards, in accordance with the Standing Orders of the Conference, the International Labour Office will prepare a *third report* containing a proposed Convention and Recommendation for work in the fishing sector. This report will be sent to all ILO member states, asking them to state within three months, after consulting with the most representative organisations of employers and workers, whether they have any amendments to suggest or comments to make. On the basis of the replies received, the Office will draw up a *final report* containing the text of the Conventions or Recommendations with any necessary amendments. These latter two reports will then serve as the basis for discussion at the 93rd Session of the International Labour Conference in June 2005, which is expected to adopt the instruments. Subject to these caveats, the ILO sets out below relevant elements in the proposed conclusions as they now stand.

Possible relevance of the proposed ILO standard to the issue of IUU fishing

The relationship between IUU fishing and conditions of work on board fishing vessels is not entirely clear. However, the nature of IUU fishing gives rise to questions concerning working conditions on board such vessels. Such operations also place fishers at risk of arrest and imprisonment. This leads to questions concerning their repatriation to their home countries.

Some provisions of the proposed conclusions prepared by the Office may be relevant to this Workshop. One proposed provision would allow port states to inspect foreign fishing vessels to ensure compliance with the standards set out in the Convention. Another provides that fishing vessels that operate internationally should be required to undergo a documented periodic inspection of living and working conditions on board the vessel. Yet another proposed non-mandatory provision states that "In its capacity as a coastal state, a member might require, when it grants licences for fishing in its exclusive economic zone, that fishing vessels comply with the standards of the Convention." This provision in particular, if retained, could contribute to action aimed at addressing IUU fishing.

Further information

For further information on the development of this standard, contact the International Labour Office (Secretariat of the ILO) at marit@ilo.org.

CHAPTER 17

IUU FISHING AND THE COST TO FLAG OF CONVENIENCE COUNTRIES[1]

Matthew Gianni, Independent Consultant on fisheries and oceans issues

The problem of IUU fishing and related infrastructure must be tackled from a number of different angles. Many of the measures debated to date have centered on taking action to deter individual vessels from engaging in IUU fishing. However, one approach worth considering might be to pursue compensation from flag of convenience states for the costs incurred by other states as a result of FOC/IUU fishing. Whether or not there is a genuine economic link between the flag state and the IUU vessels or fleets flying its flag, the flag state bears the ultimate responsibility for the activities of the vessel in relation to compliance with relevant international instruments including the conservation and management measures adopted by regional fisheries management organisations (RFMOs).

It could be argued that legitimate flag states, which are members of, participate in, and contribute to the activities of a regional fisheries management organisation, should have the right to derive long-term benefit from sustainably managed fishing in the region, commensurate with the effectiveness of conservation measures agreed by the organisation, provided they ensure that vessels under their jurisdiction abide by the rules. The conservation and management of the fisheries and the measures undertaken by a state with respect to monitoring, compliance and enforcement all come at a cost.

Conversely, a state whose vessels consistently operate in a region in contravention of the rules adopted by the relevant regional fisheries management organisation should be liable for a portion of the costs incurred by responsible flag states. While an FOC state may not be compelled to join a regional management organisation, it does have a clear duty under UNCLOS to co-operate with other states in the conservation and management of the fisheries in the region. Should it fail to do so while 'allowing', either willfully or by clear negligence, its vessels to consistently fish in the region, then the state should be liable for the costs incurred by responsible members of the RFMO associated with the failure of the FOC state to either co-operate with the regional management organisation or to exercise control over the activities of its fishing fleets operating in the area of competence of the organisation.

Costs could be measured in a number of ways. The short-term, or annualised, costs to legal operators in the fishing industry could be considered to include lost revenue resulting from lower

---

[1] This paper was submitted as a background paper to the Workshop. Paper prepared by Mathew Gianni, independent consultant on fisheries and oceans issues, Cliostraat 29-II, 1077 KB Amsterdam, Netherlands. matthewgianni@netscape.net

quotas, higher catch per unit effort costs as a result of overfishing by IUU operators, and lower prices as a result of excess supply of IUU-caught fish on the market. Costs to governments might be calculated on the basis of factors such as the expense of extra research resulting from scientific uncertainties arising from lack of sound information on the catch and biological characteristics of the species caught in IUU fisheries, the increased cost of monitoring, surveillance and enforcement at sea and port and market-based inspection schemes to combat IUU fishing, and the costs associated with dues and participation at annual meetings of an RFMO and its various committees. Longer-term costs could also be factored into the equation, in particular the loss of long-term benefits to the economy because of the lower productivity of overfished stocks as a result of IUU fishing, loss of future earnings from more sustainable fisheries, and the loss of tax revenue or income to the state.

Given the significant cost of IUU fishing to responsible governments and industry operators, what are the benefits to the states involved in issuing flags of convenience? Clearly, unscrupulous operators themselves benefit financially from the freedom to engage in IUU fishing on the high seas with the impunity conferred by the flags of convenience system. But are there economic benefits to FOC States that might argue for the legitimacy of the FOC system?

The information contained in a 2002 UN FAO report on open registries in relation to fishing suggests that the benefits derived by FOC states in flagging large-scale fishing vessels are relatively small. Based on information in the report, the total revenue derived from registering fishing vessels by 20 countries operating open registries (flags of convenience) was slightly more than USD 3 million per year in recent years.[2] The report states that the top four FOC countries – Belize, Honduras, Panama, and St. Vincent and the Grenadines – had a combined total of 1 148 large-scale fishing vessels registered to fly their flags. These same four countries generated approximately USD 2 625 000 in revenue from registration fees and related charges from the fishing vessels on their registries. They earned, on average, less than USD 2 500 per year for each fishing vessel registered to fly their flag. The report states that the figures are almost certainly underestimates of the total revenue derived from registering fishing vessels. However, even if the figures are off by 100% or 200% of gross revenue, it is clear that the income derived by FOC countries from flagging fishing vessels is still quite small.

It is further interesting to note, in the FAO report, the frequency and type of enforcement actions taken by the government of Belize against fishing vessels flying its flag operating outside of Belize waters. From the period 1997 through 2001, Belize reported that it took enforcement action 17 times against fishing vessels on its registry. In only five instances were the fishing vessels actually fined. Most of the fines levied were in the vicinity of USD 20 000 but only one of these vessels was actually reported to have paid the fine. Belize reported that the most common means of penalising an offending vessel was to delete (deflag) the vessel from the Belize registry. This, however, would have been at best a minor inconvenience for the vessels concerned. A fishing vessel can obtain a flag of convenience easily, with provisional registration being granted by some flag states within 24 hours of application. Many vessels change flags often, a phenomenon known as 'flag-hopping', taking advantage of the ease in obtaining a flag of convenience.

This history of enforcement is remarkably limited and virtually ineffective considering that several hundred large-scale fishing vessels flew the flag of Belize during the same period of time. Belize was in the top two FOC countries flagging large-scale fishing vessels in 1999 and 2001, according to an analysis prepared by Gianni and Simpson for WWF.[3] The number of large-scale

[2] Swann, J. Fishing Vessels Operating Under Open Registers and the Exercise of Flag State Responsibilities: Information and Options. FAO Fisheries Circular No. 980, Rome 2002.

[3] Gianni, M. Simpson, W. - see Chapter 6 of this report.

fishing vessels registered to Belize in 1999 and 2001 was 409 and 455 vessels respectively. The average tonnage of the fishing vessels on the registry for both years was 853 GT and 768 GT respectively. These are large vessels by fishing industry standards (the FAO reports the average tonnage of large-scale fishing vessels in 2000 was 370 GT[4]). According to the FAO report, Belize-flagged vessels were reported by RFMOs to be engaged in IUU fishing in the Atlantic, Pacific, and Indian Oceans as well as the Southern Ocean around Antarctica. To its credit, the government of Belize at least provided information to the author of the FAO report and appears to have significantly reduced the number of fishing vessels on its registry since 2001, although the number still on the registry is high. All of the other countries with open registries and substantial numbers of fishing vessel on their registries ignored the request for information by the author of the FAO report.

Clearly, states that operate flags of convenience in the fisheries sector externalise the costs of their failure to regulate 'their' fishing fleets. Other countries must pay these costs in terms of scientific uncertainty in stock assessments, reduced quotas and lost revenue for legitimate operators, and the additional costs of enforcement, among other things, as well as the depletion of fish stocks and ecosystems associated with flag of convenience fishing. The costs to legitimate operators and responsible flag states are likely to far outweigh the revenue derived by FOC states in registering large-scale fishing vessels.

An important legal question arises: Does a state have the right to enjoy the privileges of being a flag state, however little these privileges may confer to the state in terms of economic benefits, while evading most, if not all, of the responsibilities associated with being a flag state, no matter how costly this evasion of flag state responsibility may be to other states and the international community as a whole?

Given the large number of IUU fishing vessels flying flags of convenience, it seems clear that the most cost effective means of eliminating the problem of IUU fishing would be to eliminate the flag of convenience system for fishing vessels. Countries which cannot or will not exercise control over fishing vessels operating outside of their EEZs should be discouraged or prevented from registering large-scale fishing vessels (*e.g.* fishing vessels greater than or equal to 24 metres as per the international standard defined by the FAO Compliance Agreement) except under strictly defined circumstances or criteria. Ultimately, what may be needed is a clear ruling from the International Tribunal for the Law of the Sea designed to further strengthen the definition of flag state responsibility under international law and ultimately render the state practice of issuing flags of convenience for fishing vessels effectively illegal.

However, until the flag of convenience 'loophole' in international law is closed, one option available to responsible flag states may be to explore the possibility of seeking compensation from FOC states for the costs incurred by responsible states as a result of IUU/FOC fishing. It would be well worth considering a means or method to document and/or reasonably estimate the costs incurred by responsible flag states as a result of FOC fishing. On this basis, compensation could then be sought, through the available international mechanisms, from specific FOC states whose vessels are fishing in a region in contravention of the measures established by a relevant fisheries management organisation to the detriment of responsible flag states' fleets and interests.

Whether or not there is a genuine economic link between the flag state and the IUU fishing vessels or fleets flying its flag, the flag state bears responsibility for the activities of the vessels. If an

---

[4]   FAO State of World Fisheries and Aquaculture 2000. United Nations Food and Agriculture Organisation. Rome, 2001.

FOC state is faced with the prospect of paying substantial compensation to other states for its failure to regulate its fishing fleets, this could act as a disincentive to the registration of fishing vessels by the FOC state. The prospect of paying potentially large sums in compensation for the failure to exercise control over fishing vessels could potentially serve as a significant deterrent to FOC/IUU fishing in ways that could complement port state controls, market restrictions, enhanced monitoring, control and surveillance and other measures adopted thus far by states and regional fisheries management organisations. The OECD can play a role in assisting OECD members in comprehensively estimating the cost to responsible flag states of fishing by vessels flying flags of convenience.

PART IV

WAYS OF COMBATING IUU FISHING

The fourth session of the workshop focused on the various means available for deterring IUU fishing activities and assessed the costs and benefits of alternative strategies, drawing examples from governmental, industry and NGO experience. Possible loopholes in current regulatory arrangements were identified, and suggestions made for ways of dealing with them. The intention of this session was to have participants think "outside the box" and explore alternative ways to combat IUU fishing.

CHAPTER 18

ADVANCES IN PORT STATE CONTROL MEASURES

Terje Lobach, Ministry of Fisheries, Norway

Introduction

Illegal, unreported and unregulated (IUU) fishing is a major threat to sustainable fisheries management and marine biodiversity. It occurs in all fisheries, whether they are conducted within areas under national jurisdiction or on the high seas. A number of international instruments, which were developed during the 1990s for the management of world fishery resources, also address the issue of IUU fishing. Of particular importance in this regard are the 1995 UN Fish Stocks Agreement, the 1993 FAO Compliance Agreement, the Code of Conduct for Responsible Fisheries and the International Plan of Action (IPOA) on Illegal, Unreported and Unregulated (IUU) Fishing. These include both so-called hard instruments (which are legally binding on parties to the agreements) and soft instruments which serve more as guidelines and toolboxes, including some options for both states and regional management organisations (RFMOs) in addressing the issue of IUU fishing.

However, despite these agreements and plans, and despite the efforts made by global organisations, by regional bodies and by a great number of states, IUU fishing continues to persist. But the international community cannot give up the fight. Vessels engaged in IUU fishing move in and out of areas under jurisdiction of multiple states and operate within the areas of competence of several RFMOs. Thus, a key word in the combat against IUU fishing is co-operation. This could be co-operation in tracing IUU vessels, tracing owners of such vessels and tracing fish and fish products deriving from IUU fishing. Furthermore, in order to harmonise and facilitate co-operation among states and RFMOs, some minimum standards for port state measures should be developed.

Link to flag State responsibilities

If all flag States complied with their obligations concerning their fishing fleets, port State control would more or less be superfluous. But this is certainly not the case. Of particular concern is the growing trend in the use of "flags of convenience" (FOC) by fishing vessels. Flagging and re-flagging of vessels is very easy and in some cases just a few moments' work on the internet is all that is required (for example there are sites offering registration services for named States with a turn-around of 24 hours or less). Under international law, the flag State is responsible for ensuring that vessels abide with relevant rules. However, some countries are willing to sell their flag with no questions asked, in exchange for a licence fee, while exerting no control over the vessel's activities. "FOC" is a term often used in relation to states with open shipping registers. In a fisheries context, the term would have a wider application as the problem with IUU fishing stems partly from it being "convenient" to use certain specific flags to avoid being bound by conservation and management measures. In principle, states with restricted shipping registers could thus be regarded as FOC in relation to fishing. The acronym "FONC" (Flag of Non-Compliance) avoids the political sensitivities attached to the term "Flag Of Convenience" and also applies to parties and non-parties to RFMOs.

Companies and individuals typically have nationalities that differ from those of the vessels themselves, and fish deriving from IUU activities are put into international trade. It is thus absolutely necessary that agencies, international organisations and states establish both formal and informal co-operation channels. This is the only way of achieving the goal of preventing, deterring and finally eliminating IUU fishing.

The call for port State measures is closely linked to the lack of flag State responsibilities. Thus, port State measures are highly relevant for counteracting IUU fishing and some initiatives have now been taken to address the issue.

Memoranda of Understanding (MOUs) for the merchant fleet

Port State regimes have gained international acceptance in recent years as a result of numerous agreements concerning the merchant shipping fleet. Inspired by the Paris MOU (Memorandum of Understanding), which was agreed among 18 countries in 1982, several MOUs have been adopted in different regions of the world in order to trace sub-standard vessels. Such mandatory port State control is tied to internationally agreed rules and standards. The International Maritime Organisation (IMO) has played an important role in this development, and in order to ensure universal standards, IMO has developed a global strategy for operating guidelines and training of control officers.

Joint FAO/IMO working group

Recalling an agreement between IMO and FAO on matters of mutual interest, a joint FAO/IMO Working Group on IUU fishing met in 2000. The main issues examined by the group were related to flag State and port State control. Concerning port State control, in brief it was noted that the majority of fishing vessels were not covered by IMO instruments, either because fishing vessels were specifically excluded, were outside the size limitations, or the flag States are not parties to the relevant instruments. Further it was noted that it might be difficult to introduce port State inspection procedures for fisheries management purposes and fishing vessel safety within existing regional MOUs on port State control. It was also recognised that the mechanism of international or regional MOUs relating to port State control could be used as an important and effective tool for enhancing fisheries management, and addressing IUU fishing. Finally the group agreed that FAO, in co-operation with relevant organisations, should consider the need to develop measures for port State control to all matters related to the management of fisheries resources.

Possible regional strategy

By examining internationally agreed instruments like the UN Fish Stocks Agreement and the IPOA on IUU fishing, as well as measures established by several RFMOs and unilateral approaches taken by some States, port State control was found to be highly relevant for fishery conservation and management. As the existing MOUs on port State control target the standards of the vessel itself, they seem not to be the right vehicles for seeking compliance with fisheries conservation and management measures. It would therefore be worth considering taking the now widely applicable regional MOUs on merchant shipping as a model for a regional approach to fisheries.

A regional system on port State control would require common procedures for inspection, qualification requirements for inspection officers, and agreed consequences for fishing vessels found to be in non-compliance.

The underlying principle formulated in Article 23 of the UN Fish Stocks Agreement is "the right and the duty" of a port State to take non-discriminatory measures in accordance with international law, in order to "promote the effectiveness of sub-regional, regional and global conservation and management measures". Emphasis needs to be put not only on the "right", but also on the "duty" and some minimum requirements for port State control should be agreed upon.

In order to establish an appropriate system, port States should adopt harmonised mandatory obligations for control of fishing vessels. Although some RFMOs have already introduced some port State duties for their members, these apply only to activities taking place in their areas of competence, which in most cases are outside areas under national jurisdiction of the parties. Furthermore, the schemes are of course limited to members of a particular RFMO, consequently creating "Ports of Convenience" in a region.

Current schemes for some RFMOs

In 1989 port State control of fishing vessels was introduced at a regional level for the first time with the adoption of the Convention for the Prohibition of Fishing with Long Drift-nets in the South Pacific (the Wellington Convention on Drift-nets). The Convention provides for restriction of both access to the ports and the use of service facilities in the ports of parties for vessels involved in drift-net fishing.

In recent years several RFMOs have established port control obligations, in particular targeting non-parties. In order to combat IUU Fishing by non-Contracting Party vessels, in 1997 the first RFMO, the Northwest Atlantic Fisheries Organization (NAFO), already adopted the "Scheme to Promote Compliance by Non-Contracting Party Vessels with Conservation and Enforcement Measures Established by NAFO", which put certain obligations on the port States of NAFO. The Scheme presumes that a non-Contracting Party vessel that has been sighted engaging in fishing activities in the NAFO Regulatory Area (*i.e.* the area outside national jurisdiction of NAFO Parties) is undermining NAFO Conservation and Enforcement Measures. If such a vessel enters a Contracting Party port, it must be inspected. No landings or transhipments will be permitted in Contracting Party ports unless vessels can establish that certain species on board were caught outside the NAFO Regulatory Area, and that for certain other species the vessel applied the NAFO Conservation and Enforcement Measures. Contracting Parties must report the results of such port inspections to the NAFO Secretariat, all Contracting Parties and the flag State of the vessel. Similar schemes were later introduced in several other regional bodies.

Some of the schemes have later been amended to include blacklisting of IUU vessels. At its annual meeting in 2002, the Commission of the Conservation of Antarctic Marine Living Resources (CCAMLR) agreed to adopt a scheme to promote compliance with CCAMLR conservation measures by Contracting Party vessels and a scheme to promote compliance with CCAMLR conservation measures by non-Contracting Party vessels. These schemes imply that procedures were agreed upon for the establishment and maintenance of lists of fishing vessels (IUU Vessel list) found to have engaged in fishing activities in the CCAMLR-area in a manner which has diminished the effectiveness of CCAMLR-measures. Procedures for the removal of vessels from the IUU Vessel list have also been adopted. Further Contracting Parties of CCAMLR have agreed to take a number of appropriate domestic actions against vessels appearing on the IUU Vessel list, including not authorising landing or transhipment in ports.

The North East Atlantic Fisheries Commission (NEAFC) has established a similar system for non-party vessels, and NAFO is in the in the process of introducing a system of blacklisting both non-party IUU-vessels and IUU-vessels flying the flags of Contracting Parties.

The International Commission for the Conservation of Atlantic Tunas (ICCAT) has taken a different approach. ICCAT has adopted a measure concerning the establishment of a record of large-scale fishing vessels authorised to operate in the Convention area (a so-called "white list"). This implies that only vessels appearing on the list are regarded as being in conformity with applicable ICCAT-measures. Vessels that are not on the "white list" are deemed not to be authorised to fish for, retain on board, tranship or land tuna and tuna-like species. Parties to ICCAT shall take measures, under their applicable legislation, to prohibit, amongst other things, the transhipment and landing of tuna and tuna-like species by large-scale fishing vessels, which are not "white listed".

Even though parties to RFMOs have agreed to take some port measures, port control schemes which include inspection procedures, result indicators, and formats for the exchange of information are rather rare. The vast majority of RFMOs do not have in place appropriate port control schemes, though some do have quite vague references to port inspections. NAFO, for example, has established reciprocal port State control obligations. According to the relevant provision a "Contracting Party whose port is being used shall ensure that its inspector is present and that, on each occasion when catch is offloaded, an inspection takes place of the species and quantities caught". NAFO is, however, now considering strengthening the port State obligation by introducing a more comprehensive system, which includes, among other things, harmonised inspection procedures and protocols for exchange of information.

In ICCAT, parties are encouraged to enter into bilateral agreements/arrangements that allow for an inspector exchange programme designed to promote co-operation, share information and educate each party's inspectors on strategies and operations that promote compliance with ICCAT's management measures. The port inspection scheme recognises that most of the recommendations can only be enforced during off-loading and therefore found that port State enforcement is "the most fundamental and effective tool for monitoring and inspection".

It should also be mentioned that port control schemes have not been established by CCAMLR or NEAFC.

CCAMLR has, however, established a Catch Documentation Scheme for *Dissostichus* spp., which requires control by port States. The Scheme builds on the principle of flag State responsibility, but at the same time the Scheme requires that landings of *Dissostichus* Spp. at its ports and all transhipments of *Dissostichus* spp. to its vessels be accompanied by a completed catch document. The document will need to be countersigned by a port State official when the catch is landed. This signature will confirm that the catches landed agree with the details on the document.

Some states have established measures reaching further than those established by the regional fisheries management organisation to which those states are parties. States like Canada, Iceland, Norway and the United States are refusing access to port services for vessels undermining conservation and management measures on the high seas.

Application of a MOU

MOUs would have a wider application, as not all port States are parties to a RFMO, and required port measures might involve more than one RFMO. In principle, port State control should be related to *all areas* where marine capture fisheries take place. In a context of a possible MOU, such control should be related to areas within the jurisdiction of the port State, areas within the jurisdiction of another State that is Party to the MOU, and on the high seas areas managed by a relevant RFMO. Port States should thus ensure that fishing undertaken in these areas has been in conformity with established conservation and management measures.

In addition, a port State should inspect vessels flying the flag of another State where fishing activities took place within the waters of that particular flag State. This last point is particularly important when conservation and management measures concerning shared stocks have been agreed upon between two or more States. Sometimes fishing is conducted within the EEZ of a party to such arrangements, but landed in the port of another State (due to port facilities, price factors, distance from fishing grounds, etc.). In these cases it is most likely that the fishing vessels leave the waters of a coastal State without being inspected to determine whether the fishing has been conducted in accordance with applicable legislation. This is also a general issue, however, as a coastal State may seek assistance from a port State to verify that fishing in the waters of that coastal State has been in accordance with relevant legislation. This may be the only way of obtaining the information required for assessing the situation.

In doing so, it is recommended that such an approach should be linked to the existing RFMOs. Most of the conservation and management measures for high seas fishing in different regions are established by such organisations. The internationally agreed measures that vessels should comply with will therefore be those of the relevant organisation. Consequently, there is a direct link between that particular organisation and port States in the region. In order to achieve a comprehensive system within a region, the RFMOs should be encouraged to enter into agreements on mandatory port State control with port States in the region that are not parties to the relevant regional fisheries body.

RFMOs were strengthened by the entry into force of the 1995 UN Fish Stock Agreement, and the importance of their role is underlined throughout the agreement. It has also inspired coastal States and distant water fishing nations to co-operate in order to establish organisations in regions previously not covered by such bodies. Further, these organisations are responsible for establishing relevant conservation and management measures in areas under respective purviews. Thus, an inspection in port should therefore examine if the fishing vessel in question has operated contrary to any conservation and management measures established by any RFMO. It is also recommended that co-operation between regional fisheries management organisations be formalized. Such co-operation would be essential in areas where IUU fishing is the concern of two or more regional bodies. For example, the conservation and management of fish resources in the Atlantic Ocean is the responsibility of several fisheries management organisations. A comprehensive system on port State control would require that IUU fishing within the area of responsibility of one specific organisation should have consequences for port States which have agreed on mandatory measures in another region.

In principle, port State control should relate to all areas where marine capture fishing operations take place. Port States should thus ensure that fishing undertaken in these areas have been in conformity with established conservation and management measures. In summary a port State should examine whether IUU fishing has taken place in:

a) the Regulatory Area (RA) by a Contracting Party of a RFMO;

b) the RA by a non-Contracting Party of a RFMO;

c) waters under national jurisdiction of a Contracting Party by a Contracting Party of a RFMO; and

d) waters under national jurisdiction of a Contracting Party by a non-Contracting Party of a RFMO.

IUU vessels move from one region to another and are therefore not the concern of one RFMO alone. In order to establish a tight system, a MOU on port State control between such bodies could be a way forward. In that context port States should have the duty to take action against vessels having participated in IUU fishing in areas managed also by other regional bodies. RFMOs should therefore be encouraged to enter into multilateral agreements on port State control. Such co-operation would be essential in areas where IUU fishing is the concern of two or more regional bodies.

FAO expert consultation

Following the recommendations by the joint FAO/IMO Working Group and the call for harmonised port State measures in a number of international instruments, FAO has convened an Expert Consultation to review port State measures to combat IUU fishing. The Consultation agreed that regional MOUs on port State measures also for fishing vessels are highly relevant and examined all aspects of possible MOUs.

The Consultation agreed that in terms of *scope,* a MOU should apply to all vessels engaged in, or supporting, fishing activities including fishing vessels and vessels transporting fish and fishery products. Criteria for targeting specific vessels might be developed for a given MOU. For instance, vessels flying a "flag of non compliance" (FONC), or vessels having a history of non-compliance established by a RFMO can be particularly targeted.

The Consultation noted that the Parties should determine whether a MOU is binding or not. A MOU will, however, include only the minimum requirements for port State measures. The question remained open with respect to the impact of and the effect of the MOU on third parties. To encourage wider application of a MOU, the Consultation observed that some IMO instruments provide that the parties to these instruments apply the requirements in the same manner to vessels of non-parties, as may be necessary, to ensure that "no more favourable treatment" is given to such vessels.

The Consultation agreed that port States should require all foreign vessels that have engaged in fishing activities or transporting fish and fishery products to provide a *prior notice* of the intention to use a port, its landing or transhipment facilities. While failure to provide satisfactory information submitted in the prior notification might be a reason for denial of access to port, the Consultation noted that it might be advisable to allow a vessel into port in order to ascertain whether a vessel has engaged in or supported IUU fishing.

The Consultation further noted that port States might, on the basis of objective and non discriminatory criteria, *set out conditions of entry to their ports or deny access* to their ports by foreign fishing vessels that have engaged in, or supported, IUU fishing. In cases of distress and *force majeure,* vessels have a right to entry to ports under customary international law. In addition, bilateral or multilateral arrangements might be in place providing reciprocal free access to ports, as well as dealing with trade-related matters. It was also observed that denial of port access in order to combat IUU fishing might not always be appropriate in practice.

The need for *harmonised and co-ordinated approaches for inspection* was discussed in the Consultation and it received wide support. The Consultation considered that the use of a single fishing vessel numbering system could be a useful tool for the effective implementation of a MOU on port State measures. It noted that a system for numbering vessels is applied in IMO. This system is based on the Lloyds register fair-play system.

The Consultation also observed that a harmonised system of certification of fishing vessels, including the clear identification of the vessel owners and managers, could be useful to facilitate the inspection of vessels in port States.

Concerning *sanctions*, the Consultation recognised that if a vessel is found to have violated applicable legislation in waters under the jurisdiction of the port State, the latter should exercise jurisdiction as a coastal State and initiate proceedings accordingly. In other situations, port States could choose between several possible actions. With the exception of detention, arrest or other measures against crew, a port State could take other more appropriate action. Such action could include refusal to allow the landing of fish and fishery products, forfeiture of fish and fishery products, or refusal to permit a vessel to leave its port pending consultation with the flag State of the vessel.

The Consultation recognised that awareness about, and *capacity building* in, port State measures, especially in developing countries, is vital to the wide application of port State measures to prevent, deter and eliminate effectively IUU fishing.

The Consultation noted that the *exchange of information and data* would be crucial for effective implementation of port State measures to combat IUU fishing.

Following discussions and an in-depth review of the elements that might be included in regional MOUs, the Expert Consultation elaborated a draft MOU on Port State Measures to combat IUU Fishing. This could be used as a template in cases where initiatives are taken to develop regional MOUs.

COFI (FAO Committee of Fisheries) agreed in March last year that FAO should convene a Technical Consultation to Address Substantive Issues Relating to the Role of the Port State to Prevent, Deter and Eliminate Illegal, Unreported and Unregulated Fishing, which will take place at FAO headquarters in late August 2004.

Conclusion

The conclusion is thus very brief. States should recognise that a number of international agreed instruments call for the establishment of compatible measures for port State control, and participate actively in the upcoming FAO consultation to develop a MOU that can serve as a model in this regard.

DRAFT MEMORANDUM OF UNDERSTANDING ON PORT STATE MEASURES TO COMBAT ILLEGAL, UNREPORTED AND UNREGULATED FISHING[1]

The Parties to this Memorandum,

Concerned that illegal, unreported and unregulated (IUU) fishing continues to persist;

Emphasizing that effective action by port States is required to prevent, deter and eliminate IUU fishing;

Noting that the relevant international instruments call for port States to establish measures to promote the effectiveness of subregional, regional and global conservation and management measures;

Recognizing that the Code of Conduct for Responsible Fisheries and the International Plan of Action to Prevent, Deter and Eliminate Illegal, Unreported and Unregulated Fishing, promote the use of measures for port State control of fishing vessels in order to meet the objectives of the Code and the plan;

Desiring to achieve co-operation and co-ordination in fisheries-related port State control in accordance with international law;

Emphasizing the need for non-Parties and fishing entities to take action consistent with this Memorandum;

have agreed as follows:

Scope

In this Memorandum,

references to fishing vessel includes vessels transporting fish and fishery products unless otherwise provided for in the text of the Memorandum; and

references to ports include offshore terminals and other installations for landing, transhipping, refuelling or re-supplying.

---

[1]   Expert Consultation to review port state measures to combat illegal, unreported and unregulated fishing, Rome 4-6 November 2002.

Commitments

Each Party to this Memorandum undertakes to:

give effect to the provisions of the present Memorandum and the Annexes thereto, which constitute an integral part of the Memorandum;

maintain an effective system of port State control with a view to ensure that foreign fishing vessels calling at its port, comply with relevant[2] conservation and management measures;

require prior to allowing a foreign fishing vessel port access that the vessel provides notice at least xx hours in advance which includes vessel identification, the authorization(s) to fish, details of their fishing trip, quantities of fish on board and other documentation[3];

require prior to allowing a vessel transporting fish and fishery products port access that the vessel provides notice at least xx hours in advance which includes vessel identification, the transport document(s), quantities of fish and fishery products on board and other documentation[4];

where there are reasonable grounds to believe that a fishing vessel has engaged in or supported IUU fishing in waters beyond the limits of its fisheries jurisdiction, either refuse to allow the vessel to use its port for landing, transhipping, refuelling or re-supplying or to take measures such as forfeiture of fish and fishery products, as may be provided for under its national legislation;

not to allow a vessel to use its ports for landing, transhipping or processing fish if the vessel which caught the fish is entitled to fly the flag of a State that is not a contracting or collaborating party of a regional fisheries management organisation or has been identified as being engaged in fishing activities in the area of that particular regional fisheries management organisation, unless the vessel can establish that the catch was taken in a manner consistent with the conservation and management measures;

not to allow a vessel to use its ports for landing or transhipment where it has been established that the vessel has been identified by a regional fisheries management organisation as having a history of non-compliance with its conservation and management measures;[5]

designate and publicize ports to which foreign fishing vessels may be permitted admission and ensure that these ports have the capacity to conduct port inspections;

ensure that port inspections take place in accordance with Appendix 18.A;[6]

obtain in the course of such inspections, at least the information listed in Appendix 18.B; and

---

[2] The creation of a list of relevant conservation and management measures for a particular MOU might be required.

[3] The details to be provided for in a prior notice should be agreed upon for each MOU.

[4] See footnote 2

[5] The RFMO should identify such vessels through agreed procedures in a fair, transparent and non-discriminatory manner.

[6] An annual total number of inspections corresponding to at least XX % of the number of individual vessels to which the MOU applies should be agreed upon.

consult, co-operate and exchange information with other Parties in order to further the aims of this Memorandum.

Inspections

In fulfilling its commitments under this Memorandum each Party undertakes to:

carry out inspections in its ports for the purpose of monitoring compliance with relevant[7] conservation and management measures;

ensure that inspections are carried out by properly qualified persons authorised for that purpose, having regard in particular to Appendix 18.C;

ensure that prior to an inspection, inspectors shall be required to submit to the master of the vessel an appropriate identity document;

ensure that an inspector can examine all areas of the fishing vessel, the catch (whether processed or not), the nets or other gear, equipment, and any document which the inspector deems necessary to verify compliance with relevant[8] conservation and management measures; and

ensure that the master of the vessel is required to give the inspector all necessary assistance and information, produce relevant material and documents as may be required, or certify copies thereof.

Subject to appropriate arrangements with the flag State of a vessel, the inspecting port State may invite the flag State to carry out or participate in the inspection.

When exercising inspections the port State will make all possible efforts to avoid unduly delaying a vessel.

Actions

If an inspector finds that there are reasonable grounds for believing that a foreign fishing vessel has engaged in activities including, *inter alia*, the following[9];

a) fishing without a valid licence, authorization or permit issued by the flag State;

b) failing to maintain accurate records of catch and catch-related data;

c) fishing in a closed area, fishing during a closed season or without, or after attainment of, a quota;

d) directed fishing for a stock which is subject to a moratorium or for which fishing is prohibited;

e) using prohibited fishing gear;

---

[7]   See footnote 1

[8]   See footnote 1

[9]   Activities other than those listed below may be specified in procedures established by a relevant RFMO (one particular example is failure to comply with Vessel Monitoring Systems (VMS) requirements).

f) falsifying or concealing the markings, identity or registration of the vessel;

g) concealing, tampering with or disposing of evidence relating to an investigation; or

h) conducting activities which together might be regarded as seriously undermining applicable conservation and management measures then the port State shall promptly notify the flag State of the vessel and, where appropriate, the relevant coastal States and regional fisheries management organisations.[10]

The port State shall take due note of any reply or any actions imposed or taken by the flag State of the inspected vessel.[11] Unless the port State is satisfied that the flag State has taken or will take adequate action, the vessel shall not be allowed to land or tranship fish in its ports.

Information

Each Party undertakes to report on the results of its inspections under this Memorandum to the flag State of the inspected vessel, the parties to this Memorandum, and to relevant regional fisheries management organisations.

Each Party undertakes to establish a communication mechanism that allows for direct, computerized exchange of messages between relevant States, entities and institutions, with due regard to appropriate confidentiality requirements.

The information will be handled in a standardized form and in accordance with the established procedures as set out in Appendix 18.D.

---

[10]   In each region there may be reference to applicable international instruments.

[11]   It is recommended to establish a list of contact points in the relevant administration of each Party to the Memorandum.

APPENDIX 18.A.
Inspection Procedures of Foreign Fishing Vessels

Vessel identification

The inspector shall

be satisfied that the certificate of registry is valid;

be assured that the flag, the external identification number (and IMO-number if available) and the international radio call sign are correct;

examine whether the vessel has been re-flagged and if so, note the previous name(s) and flag(s);

note the port of registration, name and address of the owner (and operator if different from the owner) and the name of the master of the vessel; and

note name(s) and address(es) of previous owner(s), if any.

Authorization(s)

The inspector shall be satisfied that the authorization(s) to fish or transport fish and fishery products are compatible with the information obtained under paragraph 1 and examine the duration of the authorization(s) and their application to areas, species and fishing gear.

Other documentation

The inspector shall review all relevant documentation[12] which may include various logbooks, in particular the fishing logbook, stowage plans and drawings or descriptions of fish holds. Such holds or areas may be inspected in order to verify whether their size and composition correspond to these drawings or descriptions and whether the stowage is in accordance with the stowage plans.

Fish and fishery products

The inspector shall, to the greatest extent possible, examine whether the fish on board is harvested in accordance with the conditions set out in the authorization. In doing so, the inspector shall examine the fishing logbook, reports submitted, including those resulting from a vessel monitoring system (VMS).

If the inspector has reasonable grounds to believe that a vessel has engaged in or supported IUU fishing the inspector may review the amount and composition of all catch on board to verify whether the fish has been taken in the areas as recorded in the relevant documents.

---

[12] It is understood that documentation includes documents in electronic format.

In order to determine the quantities and species which are fresh on ice, frozen but not packed, processed, packed or in bulk, the inspector [shall/may][13] examine the fish in the hold or during the landing. In doing so, the inspector may open cartons where the fish has been pre-packed and move the fish or cartons to ascertain the integrity of fish holds.

If the vessel is discharging, the inspector shall verify the species and quantities landed. Such verification shall include presentation (product form), live weight (quantities determined from the logbook) and the conversion factor used for calculating processed weight to live weight. The inspector shall also examine any possible quantities retained on board.

Fishing gear

The inspector shall be satisfied that the fishing gear on board is in conformity with the conditions of the authorization(s). The gear [shall/may][14] also be checked to ensure that the mesh size(s) (and possible devices), length of nets, hook sizes etc. are in conformity with applicable regulations and that identification marks of the gear correspond to those authorised for the vessel.

The inspector [shall/may][15] also search the vessel for any fishing gear stowed out of sight.

Report

The result of a port inspection shall be presented to the master of the vessel and a report shall be completed, signed by the inspector and the master. The master shall be permitted the opportunity to add any comments to the report.

---

[13]   In view of certain practical problems of such inspections, this has been presented in the alternative "shall/may".

[14]   See footnote 12

[15]   See footnote 12

APPENDIX 18.B.
Results of Port Inspections

Results of port inspections shall include at least the following information:

Inspection references

Inspecting authority (name of inspecting authority or the alternate body nominated by the authority);

name of inspector;

port of inspection (place where the vessel is inspected); and

date (date the report is completed).

Vessel identification

Name of the vessel;

type of vessel;

external identification number (side number of the vessel) and IMO-number (if available) or other number as appropriate;

international Radio Call Sign;

MMSI-number (Maritime Mobile Service Identity number), if available;

flag State (State where the vessel is registered);

previous name(s) and flag(s), if any;

whether the flag State is party to a particular regional fisheries management organisation;

home port (port of registration of the vessel) and previous home ports;

vessel owner (name and address of the vessel owner);

vessel operator, responsible for using the vessel if different from the vessel owner;

name(s) and address(es) of previous owner(s), if any; and

name and certificate(s) of master.

Fishing authorization (licenses/permits)

The vessel's authorization(s) to fish;

State(s) issuing the authorization(s);

areas, scope and duration of the authorization(s);

species and fishing gear authorised; and

transshipment records and documents[16] (where applicable).

Trip information

Date trip commenced (date when the current trip started);

areas visited (entry to and exit from different areas);

ports visited (entry into and exit from different ports); and

date trip ended (date when the current trip ended).

Result of the inspection on discharge

Start and end (date) of discharge;

fish species;

presentation (product form);

live weight (quantities determined from the log book);

conversion factor (as defined by the master for the corresponding species, size and presentation);

processed weight (quantities landed by species and presentation);

equivalent live weight (quantities landed in equivalent live weight, as "product weight multiplied with the conversion factor"); and

intended destination of fish and fishery products discharged.

Quantities retained on board the vessel

Fish species;

presentation (product form);

conversion factor (as defined by the master for the corresponding species, size and presentation);

processed weight; and

equivalent live weight.

Results of gear inspection

Details of gear type inspected and attachments, if any.

---

[16] The transshipment records and documents must include the information provided for in paragraphs 1-3 of this Appendix 18.B.

APPENDIX 18.C.
Training of Port Inspection Officers[17]

Elements that shall be included in a training programme:

Overview of conservation and management measures applicable for a particular Memorandum of Understanding;

information sources, such as log books and other electronic information that may be useful for the validation of information given by the master of the vessel;

fish species identification;

catch landing monitoring, including determining conversion factors for the various species and products;

vessel boarding/inspection, hold inspections and calculation of vessel hold volumes;

gear inspections;

gathering, evaluating and preservation of evidence; and

range of measures available following the inspection.

---

[17] More extensive criteria should be developed for the qualification (*e.g.* skills and knowledge) of inspectors. The skills and knowledge listed below are minimum requirements.

APPENDIX 18.D.
Information System on Inspections

Computerized Communication between States and between States and relevant RFMOs would require the following:

Data characters;

structure for data transmission:

protocols for the transmission; and

formats for transmission including data element with a corresponding field code and a more detailed definition and explanation of the various codes.

International Agreed Codes shall be used for the Identification of the following items:

States: 3-ISO Country Code;

fish species: FAO 3-alpha code;

fishing vessels: FAO alpha code;

gear types: FAO alpha code;

devices/attachments: FAO 3-alpha code; and

ports: UN LO-code.

Data Elements Shall at least include the following:

Inspection references;

vessel identification;

fishing authorization(s) (licenses/permits);

trip information;

result of the inspection on discharge;

quantities staying on board the vessel;

result of gear inspection;

irregularities detected;

actions taken; and

information from the flag State.

CHAPTER 19

POTENTIAL LINK BETWEEN IUU FISHING AND THE STATUS OF SAFETY-RELATED INTERNATIONAL INSTRUMENTS APPLICABLE TO FISHING VESSELS AND FISHERS

Brice Martin-Castex, International Maritime Organisation (IMO)

This presentation aims at highlighting the potential link between *i)* the lack of international instruments in force addressing the safety of fishing vessels and the training of fishermen, and *ii)* the conduct of illegal fishing activities, in the current context of limited control and inspection measures applying both to maritime safety and fisheries management.

In the future, the IMO will use its specific experience to identify the substantial differences between merchant marine activities and fishing activities as far as the international legal framework is concerned.

The IMO will address the following issues in order to illustrate the areas where specific efforts are being made to improve compliance with international regulations and standards:

- harmonisation of port State control activities;
- search and rescue;
- self-assessment of flag State performance;
- Code for the implementation of [mandatory] IMO instruments;
- IMO Voluntary Member States Audit Scheme;
- non-convention ships;
- increased collaboration between flag States and port/coastal states; and
- transparency.

These areas of particular interest for the IMO will be assessed *vis-à-vis* the identified specificities of the fishing industry.

The key aspects of the recommended future activities aimed at deterring IUU fishing will be considered in the context of identified bottlenecks preventing the entry into force and implementation of international standards, highlighting the areas where increased co-operation and partnership may be needed.

CHAPTER 20

ENFORCEMENT AND SURVEILLANCE: WHAT ARE OUR TECHNICAL CAPACITIES AND HOW MUCH ARE WE WILLING TO PAY?[1]

Serge Beslier, European Commission

Introduction

The role of our political institutions, whether national or international, is to draw up rules of law that are conducive to the harmonious and sustainable development of society.

If those rules are to be effective, they must be enforced. Law without enforcement and without sanctions is non-existent.

Enforcement comes at a cost. It is therefore important for policy makers to be aware of that cost when drawing up the rules governing any economic activity. In the fishing industry, costs can be assessed in a variety of ways. The term "cost" can be defined alternatively as financial or economic.

First there is the budgetary cost. What financial resources are the public authorities ready to allocate to enforcement in a given industry? Then there is the environmental cost. What are the risks to the environment when inadequate funding is allocated to the goal of achieving sustainable fisheries management? What are the risks to endangered species and biodiversity? Finally there is the economic and social cost. What are the repercussions on stock management if these measures are not conducive to optimal yields, and what are the implications for firms and workers?

No one would dispute the need for sound enforcement, which is crucial to sustainable fishing as well as being in the interests of society at large, and firms and workers in particular.

The real issue is the cost to society and how much we are willing to pay. The current trend is towards budgetary restraint, and the OECD is the first to stress the need for spending controls. Another aspect of the issue is the cost to firms, and the subsequent implications for competitiveness and fair competition at international level.

---

[1]   Paper translated from French original.

The cost to society

Monitoring at sea is costly. This is nothing new, merely a fact that has to be faced. However, there is some uncertainty clouding the issue. This is why the European Commission attaches so much importance to the work being done by the OECD to assess the economic and social effects of IUU fishing.

The overall cost of monitoring fishing activities in the EU and its member states amounts to some EUR 300 million. That may be a somewhat conservative figure, however, as fishery surveillance is not always targeted and there is no approved method for collecting such information. The figure should be set against the value of landings by EU fishing vessels, estimated at around EUR 5.5 billion. This puts monitoring and surveillance costs at around 5% of the value of production. In the specific case of NAFO, the cost of monitoring EU vessels amounts to some €4 million, for a total of around EUR 55 million in landings (in 2002), *i.e.* over 7% of the value of production.

These two examples highlight the relatively high cost of enforcement and surveillance in this industry. It certainly exceeds fishing firms' profit margins. Consequently, the cost of this type of government action cannot be viewed solely in terms of the benefits to the sector directly concerned, but should instead be assessed in terms of the industry as a whole and – even harder – its impact on society (including environmental and other effects).

Those are just the direct costs of fisheries enforcement and surveillance. But there is also a need to assess the cost of customs inspections for fishery products. The EU market, along with the Japanese and North American markets, is one of the three major outlets for fishery products. So both the EU authorities and the customs authorities in individual member states have their part to play here. Yet it should be borne in mind that trafficking in illegal fishery products is not the only form of international crime of concern to the authorities. Others are considered to be far more of a threat to social equilibrium, including drug trafficking, people trafficking, arms dealing and money laundering, all of which take up a huge amount of resources and energy. However, all of these different forms of crime stem from the same rationale, namely unbridled globalisation and the inability of individual countries to resolve such problems alone.

It is clear that, in an open environment like the sea, international co-operation is vital. It is – not without reason – one of the pillars of the United Nations Convention on the Law of the Sea. International co-operation is not only an obligation in terms of conserving and managing fishery resources, it is also a necessity in terms of enforcement and surveillance. It has become even more vital now that budget constraints demand that the system be as cost-effective as possible.

International co-operation is all the more necessary because the economic interests of the countries concerned may not necessarily converge. The interests of a coastal state are not those of a flag State, which in turn differ from those of a port state or the state in which fishery products are actually used.

As part of its work on Common Fisheries Policy reform, the European Union looked into the efficiency of the EU system. It concluded that the separation of powers in the traditional system, with the EU wielding legislative powers and member states the executive powers, did not satisfactorily meet the need for co-operation, including co-operation between EU member states.

This is one of the reasons why the European Union is envisaging the creation of a Community Fisheries Control Agency. This would not only enhance the quality and effectiveness of the EU control system, but also give better value for money in terms of EU and member state budget expenditure. Another advantage of the new Agency would be to foster international co-operation with the introduction of a system of information exchange as it would, for instance, be a member of the MCS (monitoring, control and surveillance) network currently under development.

International co-operation necessarily involves the regional fishing organisations (RFO) too. The introduction of streamlined control schemes in all of the RFOs is bound to generate savings and improve efficiency. The EU attaches great importance to the fact that the inspection and control schemes in each RFO are tailored to the profile of the relevant fishery. One example is the process used by the Indian Ocean Tuna Commission (IOTC) to develop its own inspection and control scheme, namely by systematically analysing all known control techniques and selecting the most cost-effective for the Indian Ocean tuna fishery. The process also took account of the capacities of each contracting party, since the RFO includes among its members both developed and developing countries. This goes to show that economic assessment and monitoring tools are diverse, and their costs may vary considerably. So the question is therefore how to optimise the financial resources invested in control and achieve optimal synergy between the various types of monitoring, whether at sea, from the air, at the dockside, or at the import stage. To date, the analysis appears to have been more empirical than rational. Economic analysis may provide scope to find the most appropriate mix of monitoring tools.

The drive for a more cost-effective approach calls for the use of more readily controllable techniques, such as lists of vessels authorised to fish. This should help to counter the practice of "open registries", also known by the non-legal term of "flag registries", which deliver what are commonly called flags of convenience.

The European Union is increasingly turning to technologies that combine performance with cost-savings. The development of satellite systems has led to remarkable advances in this field. There are currently plans to make the VMS – Vessel Monitoring System – compulsory for all vessels over 15 metres in length. Satellite monitoring makes it possible to track vessels that are not necessarily fitted with transceivers, and the technology certainly holds as yet untapped development potential for the fishing industry. Satellites combined with computers are also offering scope to improve fishing and fishery surveillance, for instance with electronic log-books for real-time monitoring.

These techniques are both efficient and relatively cheap, and in any case less of a burden on the public purse than the classic at-sea monitoring or quayside-inspection techniques. An overview conducted some time ago showed VMS to be cost-effective if it could achieve a 10% cut in the cost of at-sea monitoring.

Enforcement and surveillance costs are not confined to the public purse, however, and companies have to shoulder a growing share.

The cost to firms

Fishing firms have to compete on two fronts. There is not only competition for the resource but also competition for access to the fishery product market.

The economic conditions governing access to fishery resources determine how competitive firms must be to gain access to the markets. For fishing fleets, inspection costs are a decisive factor when it comes to competitiveness.

The constraints imposed on firms by monitoring and inspection costs can be measured at two levels:

- At the firm or microeconomic level, vessel-owners must shoulder the cost of keeping log-books, declaring catches, installing VMS transceivers and using selective fishing techniques. The fact that vessels flying flags of convenience avoid these costs is bound to give them a competitive advantage. However, the factor that most seriously distorts competition may not lie there but, more importantly, in the opportunities those vessels have to avoid compliance with conservation and management rules, and in particular restrictions on catches. Vessels that do comply with such restrictions are clearly not competing on an equal footing with vessels that have no limitations on their fishing. Another point to bear in mind is that the vessels involved in IUU fishing are also the first to breach the standards on navigational safety, vessel safety, crew safety and conservation of the marine environment.

- At the macroeconomic level, that of the economic environment in which firms operate, the absence of an enforcement and surveillance policy is a major competitive advantage for vessel-owners. Whether monitoring and inspection costs are passed on to firms *via* a cost-recovery scheme (ITQs, licence-fees or a similar system) or there is a beneficial tax regime, vessel registration under a flag of convenience often goes hand in hand with company registration in a tax haven, and gives owners operating under flags of convenience a competitive advantage over their competitors operating under the flags of "civilised" countries.

If fishery management systems were watertight, control cost assessments would focus mainly on the costs that governments are willing to pay.

Paradoxically, the more resources are allocated to fisheries enforcement and surveillance, the more vessel-owners are tempted to circumvent the system.

Even if fraud is not confined to fishing by vessels flying flags of convenience, the scope for avoiding increasingly tighter controls is certainly encouraging some owners to change flags. The ease with which they can re-register their vessels is specific to the marine environment, and common to maritime transport and sea fishing. In both industries, it only takes an entry in a register for firms to relocate under a new flag, whereas land-based firms would also have to move premises.

The "convenience" issue is not confined to flags. Fishing vessels are so mobile that they can choose where to land their catches and hence where to market them and obtain the best prices. That choice may be based on legitimate economic criteria such as proximity to fishing grounds, markets for specific catches or the commercial performance of port operators. But ports may be chosen for illegal reasons, for instance the absence of inspections. So the problem is not just flags of convenience but ports of convenience. There is fairly little incentive for a country to inspect catches landed in its ports when those catches are not from stocks harvested by its own fleet. It can enjoy the economic spin-offs from its port activities, without suffering any loss or unfair competition from the predatory harvesting of its own stocks.

Because of the competitive distortion they bring to economic relations, flags of convenience and ports of convenience – by their very existence – hamper the development of enforcement and surveillance schemes by countries wishing to set up sustainable resource management systems, combined with effective control mechanisms.

An economic analysis of all these factors should provide policy makers with more insight into the implications of their decisions on the conservation and management of fish stocks, and the controls that necessarily accompany them. It should enable them to identify the most urgent areas for improvement in the international legal system and tackle the challenges facing the international community, if the formal commitments made by governments to promote sustainable development are not to remain devoid of meaning.

CHAPTER 21

WORKING TOGETHER - WHAT INDUSTRY CAN DO TO HELP

Martin Exel, COLTO, Australia

The Coalition of Legal Toothfish Operators (COLTO) is a group of legal industry members, working to assist governments, environmental organisations and other authorities to eliminate illegal, unregulated and unreported fishing for toothfish in sub-Antarctic and Antarctic waters. The unique environment of these southern oceans, together with their remoteness and hostile environment has meant that effective enforcement is extremely difficult to achieve by government actions alone.

Governments are more classically regarded as "responsible" for the enforcement of fisheries management regimes, including surveillance and compliance of rules and regulations. It became very clear to legal operators that IUU fishing was able to deplete stocks of toothfish fisheries to non-commercially viable levels faster than governments could eliminate or control IUU operations. It was essential that industry help government agencies to "clean up" IUU fishing, or there would be no future for the legal toothfish industry.

The main driver that made legal operators decide to work together against IUU fishing operations was the vital need to provide for toothfish stock sustainability into the future. Management measures in the Commission for the Conservation of Marine Living Resources (CCAMLR) are amongst the most precautionary in the World, and IUU fishing was directly jeopardising a number of stocks in the CCAMLR region. IUU fishing was therefore, in turn, directly reducing available quantities of fish to legal operators, as well as making legal fishing non-viable.

COLTO was launched in May 2003. It comprised 27 companies from 10 separate countries, and is fully industry funded. Over the course of its first eleven months of operation, COLTO has:

- set up a website at www.colto.org with details on IUU vessels, as well as recent press articles or reports;

- had a global "Wanted" campaign, with rewards offered up to USD 100 000 for information leading to the conviction of illegal operators;

- created a database of information on vessel movements, product unloading, individuals and companies involved in IUU fishing;

- participated as an official observer at CCAMLR in November 2003;

- produced a public report, titled "Rogues Gallery" summarising IUU operations;

- provided information either on request or informally to governments and government agencies;

- provided analyses of vessels for identification where names and/or flags have been changed;

- worked to encourage publicity against IUU fishing, and to support governments in their actions against IUU operators.

The concept of industry working closely with governments has been mooted often, and works in a number of smaller fisheries. The real test for COLTO was the ability to bring together varied operators from different countries with differing views on IUU approaches and make a positive contribution. To date, the impacts have been positive and, with sufficient energy and responsiveness from authorities to information provided by industry, COLTO will continue to provide a central role of linkage between legal operators, conservation groups, and governments.

Industry members are motivated primarily by the maintenance of their livelihoods – and that is only going to come with access-right security, well-managed fisheries that are ecologically sustainable, clarity of management measures, and effective enforcement and compliance arrangements in the fishery. This, in the case of toothfish, is going to necessitate new concepts of management for the RFMO to control more effectively not just the high seas areas, but also those States party to CCAMLR who are not as effectively implementing their flag State responsibilities over vessels and IUU operators.

The weaknesses in the system that need to be addressed include:

- the need for government agencies to develop effective and rapid mechanisms to exchange information and data on vessels that are identified as IUU;

- CCAMLR to consider the regular publication of "white lists" of legitimate operators in addition to the existing black list;

- CCAMLR to develop effective access-right security management arrangements for high seas areas, as opposed to competitive TAC arrangements;

- consideration of changes on the current IUU loopholes that are facilitated under UNCLOS, such as the "freedom of the high seas" which enables re-flagging to non-party States by IUU boats;

- development and/or enhancement of market-based mechanisms such as paper trails for trade in toothfish which can be used to assist the identification of legal catches.

These challenges are not unique to CCAMLR or toothfish fisheries, and many agencies and governments have been working to achieve the best results possible. The biggest challenge is to achieve the desired results before IUU fishing undermines management measures to the extent that legal fishing is no longer viable, and stocks are reduced to critically low levels.

CHAPTER 22

PRIVATE INITIATIVES: A POSSIBLE WAY FORWARD?

Hiroya Sano and Yuichiro Harada, OPRT, Japan

Introduction

We welcome this opportunity to present the views of the Organization for the Promotion of Responsible Tuna Fisheries (OPRT) on the question of whether private initiatives are a way forward to address the problem of illegal, unreported, and unregulated (IUU) fishing activities.

From the outset, one of OPRT's main goals was to combat IUU tuna fishing. This paper recounts OPRT's experience and describes how the organization has been dealing with the problem of IUU tuna fishing.

The establishment of OPRT is in itself the result of private initiatives. Since the introduction of the Positive List Scheme on a global scale towards the end of last year, tuna caught by IUU tuna longline fishing vessels can no longer be traded in international markets. This ensures that in the case of tuna caught by large-scale longliners, IUU fishing cannot survive. The tuna longline fishing industry itself has become aware that it is necessary to promote responsible fisheries, show the legitimacy of their fishing activities to the international community and make efforts to increase the transparency of their fishing operations. It has actually been a long, difficult journey for private stakeholders to finally come together to this end. Private stakeholders took every possible action available to eliminate IUU tuna fishing. This paper will focus on just a few of the most important private initiatives taken.

OPRT's private initiatives were solely targeted to IUU tuna longline fishing activities, and may not be applicable to the wider IUU fishing problem. However, there are many lessons to be learned, and these may contribute to the ongoing effort to combat other types of IUU fishing operations.

What is IUU tuna longline fishing?

In order to provide more in-depth understanding of the problems specific to IUU tuna longline fishing, some of the basic facts are outlined below:

a) A large-scale tuna longline fishing vessel is defined as a fishing vessel of over 24 meters in length, equipped with longline gears and a super-freezing capacity of minus 60 degrees

Celsius in order to keep the catch fresh. These vessels are highly mobile, and are able to operate in the Pacific, the Atlantic and the Indian Ocean in pursuit of tunas throughout the entire year. They are subject to the Positive List Scheme.

b) The motivation for IUU tuna longline fishing vessel operations is their ability to sell their catch to the Japanese sashimi market. The nature of the Japanese sashimi market is as follows:

- First, high-quality tuna is consumed in fresh and raw form in the traditional Japanese dishes "sashimi" and "sushi."

- Second, the Japanese tuna market demand is large and stable. About one third of the world's tunas catches are consumed in Japan, mainly as "sashimi" and "sushi".

- Third, prices in Japan's tuna market are very high. Prices differ by species, but they are 10 to 30 times higher than in other markets, including the canned tuna market.

- Fourth, Japan's tuna market is *the* international market. The number of countries exporting tuna to Japan has more than doubled over the last 15 years, and about 70 countries now export tuna to Japan.

Why was OPRT established?

OPRT was established on December 8, 2000 in Tokyo, Japan. The letter of intent for its establishment by the founding parties states that OPRT aims to contribute to the development of tuna fisheries in accordance with international and social responsibility, to promote the sustainable use of tuna resources through measures to reinforce the conservation and management of tuna, to foster healthy tuna markets, and to further international co-operation among fishermen.

The initial members were the tuna longline fishing industries in Japan and Chinese Taipei along with organisations of traders, distributors and consumers in Japan. Membership gradually grew to include like-minded tuna longline fishing industries in Korea, Indonesia, the Philippines, the Peoples Republic of China, and Ecuador. As of the end of March 2004, 1 460 large-scale tuna longline fishing vessels were registered with OPRT. This number includes almost all of the large-scale tuna longline fishing vessels around the world.

The initiative to establish OPRT originally came from the Japanese tuna fishing industry. The organisation's main goal is to eliminate IUU tuna longline fishing vessels. There was a clear motive for the industry to take such an initiative. The Japanese government had urged the industry to reduce the number of vessels in response to the United Nations FAO's International Plan of Action for the Management of Fishing Capacity, adopted in February 1999. The Plan stated that the world's tuna resources are being excessively exploited and indicated the need for urgent action to reduce the world's large-scale tuna longline fishing fleets by 20 to 30%. With financial support from the Japanese government, the Japanese industry scrapped 132 large-scale tuna longline fishing vessels, equivalent to 20% of its total fleet, in the expectation that other tuna longline fishing industries would also reduce their fleets. The goal was to stop the excessive exploitation of tuna and ensure the sustainability of tuna resources.

Given this development, the Japanese tuna fishing industry was greatly concerned by the fishing activities of flag-of-convenience tuna fishing vessels (also referred to as IUU vessels) that operate without adhering to international tuna fishery management measures. There were reportedly 250 such vessels, and an analysis of trade statistics showed that they exported all of their catch to the Japanese market. The industry realised that the benefits to tuna resource sustainability through the reduction of its fleet would be nullified unless IUU tuna longline fishing activities were eliminated.

Since it was known that the effective source of IUU fishing was Chinese Taipei vessel owners who conducted IUU tuna longline fishing using Japanese second-hand tuna longline fishing vessels, the Japanese tuna industry proposed consultations with the tuna fishing industry of Chinese Taipei. These consultations continued for almost two years, culminating in a Joint Action Plan to eliminate IUU tuna fishing vessels. The plan consists of projects to scrap Japanese secondhand IUU tuna longline fishing vessels, and also to re-register Chinese Taipei built fishing vessels to the Chinese Taipei registry. It was necessary to find financial compensation for the vessels to be scrapped. With the support of the Japanese government, a compensation fund of JPY 3.2 billion (USD 30 million) was arranged by the Japanese and Chinese Taipei industries to implement the scrapping of these IUU tuna longline fishing vessels, and the OPRT was established to implement the scrapping project. Forty-three vessels were scrapped by OPRT during the 3-year project period, from 2001 to 2003.

Why did IUU tuna fishing vessel owners accept the scrapping of their vessels?

It was not easy to get IUU tuna fishing vessel owners to agree to scrap their vessels. The Japanese tuna fishing industry pointed out that the elimination of IUU tuna fishing vessels would eventually benefit both industries by ensuring sustainable tuna resources. This argument seemed to be partially understood, but it was not in itself successful.

A carrot and stick approach was necessary. The carrot was the compensation fund, and the stick was an anti-IUU tuna fishing vessel campaign. The level of compensation was the subject of lengthy negotiations between the two industries, but agreement on the scrapping project over a 3-year period was finally reached. During the consultations, the Japanese tuna fishing industry carried out an extensive, continuous campaign to raise public and government awareness of the problem of IUU tuna longline fishing, emphasising the fact that Japan is virtually the only market in the world for products from IUU tuna longline fishing operations. The industry asked the Japanese government and Diet Members to take appropriate measures to eliminate IUU tuna longline fishing vessels, emphasising that Japan is responsible for ensuring the sustainable tuna resources because it is not only a major tuna fishing nation but also one of the largest tuna consuming nations. The Japanese market was, in effect, providing the economic incentive for IUU tuna fishing activities. The industry therefore also asked importers, distributors, and the public to refrain from buying IUU tuna. In order to increase public awareness of the problem, meetings, seminars, TV interviews, etc. were conducted. Also, trade information on tuna caught by IUU tuna longline fishing vessels was continuously monitored.

The Japanese tuna industry also took the problem to the international community. The issue was brought before the International Coalition of Fisheries Associations (ICFA), a non-governmental organisation founded in 1988 with membership open to national fishery organisations. ICFA brings together leaders from the world's seafood industry to ensure the health of the private sector involved in commercial fisheries and to provide a united voice for presentation in international forums.

ICFA is actually another example of a private initiative that is working to combat IUU fishing. ICFA has been making strong efforts to support the elimination of IUU fishing through participation in relevant international forums, such as FAO's Committee on Fisheries. ICFA has distributed information which includes the problem of IUU fishing, and has adopted a resolution supporting OPRT's activities against IUU fishing. ICFA's other efforts to combat IUU fishing include educational activities through appeals on its website and press releases. Additionally, ICFA members have worked with their national governments to seek support on national and international levels to fight against IUU fishing. An ICFA resolution passed at its meeting in 2003 calls upon the WTO to widen its consideration of the application of trade measures to encourage compliance with multilateral environmental arrangements, in particular regional fisheries management organisations, including the application of such measures to non-parties to these arrangements or agreements. It is obvious that unless these trade measures are widely applied, IUU fishing can escape this sanction simply by being associated with a flag government that does not belong to the relevant organisation. This matter therefore needs to be addressed and supported by governments in the WTO discussion.

Finally, the international community became aware of the IUU tuna longline fishing problem, and concern was expressed at various levels. Pressure on IUU tuna longline fishing vessel owners and Chinese Taipei increased, leading to an agreement to abandon IUU tuna longline fishing.

It is estimated that there are still about 25 IUU large-scale tuna longline fishing vessels, but they are reportedly no longer used as tuna fishing vessels because of their age.

Lessons from taking action

Having conducted such a campaign, it became apparent that the elimination of IUU fishing activities would not be possible simply through the efforts of tuna fishermen, but that it would be necessary to have the co-operation of all stakeholders. However, it was not easy to get the different sectors of stakeholders to understand and support OPRT's efforts to tackle IUU fishing. In reality, traders and distributors were somewhat reluctant to support the elimination of IUU tuna fishing because, in all honesty, they would be losing a profitable business when they could no longer purchase fish from flag-of convenience operations. Similarly, it was difficult for consumers to understand and support the initiative. They buy tuna based on quality and price, and are not aware of, or overly interested in, conservation and management issues.

This matter of different interests among sectors was also true within the producing sectors of different OPRT members. The main interest of many producers from countries other than Japan was to maintain their access to the lucrative Japanese tuna market. The point is that it was a difficult and lengthy process to convince producers from all the countries concerned to join OPRT and work together to promote responsible tuna fisheries.

We also learned that the activities of IUU tuna fishing vessels must be closely monitored in a timely manner. This was achieved by the careful monitoring and analysing of import data by OPRT. Since all the IUU tuna products were exported only to the Japanese market, the activities of each IUU tuna longline fishing vessel became transparent through the data and its analysis. It was found that IUU fishing vessels changed the names of vessels and changed registration countries in order to circumvent sanctions imposed by regional fishery management organisations like ICCAT.

In addition, IUU vessels transhipped their products to legally licensed vessels. This began when the Japanese government required importers to submit records of the fishing vessel's previous national registry in order to determine whether the vessel had operated as an IUU fishing vessel. These cases were also detected by the analysis of trade information. For example, several legally licensed vessels suddenly increased their exports by 3 to 4 times in one year. Practically, this was not possible, given the limited capacity of the fishing vessel. Through such findings, the urgent need to introduce the Positive List Scheme was recognised and supported by OPRT members. Action was taken to urge national governments and international management organisations to adopt such a scheme.

In fact, most Chinese Taipei tuna longline fishing operators had been operating both legitimate fishing vessels and IUU fishing vessels. The involvement of these Chinese Taipei fishing operators in

OPRT's private initiatives as OPRT members was highly effective in promoting the elimination of IUU tuna longline fishing. Chinese Taipei fishing vessel operators became aware of the need and the moral obligation to conduct fishing in a responsible manner, through direct communication on a private level, such as the case of consultations concerning the introduction of the Positive List Scheme.

Conclusion

Private initiatives play a very important role in the fight against illegal fishing but cannot, by themselves, be successful. Private initiatives must be part of a broad mosaic most of which is composed of government and international elements. By participating in private initiatives by OPRT, traders and dealers who had dealt with tuna harvested by IUU fishing were forced to change their business practices. Actions by the private sector cannot flourish unless they operate within a legal framework and international rules supported by governments.

It is also true that any single segment of private industry cannot be successful in dealing with IUU fishing. For example, initiatives by fishermen or vessel owners must be accompanied by initiatives involving buyers, traders, distributors, processors, as well as consumers. All sectors must work together. And in an increasingly globalised world, where the economies of many nations impact each other and where the industries of different nations often have close relations and connections, the private sectors of all nations involved must work together. This is especially true when dealing with highly migratory species such as tuna, which are fished in all of the world's oceans and on the high seas, and found in the markets of many countries.

While the private sector can do much to discourage IUU fishing, actions by the private sector cannot flourish unless they operate within a legal framework and in an atmosphere of opposition to IUU activities provided by international rules supported by governments. Otherwise, private initiatives can only be voluntary, and while there might be a place for voluntary action, it cannot be expected to be overly effective.

It is therefore extremely important that efforts continue at the international level to combat IUU fishing. This is indispensable if the private sector is to play its part. The FAO International Plan of Action and the actions taken by the international tuna management organisations are crucial to the success of the overall effort.

Ongoing private initiatives have contributed much to the elimination of IUU fishing in the case of tuna. As mentioned, these private initiatives started with tuna fishers who had suffered severe financial losses in their businesses, many of which had been built up over a number of years. This hard experience was the catalyst for their strong motivation to eliminate IUU tuna fishing vessels.

Future of OPRT

OPRT continues to monitor trade information to ensure that IUU tuna fishing activities do not reappear. OPRT will also extend the scope of its work ensuring responsible tuna fisheries by addressing other issues, such as the control of fishing capacity of large-scale tuna longline fishing vessels, the incidental catch of non-target species such as sea birds, sharks and sea turtles, and the promotion of responsible purse seine fisheries. OPRT is now particularly concerned with the rapid increase of large-scale purse seiners and their excessive fishing pressure on tuna resources. This increase appears to have resulted from the termination of IUU tuna longline fishing activities. Vessel construction funds seemed to turn to the construction of large-scale purse seiners. The effectiveness of OPRTs past efforts to ensure the sustainability of tuna resources may be nullified by the uncontrolled expansion of large scale purse seiners. In order to ensure the sustainability of tuna resources, the problem of the rapid expansion of large-scale purse seiners needs to be addressed. OPRT hopes that the European Union and the United States will take the initiative in this matter, as the major markets for purse seiners. Our experience in dealing with IUU tuna longline fishing activities shows that monitoring trade information is an important measure.

It is a difficult task to eliminate IUU fishing. It requires unrelenting efforts from a wide range of players, but if everyone concerned works together in a co-operative manner, we will ultimately be successful.

CHAPTER 23

PROMOTING CORPORATE RESPONSIBILITY: THE OECD GUIDELINES FOR MULTINATIONAL ENTERPRISES

Kathryn Gordon, OECD

The Workshop organisers asked me to present the OECD's core corporate responsibility instrument – the OECD Guidelines for Multinational Enterprises. These Guidelines are a comprehensive code of conduct for international business with a distinctive, government-backed follow-up mechanism. The OECD Investment Committee has oversight responsibility for this instrument, but it often works in partnership with other policy communities. It would therefore be useful to reflect on a possible role for the OECD Guidelines in the broader approach to controlling IUU fishing.

It would seem that some of the corporate responsibility issues currently being dealt with under the Guidelines resemble those encountered in IUU fishing. For example, the Guidelines are currently being used to look at illegal or unethical exploitation of natural resources in the Democratic Republic of Congo, where the OECD Investment Committee is following up on a request for co-operation made by the UN Security Council as part of its peace-making efforts in that troubled country.

In reviewing some of the papers prepared for this workshop, I noted at least four common features linking the two resource-exploitation problems:

i) Significant shortcomings in transparency, disclosure and reporting. When Upton and Vitalis state that the IUU fish harvest is an "unknown percentage of an ill-defined resource" they could just as easily have been talking about many forms of mineral exploitation in the DRC. I also note that the many product-tracking and reporting schemes in both sectors – fisheries and, for example, diamonds – have shown both their potential and their practical limitations in their market and policy contexts.

ii) Problems straddling many areas of corporate responsibility. In both sectors, problems of corporate responsibility manifest themselves in multiple and inter-connected ways. Illegal exploitation of minerals in the DRC is associated with a whole host of problems – corruption, environmental mismanagement, money laundering, non-compliance with laws of all sorts, serious violations of human rights, including such basic labour standards as freedom from forced labour and reasonable guarantees of occupational health and safety. For the fisheries sector this multiplicity of CR problems is shown quite clearly in Jon Whitlow's paper on the social dimension of IUU fishing, while environmental, corruption and other compliance problems are described in some of the other papers prepared for this workshop.

iii)   Interaction between corporate irresponsibility and government irresponsibility. This is a long-standing theme of the Investment Committee's work on corporate responsibility and I see that it is also relevant for IUU fishing. The Investment Committee has seen many instances where policy environments are not only weak, but deliberately weak – weak policy frameworks reflect not only a genuine lack of capacity in some countries, but also deliberate opportunism, guile and overt wrongdoing on the part of some public officials. This inseparable relationship between wrongdoing in the public and private sectors shows up clearly in David Bolton's paper on dealing with bad actors of ocean fisheries.

iv)   Comparative ineffectiveness of formal law enforcement and deterrence in coming to grips with the wrongdoing. This is shown *inter alia* in the Agnew-Barnes paper on economic aspects and drivers of IUU fishing. Their paper's description of complex company ownership structures and laundering of products could equally apply to illegal exploitation of mineral resources in the DRC. I should stress that while deterrence is not especially effective – in preventing either illegal exploitation of Africa's mineral wealth or IUU fishing – it nevertheless has an essential role to play in a durable solution to both problems.

All in all, then, what we have before us is a pretty dreary picture – there are enormous and intractable problems of illegal resource exploitation. The purpose of this presentation is to suggest that the Workshop might want to consider what contribution, if any, the OECD Guidelines could make to the broader strategy for fighting IUU fishing, just as the UN Security Council has asked the Guidelines institutions to assist it in looking into illegal exploitation of natural resources in the DRC.

First let me tell you a bit more about the instrument. The OECD Guidelines for Multinational Enterprises express the shared views of 38 adhering governments on ethical business conduct.

Key features of the Guidelines are[1]:

- They contain voluntary recommendations to multinational enterprises (MNEs) in such areas as human rights, labour and environmental standards, corporate governance, disclosure of information and transparency, anti-corruption, taxation, and consumer protection. Thus, they provide a means of looking at the wide range of corporate responsibility issues presented by IUU fishing.

- While the observance of the Guidelines is voluntary for companies, the 38 adhering governments sign a binding commitment to promote them among multinational enterprises operating in or from their territories. Thus, the Guidelines embody a unique combination of voluntary and binding elements.

- Since the Guidelines are the only comprehensive corporate responsibility instrument backed up by an inter-governmental implementation process, their comparative advantage lies in dealing with issues located at the intersection of corporate responsibility and government responsibility. I would suggest that IUU fishing lies precisely at this intersection – it involves both business and government actors, a very challenging enforcement environment and

---

[1]   For fuller information on the Guidelines, see www.oecd.org/daf/investment/guidelines/.

somewhat patchy policy framework. This might well be a situation in which to test government use of the Guidelines to complement "harder" law enforcement efforts.

- The values expressed in the Guidelines are so fundamental to the OECD that countries are required to sign up to them when they become OECD members. In some sense, the Guidelines help to define what it means to be a member of the OECD. The OECD Investment Committee is the official guardian of the instrument, but other committees have worked or plan to help the Investment Committee explore what the Guidelines mean for companies operating in their policy area. Recent examples include the Environment, Agriculture, Development Assistance and the Labour and Social Affairs Committees. There would seem to be no reason that the Fisheries Committee would not do the same thing. In this way, the Fisheries Committee could contribute to a broader, OECD-wide exploration of the meaning of responsible business conduct in today's globalizing world.

- The most visible sign of adhering governments' commitment to the Guidelines is their participation in the instrument's distinctive follow-up mechanisms. These include the operations of so-called National Contact Points (NCPs), which are government offices charged with promoting the Guidelines and handling enquiries in the national context. The fisheries policy community is in a good position to engage with the National Contact Points on IUU fishing – indeed, flags of convenience and occupational health and safety on the high seas has already come to NCPs attention in a number of contexts. One role of the fisheries community could be to ensure that the important issue of IUU is given the attention it deserves.

The purpose of Guidelines-related activities is to encourage companies to act responsibly. The Guidelines implementation procedures create a number of channels for doing this. We have seen that in dealing with the small mining companies active in the Democratic Republic of Congo, the National Contact Points have formally engaged with companies that had become accustomed to the lack of monitoring and accountability that comes from operating in one of the least transparent countries in the world. These companies have seen that they are indeed being watched – the Guidelines provide one of the few means by which governments can impress upon these companies that they too are subject to external scrutiny. While this is probably a small thing in relation to the magnitude of the DRC's problems, it is nevertheless an important step in the right direction. If the Fisheries Committee is interested in working with the Guidelines institutions the exact modalities of such co-operation would have to be explored.

The goal of the Guidelines is to help governments, business, labour unions and NGOs to align private business initiatives with public policy goals so that business works in greater harmony with surrounding societies. I think that Mr. Hiroya Sano got the positioning of this sort of co-operation right - he stated that "Private initiatives play a very important role in the fight against illegal fishing but cannot, by themselves, be successful. Private initiatives must be part of a broad mosaic most of which is composed of government and international elements. Actions by the private sector cannot flourish unless they operate within a legal framework and international rules supported by governments." I would invite the members of this Workshop to consider whether it is not worth finding ways for governments and other actors to engage with companies to try to find ways to make the fight against IUU fishing more effective.

CHAPTER 24

WHAT ROLE FOR RFMOS?

Denzil G.M. Miller of CCAMLR, Tasmania, Australia [1]

This presentation focuses on the role that Regional Fishing Management Organisations (RFMOs) could be called on to play in global efforts to combat Illegal, Unreported and Unregulated (IUU) fishing. The key elements entail the need to improve operational efficiency and to mobilise political will.

Two possible approaches open to RFMOs are identified:

a) Ignore IUU fishing until stocks become self-regulating (*i.e.* fishing is no longer sustainable); or

b) Improve current, and develop new, initiatives to combat IUU fishing.

Option a) is dismissed since it is contrary to current "best practice" and is certainly not sanctioned by international law. Option b) is examined in more detail.

Effective coastal State enforcement is seen as essential to option (b). However, it is relatively expensive and generally not fully applicable to high seas areas, given the nature of these fisheries and their geographic extent. A series of specific considerations are elaborated in Table 7.1, (see Chapter 7) and an additional five are also discussed. These five considerations relate to potential synergies, and/or contradictions, concerning trade measures, port-flag State modalities, institutional competencies and the role of related measures.

Finally, the elements of "political will" necessary to combat IUU fishing are discussed and it is suggested that there may be merit in considering development of an international fisheries policing organisation (FISHPOL), following the precedent of INTERPOL and building on the current Monitoring Control and Surveillance Network (MCS Network).

---

[1]   (email: denzil@ccamlr.org).Many of the points raised in this summary are discussed in more detail in Chapter 7.

Box 24.1: Some Issues to be Addressed in Improving RFMO Enforcement

Resolve Jurisdictional Issues [Flag/Coastal State]
[Avoid "Creeping" Expansion of Coastal State Rights]

Operationalize Key *LOSC* Provisions
[Especially Articles 116-119]

Improve Institutional Enforcement & Co-operation

Improve Links between Vessels & Flags [FOC]

Improve Compatibility between National/International Measures

Elaborate "Nationals" [Beneficiaries] Responsibilities/Obligations
[*e.g.* Following *LOSC* 116-119 & *UNFSA* Article 10.(l)]

Standardise Sanctions
[*e.g.* As per *SADC Fisheries Protocol* Article 8.4.(b)]

Address *NCP* Role in *RFMOs* [*UNFSA* Article 17]

Promote Cult of Responsible Fishing Activity
[As per *FAO Code of Conduct*]

Implement FAO IPOA-IUU

Build Regional/National Enforcement Capacity
[As per *UNFSA Articles* 24-26]

CHAPTER 25

THE DEVELOPMENT AND ENFORCEMENT OF NATIONAL PLANS OF ACTION: THE SPANISH CASE

Ignacio Escobar, Secretaria General de Pesca Maritima, Spain

Introduction

In response to the International Plan of Action (IPOA-IUU), adopted by the International Community in FAO in 2001, Spain developed its National Plan of Action to prevent, deter and eliminate illegal, unreported and unregulated fishing (NPOA) in November 2002.

Although the national plan of action only dates from 2002, Spain has been implementing an IUU control scheme since the year 2000, so we have accumulated four years' experience in dealing with different cases of IUU fishing.

Combating IUU fishing is a complex task that requires a global approach, an idea which was borne in mind by Spain when developing its national plan of action. Indeed, our plan includes measures concerning state responsibilities, flag State responsibilities, coastal state measures, port State measures, internationally agreed market-related measures, scientific research, co-operation with regional fisheries management organisations, and the special requirements of developing countries.

Legal and administrative instruments

Spain's main body of legislation dealing with fishing activities is Act 3/2001 on Marine Fisheries, which is applied to all national vessels wherever they are fishing, and to third-country vessels operating in waters under Spanish sovereignty or jurisdiction. It comprises a system of offences and penalties in the area of marine fishing (both within the EEZ and on the high seas), management of the fishing sector and trade of fishing products. Port State control is considered as the basic and essential tool to deal with IUU fishing.

Other legal instruments to combat IUU fishing are as follows:

- Royal Decree 1797/1999, on the monitoring of fishing operations by vessels of third Countries in waters under Spanish sovereignty or jurisdiction.
- Royal Decree 1134/2002, on the application of penalties to Spanish nationals employed on flag-of-convenience vessels.
- Royal Decree 176/2003, regulating control and inspection of fishing activities.

- Ministerial Order of 12 November 1988, concerning a satellite-based vessel monitoring system.
- Royal Decree 2287/1998, which defines the criteria and conditions of interventions with a structural purpose in the fisheries sector.
- Royal Decree 601/1999, regulating the Official Register of Fisheries Companies in Third Countries.
- Royal Decree 3448/2000, with the basic regulations for structural support in the fisheries sector.

Enforcement

Spain strongly believes that one of the most efficient ways to combat IUU fishing is through cutting off access to markets. This is why Spain has concentrated its efforts in this area, which implies stringent port and customs controls.

Through the European Community, Spain has fostered the adoption of binding instruments in the main RFMOs concerning monitoring, control and surveillance schemes, as well as catch documentation schemes and countermeasures against countries or territories that engage in IUU fishing or do not exercise control over their fleets. In any case, the adoption of the so-called "positive" lists of vessels and bluefin, bigeye, swordfish and toothfish statistical documents have proved to be an effective tool in combating IUU fishing. The landing, transhipment and import of fish products from third countries are subject to other systematic controls. Both Customs and Fisheries Control must grant dual approval before the entry of products is permitted.

Identification of vessels has emerged as another major issue. Foreign vessels coming into Spanish ports are subject to being photographed, and copies of hold plans are made.

Another difficult point is the ownership of vessels. Spain has started a project aimed at determining the actual links between the vessel owners. A team of economists, lawyers and fisheries technicians has been created, but there is a need for increased co-operation among interested countries.

Vessel laundering is another way to avoid controls, and this is why Spain has adopted regulations to prevent flag-hopping. It is extremely difficult to determine the actual ownership of a vessel when it is covered by off-shore companies in tax-havens.

Following the adoption of Royal Decree 1134/2002 on the application of penalties to Spanish nationals employed on flag-of-convenience vessels, a number of actions have been taken, especially in the field of co-operation with arresting States. It must, however, be pointed out that this legal instrument has a subsidiary nature, and is only applicable in such cases where the flag country or another country involved has not punished the infringing vessels.

One of the main obstacles Spain has encountered when trying to implement this Decree is the lack of official co-operation with both the flag State and the country that detained the IUU vessel. It is very difficult to start any punitive procedures without having official supporting evidence. Although considerable information on IUU activities or specific vessels can be found in press articles or NGO documentation, it does not guarantee success in a court of law or administrative procedure.

In any case, the most important aspect of combating IUU fishing is whether or not there is the political will and the subsequent allocation of adequate human and financial means to tackle the issue.

Co-operation with other countries has also proved to be an excellent tool.

CHAPTER 26

OECD INTRUMENTS AND IUU FISHING[1]

Ursula A. Wynhoven, Consultant

Introduction and Executive Summary

The context for this paper is the problem of illegal, unregulated and unreported (IUU) fishing, its impact on world fisheries, and the associated serious economic, environmental and social consequences. The number and complexity of the factors driving IUU activities demands a multidisciplinary and multifaceted response.[2] One avenue of investigation is to examine the instruments and follow-up mechanisms that already exist to determine their potential contribution to a solution. One category of existing instruments, and the focus of this paper, is the OECD's investment instruments.

This paper focuses on the OECD's investment instruments with a view to assessing their utility in combating IUU fishing. The instruments are: the OECD Codes on Liberalisation, the OECD Declaration and Decisions on International Investment and Multinational Enterprises (including the OECD Guidelines for Multinational Enterprises), and the Convention on Combating Bribery of Foreign Public Officials in International Business Transactions.

The paper presents a brief overview of the nature and scope of each instrument and an analysis of how the instrument could be used in a way that could contribute to the fight against IUU fishing and/or how the instrument could present an obstacle. Though none of them explicitly address the problem of IUU fishing, their potential contribution is significant. Some of the more promising possibilities are:

---

[1]   This paper was submitted as a background document to the workshop. The author wishes to thank Kathryn Gordon and Eva Thiel, both of the OECD's Directorate for Financial, Fiscal and Enterprise Affairs, for their guidance on the accuracy of facts in this paper, especially the sections on the OECD Guidelines for Multinational Enterprises and the OECD Codes of Liberalisation. The opinions expressed, and any mistakes, are the author's own.

[2]   A description and analysis of the economic and social drivers of IUU, including market control, price distortion, effect of the global economy and world fishing opportunities, international regulations, fishing agreements, re-flagging, national fisheries management policy (including subsidies, excess capacity and surveillance activities) is contained D.J. Agnew and C.T. Barnes, "The Economic and Social Effects of IUU/FOC Fishing," February 2003.

- Using the OECD Guidelines on Multinational Enterprises – a set of recommendations on good corporate behaviour by 38 governments to the multinational enterprises operating in or from their territories – to start a dialogue about corporate responsibility for IUU fishing and/or to raise awareness generally of IUU activities as a corporate responsibility problem. Specific recommendations in the Guidelines could be used to encourage enterprises to, among other things, respect the environment, disclose more information about their activities and corporate structure, provide protection for whistleblowers, apply pressure to their suppliers and other business partners to act more responsibly, not engage in bribery, refrain from seeking or accepting exemptions not contemplated in relevant statutory or regulatory frameworks, use fair marketing and advertising practices etc.

- Bringing to the attention of the relevant adhering country National Contact Point (NCP) situations of alleged corporate failure to observe the OECD Guidelines on Multinational Enterprises in connection with IUU and/or related activities. This would then invoke a set of procedures – described in Box 2 below – pursuant to which the NCP would deal with the situation.

- Referencing the OECD Declaration and Decisions on International Investment and Multinational Enterprises in encouraging adhering countries to address the impact of investment incentives and disincentives on the drivers of IUU.

- Prosecuting, under national legislation implementing the OECD Convention on Combating Bribery of Foreign Public Officials in International Business Transactions, persons who have bribed a foreign public official in connection with IUU or related activities. The Convention makes it a crime to offer, promise or give a bribe to a foreign public official in order to obtain or retain international business deals.

- Using the OECD Convention on Combating Bribery of Foreign Public Officials in International Business Transactions, and domestic legislation implementing it, as a demonstration of the legal possibility of holding a country's own nationals responsible for their conduct engaged in abroad.

- Using peer review mechanisms – of which there are many examples currently in use across the OECD – as a model for devising a peer review mechanism for country efforts to deal with IUU fishing.

However, the non-discrimination and national treatment principles embodied in the OECD's investment instruments may also present some potential obstacles:

- National treatment is the commitment by a country to treat enterprises operating on its territory, but controlled by the national of another country, no less favourably than domestic enterprises in similar situations. This may mean that measures aimed at curbing IUU fishing should not single out foreign-controlled enterprises or vessels (either in general or the enterprises or vessels of particular countries) for less favourable treatment than is accorded national enterprises or vessels in gaining access to the country's fisheries. This is despite the fact that the enterprises and vessels flying flags of particular countries may be statistically more likely to be involved in IUU activities.

- The OECD Codes of Liberalisation require OECD countries to move progressively towards open markets and liberalisation, whereas some measures aimed at discouraging IUU fishing and related activities could, at least theoretically, be construed as moving in the opposite direction. Examples of such measures might be a law prohibiting nationals from registering their ships in another country, and boycotts or blacklists of foreign vessels or the entities owning them that cause Foreign Service providers difficulties in entering the market of the country imposing the sanctions.

These possible contributions and obstacles are discussed in more detail below. The paper concludes with some suggested future actions that the OECD Fisheries Committee and others may wish to take.

The OECD's investment instruments

The OECD has a number of legal instruments on international investment and trade in services.[3] They are:

- The OECD Codes of Liberalisation;
- The Declaration and Decisions on International Investment and Multinational Enterprises (which incorporates the OECD Guidelines on Multinational Enterprises);
- The Convention on Combating Bribery of Foreign Officials in International Business Transactions.

Together, these instruments establish the "rules of the game" for adhering countries and multinational enterprises based in or operating in their countries for capital movements, international investment and trade in services. These are instruments to which all member countries of the OECD must adhere. In addition, some of these instruments are open for adherence by non-member countries.

None of the instruments discussed in this paper expressly deal with IUU fishing. Nevertheless, they can make a contribution in the fight against it. This part of the paper explores the nature and scope of each instrument and presents an analysis of how the instrument could be used in this context and/or possible obstacles to be overcome. Each instrument is introduced with a brief overview explaining what it is, to whom it applies and how it is implemented. A copy of each of the instruments is included in the appendices.

OECD Codes of Liberalisation

What they are

The Codes of Liberalisation is a collective term for two separate instruments. the Code of Liberalisation of Capital Movements; and the Code of Liberalisation of Current Invisible Operations.

---

[3]   Another OECD legal instrument on international investment was negotiated between 1995 and 1998. However, negotiations were abandoned in December 1998 after a six month hiatus, during which no official meetings of negotiators took place. The instrument, which was called the draft Multilateral Agreement on Investment (MAI), was to be a "free standing international treaty, open to all OECD Members and the European Communities, and to accession by non-OECD Member Countries" and its proposed objective was to "provide a broad multilateral framework for international investment with high standards for the liberalisation of investment regimes and investment protection and with effective dispute settlement procedures." A key reason for the cessation of negotiations was that public interest groups were concerned about the impact of globalization on labour and human rights, the environment and consumer and development issues. See OECD Codes of Liberalisation of Capital Movements and Current Invisible Operations: Users' Guide, OECD, April 2003, p. 13.

They prescribe progressive, non-discriminatory liberalisation of capital movements, the right of establishment and current invisible transactions (mainly services).[4] Both were formally adopted in 1961 and have been revised and expanded in scope a number of times. Though not a treaty or an international agreement, they are, nevertheless, legally binding rules of behaviour for the governments of OECD countries.

The Codes are very similar, sharing many provisions in common. The main difference is that one Code concerns capital movements and the other invisible transactions and transfers. The structure of each Code is as follows: In Article 1 members commit to eliminating, between one another, restrictions on capital and current account operations. The remainder of each Code sets out the framework for working towards this objective. Each Code has two principal annexes: a list of economic activities covered, and a list of countries' current reservations.

In the Code of Liberalisation of Capital Movements, 16 categories of economic activities are covered, including direct investment, liquidation of direct investment, credits directly linked with international transactions or with the rendering of international services, and a large number of other short- and long-term capital movements.[5] Both capital inflows and capital outflows are covered, as are actions initiated by non-residents in the country concerned and actions abroad initiated by non-residents.[6] For example, it addresses direct investment – including creating, acquiring or participating in a new or existing business – both in the country concerned by non-residents, or abroad by residents.

The coverage of the Code of Liberalisation of Current Invisible Operations is more limited. It is concerned with liberalisation of cross-border trade in services, namely, the supply of services to residents by non-resident service-providers and *vice versa*. A variety of sectors is covered, including banking and financial services, insurance, professional services, maritime[7] and road transport, and travel and tourism.

The kinds of restrictions that members are expected to eliminate progressively are laws, decrees, regulations, policies and practices taken by authorities that may restrict the conclusion or execution of economic activities covered by the Codes.[8] Non-discrimination is a key principle of the Codes: OECD members are expected to grant the benefits of open markets to residents of all other OECD member countries.[9] A measure is a restriction if it discriminates between residents and non-residents. Although

---

[4] The Codes are actually Decisions of the OECD Council, which are legally binding on OECD member governments. See OECD Codes of Liberalisation of Capital Movements and Current Invisible Operations: Users' Guide, April 2003, p. 6.

[5] Annex A, OECD Code of Liberalisation of Capital Movements.

[6] OECD Codes of Liberalisation of Capital Movements and Current Invisible Operations: Users' Guide, OECD April 2003, p. 22.

[7] The Code of Liberalisation of Current Invisible Operations covers maritime freights (including chartering, harbour expenses, disbursements for fishing vessels, etc.), maritime transport (including bunkering and provisioning, maintenance, repairs, expenses for crews, etc) and other items that have a direct or indirect bearing on international maritime transport. It is intended to give residents of one member country the unrestricted opportunity to avail themselves of, and pay for, all services in connection with international maritime transport that are offered by residents of any other member country. See Notes to Annex A of the Code of Liberalisation of Current Invisible Operations, Note 1.

[8] OECD Codes of Liberalisation of Capital Movements and Current Invisible Operations: Users' Guide, OECD April 2003, p. 17.

[9] Article 9 of the Codes.

"resident" is not synonymous with "national," nationality requirements are generally considered incompatible with the Codes.[10] Non-discrimination is not to be confused with preferential treatment, which non-residents are not entitled to.[11] Non-residents are also subject to the same general regulations as residents.[12] The treatment of residents and non-residents need not be identical as long as it is equivalent.[13] Reservations are generally not permitted to the non-discrimination principle.[14]

A copy of the Codes can be found in Appendices 1 and 2 respectively.

To whom they apply

The Codes create rights and obligations for OECD member countries only.[15] However, members have agreed to use their best offices to extend the benefits of liberalisation to all members of the IMF.[16] Moreover, the adoption of the General Agreement on Trade in Services (GATS) has meant that a number of the economic activities covered by the Codes, especially establishment and cross-border trade in services, are now also subject to the liberalisation obligations in the GATS.[17] The result is that where OECD members have committed themselves to non-discrimination between GATS members, liberalisation benefits under the Codes overlapping with those in the GATS are also to be extended to all GATS signatories.[18]

How they are implemented

The Codes ask OECD member countries to implement their obligations through necessary measures at the national level. Non-conforming measures are required to be listed in country reservations lodged under the Codes, and members are expected to progress at their own pace towards open markets, that is, the full abolition of restrictions.

Further implementation or follow-up occurs through policy reviews and country examinations, which rely on peer pressure to encourage unilateral liberalisation. These take place in the context of the OECD Committee on Capital Movements and Invisible Transactions (CMIT), which is the forum where member countries meet to discuss application and implementation of the Codes. There are no direct sanctions involved in the compliance review process. Nevertheless, peer review and peer

---

[10]   OECD Codes of Liberalisation of Capital Movements and Current Invisible Operations: Users' Guide, OECD April 2003, p. 18.

[11]   OECD Codes of Liberalisation of Capital Movements and Current Invisible Operations: Users' Guide, OECD April 2003, p. 18.

[12]   *Ibid* at 22.

[13]   *Ibid* at 18.

[14]   Note, however, that Article 10 provides (paraphrasing) that Members that are part of a special customs or monetary system, such as the European Community, are permitted to liberalise more rapidly or widely among themselves, or, in other words, to maintain more restrictions in relation to other members that are not in their special customs or monetary system

[15]   In addition, the entities subject to the liberalisation obligation are government authorities, not private entities. *Ibid.* at 16.

[16]   Article 1(d) of the Codes.

[17]   OECD Codes of Liberalisation of Capital Movements and Current Invisible Operations: Users' Guide, OECD April 2003.

[18]   *Idem.*

pressure in a multilateral setting have provided strong incentives for authorities to undertake policy adjustments through "benchmarking" regulations and administrative procedures against those adopted and enforced by peer members.[19]

Neither individuals nor enterprises can directly invoke rights to invest abroad, move funds or provide cross-border services under the Codes. Their complaints can only be raised through their own (OECD) governments, which could then raise a case under the Codes with the CMIT.

How they could be used to tackle IUU fishing

The Codes are aimed at liberalisation – in the sense of removing unnecessary barriers to the free circulation of capital and services. As such, it is difficult to see what contribution they could make in the fight against IUU fishing, which seems to need more restrictions rather than less. For this reason, the Codes could actually present an obstacle in searching for new ways to deal with IUU fishing insofar as they have the potential to restrict the ability of a country to take certain measures to deter IUU fishing and related activities. However, most, perhaps all, of these obstacles could be avoided if law makers consult with their international investment colleagues to ensure that the proposed measures will not contravene the country's liberalisation commitments.

Peer reviews that are conducted as part of the Codes implementation process and/or elsewhere at the OECD could also present an interesting model for dealing with IUU fishing. Having used them since its inception, the OECD has developed a comparative advantage in conducting peer reviews – the assessment of the policies and performance of a country by other countries with a view to improving the first country's policies and helping it comply with established standards and principles.[20] A recent analysis of peer review processes at the OECD observed that there was no other international organisation in which the practice of peer review has been so extensively developed.[21] The analysis also articulated a (best practice) model based on the different peer review mechanisms in operation across the OECD. This model and the OECD's expertise in this area could perhaps be used to help construct a peer review process for country efforts to deal with IUU fishing.

Potential limitations to their use in combating IUU fishing

At least in theory, the Codes have the potential to constrain a member country's ability to introduce and maintain measures to deal with IUU fishing, especially where those measures distinguish between residents and non-residents or between nationals and foreigners. For example, a country's efforts aimed at actively discouraging insurance, banking, and shipping industries and other related sectors from providing products and services to vessels and companies from certain flag of convenience countries, closing ports to them or refusing to grant them licences and approvals on a blanket basis because of their flag, could be inconsistent with its obligations under the Codes to progressively liberalise. Similarly, measures aimed at restricting their own residents from registering their fishing vessels in other states or being involved in the fishing industry in other states could also implicate the Codes. In practical terms, however, the impact of the Codes on measures designed to tackle IUU fishing may be minimal because of the way concepts in the Codes have been interpreted, exceptions in the Codes themselves, and reservations that countries have lodged.

---

[19]   "Successful Capital Movements Liberalisation: A Question of Governance – Recent OECD Experience" in *International Investment Perspectives*, No. 1 2002, p. 118.

[20]   Peer Review: A Tool for Co-operation and Change – An Analysis of an OECD Working Method, OECD, 2003 (SG/LEG(2002)1).

[21]   Ibid at 7.

The coverage of the Codes and the way they have been interpreted. Measures aimed at curbing IUU fishing and related activities will not necessarily fall foul of the Codes. For example, the Codes do not cover domestic transactions. Thus, if there is no international element, in the sense of involving residents of more than one OECD country, the Codes will not apply. As already described, the Codes also only address certain kinds of international transactions. If an anti-IUU fishing measure has no impact on one of these transactions, the Codes will not be relevant. In addition, if the measure does not discriminate it will not be inconsistent with the obligation of non-discrimination.[22] For example, encouraging service providers not to maintain business relations that they may have with vessels identified as engaging in IUU fishing does not discriminate and would be unlikely to contradict liberalisation obligations under the Codes.

Notwithstanding the potential for licensing requirements and other domestic regulations to affect operations under both Codes, the CMIT has generally considered that such measures do not constitute restrictions under the Codes as long as they are applied in a non-discriminatory manner.[23] This means that governments can be relatively confident that non-discriminatory licensing requirements and domestic regulations can be used in dealing with IUU and related activities without violating their obligations under the Codes. The International Plan of Action to Prevent, Deter and Eliminate Illegal, Unreported and Unregulated Fishing also recognises the importance of non-discrimination, providing that "The IPOA should be developed and applied without discrimination in form or in fact against any State or its fishing vessels."[24]

Liberalisation obligations do not generally apply to subsidies or the conditions attached to them, or the levying of taxes, duties and other charges. However, if they have the effect of frustrating liberalisation, suggestions may be made to encourage their removal or modification.[25]

Exceptions in the Codes. The Codes also have a number of exceptions built in that preserve for countries a degree of latitude in taking actions that might otherwise fall within the scope of the Codes. There are exceptions for action that the member considers necessary for the maintenance of public order or the protection of public health, morals and safety; the protection of its essential security interests; and the fulfilment of its obligations relating to international peace and security.[26] These exceptions allow members to introduce, reintroduce or maintain restrictions that are not covered by

---

[22]   The importance of non-discrimination is also emphasised in the European Commission's Community action plan for the eradication of illegal, unreported and unregulated fishing, 28 May 2002.

[23]   OECD Codes of Liberalisation of Capital Movements and Current Invisible Operations: Users' Guide, OECD, April 2003, p. 17. Article 16 of the Codes is concerned with situations where domestic regulations do not discriminate directly (*i.e.* on the face of the regulation), but nevertheless have an unreasonable discriminatory effect. In such situations, an affected member country can refer the situation to the OECD, which will make a determination as to whether the regulation does indeed have the effect of frustrating its measures of liberalisation and, if so, will make suggestions about removing or modifying it. An example of a regulation that had a discriminatory effect on or prejudiced foreigners would be any regulation that was much more difficult for foreigners to comply with than nationals.

[24]   See 9.6 Non-discrimination.

[25]   OECD Codes of Liberalisation of Capital Movements and Current Invisible Operations: Users' Guide, OECD, April 2003, p. 46.

[26]   Article 3 of the Codes.

reservations to the Code and to exempt them from the principle of progressive liberalisation.[27] More comfort would be provided, though, if there was also an exception for measures protecting the environment.[28] Another exception concerns obligations in existing multilateral international agreements.[29] However, this exception does not apply to obligations in agreements concluded after the adoption of the Codes.[30] Yet another important exception concerns law enforcement, specifically, the powers of members to verify the authenticity of transactions and transfers and to take any measures required to prevent the evasion of their laws or regulations.[31] A note by the Chairman of the Negotiating Group on the Multilateral Agreement on Investment (MAI) concluded that this exception presumably includes laws and regulations concerning the environment.[32] Members that are part of a special customs or monetary system, such as the European Community, are also permitted to liberalise more rapidly or widely among themselves, or, in other words, to maintain more restrictions in relation to other members that are not in their special customs or monetary system.[33]

Reservations to the Codes. By lodging a reservation, a member retains the right to maintain restrictions, as specific in the reservation, on the economic activity concerned.[34] Many OECD countries have reserved to themselves the right and power to regulate such things as ownership or registration of their flag vessels by non-residents,[35] and the ownership of a business engaged in commercial fishing or investment in fishing and/or primary fish processing by non-residents.[36] This means that the countries concerned would not be violating their obligations under the Codes if they maintain restrictions that are consistent with these reservations. Table 26.1 sets out some of these reservations.[37] There are, however, almost no reservations concerning outward direct investment, that

---

[27]   OECD Codes of Liberalisation of Capital Movements and Current Invisible Operations: Users' Guide, OECD April 2003, p. 24. Nevertheless, in recent years, members have been encouraged to lodge reservations when introducing restrictions for national security concerns. Idem.

[28]   Note that even the General Agreement on Tariffs and Trade has an exception for measures "necessary to protect human, animal, or plant life or health" (Article XX(b)). This provision has attracted a lot of controversy because of its narrow scope/it has been interpreted narrowly. For example, a United States embargo on tuna products was ruled impermissible notwithstanding its stated aim was to protect dolphins. See GATT Dispute Settlement Panel Report on United States Restrictions on Impacts of Tuna, Aug. 16, 1991, 30 I.L.M. 1594 (1991) and June 16, 1994, 33 I.L.M. 839 (1994).

[29]   Article 4 of the Codes.

[30]   OECD Codes of Liberalisation of Capital Movements and Current Invisible Operations: Users' Guide, OECD April 2003, p. 25.

[31]   Article 5 of the Codes.

[32]   MAI and the Environment, Note by the Chairman, Negotiating Group on the Multilateral Agreement on Investment (MAI), 9 October 1996 (DAFFE/MAI(96)30).

[33]   Article 10 of the Codes.

[34]   The CMIT has taken special pains to ensure that the language of each reservation is as specific and narrow as possible so as to promote transparency and discourage backwards sliding.

[35]   See, for example, the reservations lodged by Australia, Austria, Belgium, Finland, France, Germany, Greece, Iceland, Ireland, Italy, Netherlands, New Zealand, Norway, Poland, Portugal, Sweden, Switzerland, and the United Kingdom to the OECD Code of Liberalisation of Capital Movements (Annex B).

[36]   See, for example, the reservation lodged by Canada, Denmark, Iceland, Japan, Korea, Mexico, New Zealand, Norway and Sweden to the OECD Code of Liberalisation of Capital Movements (Annex B).

[37]   Table 26.1 shows only those reservations that relate to fishing, shipping and related activities.

is, abroad by residents.[38] In other words, restrictions on outward direct investment by residents will generally be inconsistent with the Codes and not allowed. In certain limited circumstances, it is possible to introduce new reservations.[39] However, even where it is possible to do so, lodging a reservation entails providing the reasons for doing so and submitting to an initial examination as well as subsequent periodic examinations.[40] Part of these examinations will involve the application of strong peer pressure to encourage the country concerned to justify its reservations, to narrow them, to look for other non-restrictive ways of achieving the same (legitimate) objectives, and to move progressively towards eliminating them. Adding, limiting or withdrawing a reservation requires a decision of the OECD Council.[41]

Declaration and decisions on international investment and multinational enterprises

What they are

The Declaration on International Investment and Multinational Enterprises (the "Declaration") is a policy commitment to improve the investment climate, encourage the positive contribution multinational enterprises can make to economic and social progress, and minimise and resolve difficulties that may arise from their operations. The Declaration is comprised of four elements, each of which is supported by a Decision of the OECD Council on follow-up procedures:

- The Guidelines for Multinational Enterprises, which are a set of recommendations by governments to multinational enterprises on responsible corporate conduct;

- National Treatment requiring that member countries accord to foreign-controlled enterprises on their territories no less favourable treatment than that accorded in like situations to domestic enterprises;

- Conflicting requirements obliging members to co-operate so as to avoid or minimise the imposition of conflicting requirements on multinational enterprises;

- International investment incentives and disincentives in relation to which members recognise the need to give due weight to the interest of members affected by laws and practices in this field and endeavour to make measures as transparent as possible.

The second, third and fourth elements are dealt with in this section. The first – the OECD Guidelines for Multinational Enterprises –is considered separately later in this paper because of its special focus on corporate responsibility – which is of key importance in the discussion of ways to tackle IUU activities.

---

[38] A notable exception appears to be Japan, which maintains an exception concerning direct investment abroad by residents applying only to investment in an enterprise engaged in fishing regulated by international treaties to which Japan is a party or fishing operations coming under the Japanese Fisheries Law.

[39] The circumstances under which a new reservation can be lodged are set out in Article 2 of the Codes.

[40] OECD Codes of Liberalisation of Capital Movements and Current Invisible Operations: Users' Guide, OECD April 2003, p. 23.

[41] *Idem.*

Table 26.1. Examples of Reservations Lodged Under the OECD Code of Liberalisation of Capital Movements

Country	Reservation
Australia	Ownership of Australian flag vessels, except through an enterprise incorporated in Australia.
Austria	Acquisition of 25% or more in ships registered in Austria.
Belgium	Acquisition of Belgian flag vessels by shipping companies not having their principal office in Belgium.
Canada	Fish harvesting.
Denmark	Ownership of Danish flag vessels by non-EC residents except through an enterprise incorporated in Denmark. Ownership by non-EC residents of one-third or more of a business engaged in commercial fishing.
Finland	Ownership of Finnish flag vessels, including fishing vessels, except through an enterprise incorporated in Finland.
France	Ownership after acquisition of more than 50% of a French flag vessel, unless the vessel concerned is entirely owned by enterprises having their principal office in France.
Germany	Acquisition of a German flag vessel, except through an enterprise incorporated in Germany.
Greece	Ownership of more than 49% of the capital of a Greek flag vessel for fishing purposes.
Iceland	Investment in fishing and primary fish processing (excluding retail packaging and later stages of preparation of fish products for distribution and consumption). Ownership of Icelandic flag vessels, except through an enterprise incorporated in Iceland.
Ireland	Acquisition by non-EC nationals of sea fishing vessels registered in Ireland. Foreign acquisition of shipping vessels registered in Ireland is subject to a reciprocity requirement.
Italy	Purchase by foreigners other than EC residents of a majority interest in Italian flag vessels or of a controlling interest in ship owning companies having their headquarters in Italy. Purchase of Italian flag vessels used to fish in Italian territorial waters.
Japan	Investment in primary industry related to fisheries. (Abroad by residents) Investments in an enterprise engaged in fishing regulated by international treaties to which Japan is a party or fishing operations coming under the Japanese Fisheries Law.

Table 26.1. (cont.) Examples of Reservations Lodged Under the OECD Code of Liberalisation of Capital Movements

Country	Reservation
Korea	Fishing in internal waters, the territorial sea and the EEZ if foreign investors hold 50% or more of the share capital.
Mexico	Investment exceeding a total of 49% in fishing, other than aquaculture, in coastal and fresh waters or in the EEZ.
Netherlands	Ownership of Netherlands flag vessels, unless the investment is made by shipping companies incorporated under Netherlands law, established in the Kingdom and having their actual place of management in the Netherlands.
New Zealand	Acquisition, regardless of dollar value, of 25% or more of any class of shares or voting power in a New Zealand company engaged in commercial fishing. Acquisition, regardless of dollar value, of assets used, or proposed to be used, in a business engaged in commercial fishing.
Norway	Ownership of Norwegian flag vessels, except a) through a partnership or joint stock company where Norwegian citizens own at least 60% of the capital, b) by registering the vessel in the Norwegian International Ship Register under the applicable conditions. Investment in a registered fishing vessel bringing foreign ownership of the vessel above 40%.
Poland	Investment in a registered vessel, except through an enterprise incorporated in Poland.
Portugal	Ownership of Portuguese flag vessels other than through an enterprise incorporated in Portugal.
Sweden	Acquisition of 50% or more of Swedish flag vessels, except through an enterprise incorporated in Sweden. Establishment of, or acquisition of 50% or more of shares in, firms engaged in commercial fishing activities in Swedish waters, unless an authorisation is granted.
Switzerland	The registration of a ship in Switzerland serving two points on the Rhine.
United Kingdom	Acquisition of United Kingdom flag vessels, except through an enterprise incorporated in the United Kingdom.
United States	Fishing in the "Exclusive Economic Zone", and deepwater ports, except through an enterprise incorporated in the United States.

To whom they apply

The Declaration binds all countries that have adhered to it. At present, there are 38 adherents (OECD countries and 8 others).[42] Other countries that are willing and able to meet the Declaration's requirements are also welcome to apply to adhere. Although the Declaration itself is binding, national treatment of foreign controlled enterprises on their territories is only a voluntary commitment. Nevertheless, in 1988 there was a unanimous pledge not to introduce new exceptions to national treatment.

---

[42]   The eight non-member countries that have adhered are: Argentina, Brazil, Chile, Estonia, Israel, Latvia, Lithuania, and Slovenia. The application of Singapore is currently being considered. Though not able to be an adherent itself, the European Commission is also an active participant in the administration of the Guidelines.

How they are implemented

Each of the four elements of the Declaration has its own distinct follow-up procedure. Implementation of the Guidelines and their unique follow-up procedures are discussed in the section devoted to the Guidelines. A description of the Declaration's treatment of the implementation and follow up of the national treatment, conflicting requirements and investment incentives, is provided below.

National treatment

The national treatment instrument consists of a declaration of principle in the Declaration and a procedural OECD Council Decision (December 1991), which requires adhering countries to notify their exceptions to national treatment and establishes follow-up procedures. The exceptions are subject to periodic examination by the Committee on International Investment and Multinational Enterprises (CIME), which, in turn, results in a decision of the OECD Council and proposals for action for the country concerned. Exceptions are usually confined to certain key sectors such as fisheries, or transport and communications, and are generally limited in scope.[43] Exceptions are limited or removed either by unilateral action by the country itself, or as a result of the examinations.

Conflicting requirements

The instrument on conflicting requirements also consists of a declaration of principle in the Declaration and an OECD Council Decision on procedures (June 1991). Under the Declaration, governments of adhering countries have committed themselves to co-operate with a view to avoiding or minimising the imposition of conflicting requirements on multinational enterprises and to take into account the General Considerations and Practical Approaches.[44] The aim of avoiding or minimising the imposition of conflicting requirements by governments is implemented through promoting co-operation among member countries, bilaterally and/or in the context of the CIME.[45]

Investment incentives and disincentives

The instrument on international incentives and disincentives is comprised of a declaration of principle in the Declaration and an OECD Council Decision (May 1984). Recognising that adhering countries may be adversely affected by investment incentives and disincentives provided by other adhering countries, the provisions ask that such measures be made as transparent as possible and encourage effective co-operation between adhering countries. The instrument prescribes consultations and review procedures, and asks that adhering countries provide information about their policies and participate in studies on investment incentives and disincentives. The consultations take place in the CIME at the request of an adhering country that considers that its interests may be adversely affected, with the objective of examining the possibility of reducing the adverse effects to a minimum.

---

[43] The exceptions are found in Annex A to the Third Revised Decision of the Council on National Treatment, December 1991. Countries that have notified exceptions in connection with the fishing sector include: Australia, Austria, Brazil, Canada, Chile, Greece, Iceland, Ireland, Italy, Japan, Korea, Lithuania, Mexico, New Zealand, Norway, Sweden and United States. These exceptions limit such things as access to fisheries by foreign flag vessels or foreign-controlled enterprises, and/or prescribe limits on ownership of registered fishing vessels by foreign-controlled enterprises.

[44] General Considerations and Practical Approaches is a document setting out a process for co-operation between countries on the subject of conflicting requirements.

[45] Decision of the OECD Council, June 1991.

How they could be used to tackle IUU fishing

Conflicting requirements

The conflicting requirements provisions offer a non-adversarial, co-operative mechanism for dealing with inconsistent requirements on multinational enterprises. The contribution that such provisions can make to the fight against IUU fishing is not immediately apparent. However, if, for example, measures aimed at curbing IUU activities resulted in different adhering governments imposing conflicting obligations on multinational enterprises, then the non-adversarial, co-operative approach might be helpful in responding to the conflict.

International investment incentives and disincentives

The OECD and others, including the WWF, have pointed to the role of various types of subsidies in keeping illegal fishing vessels in operation and/or encouraging the export of excess capacity to other areas of the world.[46] The WWF has also called for the urgent control of subsidies that assist in driving IUU fishing.[47] The International Plan of Action to Prevent, Deter and Eliminate Illegal, Unreported and Unregulated Fishing also calls on States to avoid conferring economic support, including subsidies, to companies, vessels or persons that are involved in IUU fishing.[48]

Part IV of the Declaration – concerning international investment incentives and disincentives – contributes to the fight against IUU fishing by recognising that adhering countries need to strengthen their co-operation in the field of international direct investment, as well as pay attention to the interests of adhering governments affected by specific laws, regulations and administrative practices providing official incentives and disincentives to international direct investment. Under the Declaration, adhering governments also commit to try to make such incentives and disincentives as transparent as possible so that their importance and purpose can be ascertained and information on them can be readily available. Promoting transparency of such measures and encouraging dialogue and co-operation between adhering countries about them, could provide another entry point for discussions about the incentives and disincentives (including subsidies) that may fuel IUU activities.

Potential limitations to their use in combating IUU fishing

National treatment

Concerns about IUU fishing and scarce enforcement resources might prompt some countries to consider restricting access to their fisheries by foreign-controlled enterprises or vessels, either generally or by nationals from certain countries and/or vessels flying flags from countries that are

---

[46]   See, for example, S.J. Cripps, A. Oliver and J. Cator, "International aspects of the control and eradication of IUU fishing – an NGO's perspective," Fisheries Monitoring, Control and Surveillance, Brussels 24-27 October 2000; "The Environmental, Economic and Social Issues and Effects of IUU/FOC Fishing Activities in the High Seas", OECD, 14 February 2003 (AGR/FI(2003)5); D.J. Agnew and C.T. Barnes "The Economic and Social Effects of IUU/FOC Fishing," February 2003.

[47]   S.J. Cripps, A. Oliver and J. Cator, "International aspects of the control and eradication of IUU fishing – an NGO's perspective," Fisheries Monitoring, Control and Surveillance, Brussels 24-27 October 2000.

[48]   Paragraph 23 (Economic Incentives).

well-known for their open registries, low standards and/or lax enforcement.[49] Alternatively, or in addition, some countries might seek to concentrate their scarce IUU fishing detection and enforcement resources on, or otherwise to treat less favourably, foreign-controlled vessels or enterprises from certain countries. Along similar lines, the International Coalition of Fisheries Associations has adopted a resolution calling on governments and the private sector to prevent flag of convenience vessels from gaining access to international markets; freighter companies to refrain from transporting any fish caught by flag of convenience fishing vessels; and trading and distribution companies to refrain from dealing in fish caught by flag of convenience vessels.[50] Others have also called for discouraging or preventing certain countries that cannot or will not exercise control over fishing vessels operating outside of their EEZs from registering large-scale fishing vessels, and for the closing of ports to flag of convenience fishing vessels.[51]

The requirement of according national treatment to foreign-owned or -controlled enterprises might present an obstacle to measures aimed at singling out foreign enterprises and vessels for worse treatment because they are foreign or because of the flag they are flying, particularly in the absence of proof that the vessels concerned had actually been engaged in IUU fishing or intended to do so.[52] Such a concern may, however, be mitigated, to some extent, by the fact that the obligation to accord national treatment is subject to adhering countries' needs to maintain public order, to protect their essential security interests and to fulfil commitments relating to international peace and security. Importantly, a number of countries have notified fishing-related exceptions to the national treatment principle. These exceptions limit such things as access to fisheries by foreign-flag vessels or foreign-controlled enterprises, and/or prescribe limits on ownership of registered fishing vessels by foreign-controlled enterprises. Table 26.2 lists the fishing-related exceptions to national treatment. Moreover, the obligation to accord national treatment is less strong with respect to non-adhering countries, which are more likely to be flags of convenience countries: adhering countries only "consider applying" national treatment to countries other than adhering countries. Lastly, the Declaration expressly states that it "does not deal with the right of adhering governments to regulate the entry of foreign investment or the conditions of establishment of foreign enterprises." Thus, if the measures being considered fell into this category, they would fall outside the national treatment instrument.

---

[49]   Some flag states are of particular concern for the management of fisheries. See, for example, M. Gianni (for WWF), "Recommendations to OECD Countries on Measures to Prevent and Eliminate the Problem of Illegal, Unreported and Unregulated Fishing," February 2003, pp. 2-3.

[50]   "ICFA Calls for Elimination of Flag-of-Convenience (FOC) Fishing Vessels," press release, International Coalition of Fisheries Associations, 5 January 2000, available at http://www.icfa.net/?a=Press%20Releases&item=158.

[51]   See, for example, M. Gianni (for WWF), "Recommendations to OECD Countries on Measures to Prevent and Eliminate the Problem of Illegal, Unreported and Unregulated Fishing," February 2003, p. 6.

[52]   Beyond the desire to engage in IUU fishing, there are a host of other reasons, some legitimate, some not, why a vessel may have a particular flag.

Table 26.2. Fishing-related Exceptions to the National Treatment Principle

Adhering country	Exception
Australia	(Western Australia only) Foreign ownership in rock lobster processing is limited to 20%; restrictions are placed on non-residents becoming directors or office bearers in corporations undertaking rock lobster processing.
Austria	Requirements to obtain the national flag: citizenship, residence in Austria, and more than 75% local ownership. The flag is required for registration of vessels.
Brazil	Exploitation of internal waters, areas within the territorial sea and some other activities are reserved to native-born Brazilians or persons who have naturalised citizenship or must be undertaken by firms registered in Brazil. Foreign vessels need authorisation from the Ministry of Agriculture to develop fishing activities.
Canada	There is no limit on foreign ownership of fish processing companies that do not hold fishing licences. Canadian fish processing companies which have more than 49% foreign ownership are not permitted to hold Canadian commercial fishing licences. Fish harvesting firms with foreign participation are subject to the same rules and policies as wholly Canadian-owned firms (*e.g.* Canadian registry and Canadian crews for licensed fishing vessels). (British Columbia only) Nationality requirement to obtain a fish buyer's license.
Chile	Ownership of Chilean fishing vessels is limited to Chilean individuals or Chilean majority-owned corporations with principal domicile and real effective seat in Chile. However, an owner of a fishing vessel registered in Chile prior to 30 June 1990 is not subject to the nationality requirements. Fishing vessels specifically authorised by the maritime authorities, pursuant to powers conferred by law in cases of reciprocity granted to Chilean vessels by other States, may be exempted from the above-mentioned requirements on equivalent terms provided to Chilean vessels by that State.
Greece	Non-EC ownership of Greek flag vessels including fishing vessels is limited to 49%.
Iceland	Foreign investment in primary fish processing (*i.e.* excluding retail packaging and later stages of preparation of fish products for distribution and consumption) is prohibited. No foreign ownership limitations apply to further fish processing.
Ireland	Registration of fishing vessels requires ownership by citizens or companies from an EC Member State and a license to fish within Irish fishing limits. The acquisition by non-EC nationals of sea fishing vessels registered in Ireland may be restricted.
Italy	Fishing in territorial waters is reserved to nationals.
Japan	Foreign-controlled enterprises may be restricted from engaging in fisheries.
Korea	Enterprises with foreign participation require authorisation to be engaged in commercial fishing in internal waters, the territorial sea and the EEZ.
Lithuania	Access to Lithuania's waters is only possible for vessels with a Lithuanian flag and registered in Lithuania or for foreign country vessels on the basis of bilateral and multilateral agreements.

Table 26.2. (cont.) Fishing-related Exceptions to the National Treatment Principle

Adhering country	Exception
Mexico	Foreign investment is permitted up to 49% in fishing in coastal and fresh waters of in the EEZ and up to 100% in aquaculture.
New Zealand	Purchase of fishing quota is restricted to enterprises where 75% of more of the voting rights are held by New Zealand residents.
Norway	As a general rule, processing, packing or re-loading fish, crustaceans and mollusc or parts and products of these, is not allowed on a foreign-controlled vessel inside the fishing limits or the Norwegian EEZ.
	To obtain ownership (including part) of a registered fishing vessel, a 60% Norwegian ownership is required.
	Foreign-controlled enterprises may not fish with trawls from Norwegian vessels.
Sweden	A legal entity, owned up to 50% or more by foreign citizens, is subject to permission for having the right to pursue commercial fishing activities in Swedish waters without holding a private fishing right.
United States	Foreign-controlled enterprises may not engage in certain fishing operations involving coastwise trade. In addition, foreigners may not hold more than a minority of shares comprising ownership in companies owning vessels which operate in US fisheries. Also, corporate organisation requirements pertain to the registration of flag vessels for fishing in the US EEZ.
	Foreign-flag vessels may not fish or process fish in the 200 nautical mile US EEZ except under the terms of a Governing International Fisheries Agreement (GIFA), or other agreement consistent with US law.

OECD Guidelines for Multinational Enterprises

What they are

Though it forms part – Annex 1 – of the previously mentioned instrument (the Declaration), the OECD Guidelines for Multinational Enterprises (the Guidelines), are discussed separately here. The Guidelines are regarded as one of the world's foremost corporate responsibility instruments and are becoming an important international benchmark for corporate responsibility. They aim to promote the positive contributions multinational enterprises can make to economic, environmental and social progress. They contain ten short chapters of voluntary[53] principles and standards for responsible business conduct addressing such areas as human rights, disclosure of information, anti-corruption, taxation, labour relations, environment, and consumer protection. The principles and standards are in the form of recommendations by the 38 countries that have adhered to them to the multinational enterprises operating in or from their territories. They express the shared values of the governments of countries that are the source of most of the world's direct investment flows and home to most multinational enterprises. Box 1 presents a brief overview of the main Guidelines recommendations.

---

[53]   Observance of the Guidelines is voluntary in the sense of not being legally enforceable (see I.1 of the Guidelines). However, as was pointed out in Corporate Responsibility: Private Initiatives and Public Goals, OECD 2001, p. 12, there are often powerful pressures acting on firms engaged with voluntary corporate responsibility initiatives.

Box 26.1. Main Recommendations of the OECD Guidelines for Multinational Enterprises[54]

The Preface situates the Guidelines in a globalizing world. The common aim of the governments adhering to the Guidelines is to encourage the positive contributions that multinational enterprises can make to economic, environmental and social progress, and to minimise the difficulties to which their various operations may give rise.

I. Concepts and Principles: sets out the principles that underlie the Guidelines, such as their voluntary character, their application worldwide and the fact that they reflect good practice for all enterprises.

II. General Policies: contains the first specific recommendations, including provisions on human rights, sustainable development, supply chain responsibility, and local capacity building, and more generally calls on enterprises to take full account of established policies in the countries in which they operate.

III. Disclosure: recommends disclosure on all material matters regarding the enterprise, such as its performance and ownership, and encourages communication in areas where reporting standards are still emerging, such as social, environmental and risk reporting.

IV. Employment and Industrial Relations: addresses major aspects of corporate behaviour in this area, including child and forced labour, non-discrimination and the right to *bona fide* employee representation and constructive negotiations.

V. Environment: encourages enterprises to raise their performance in protecting the environment, including performance with respect to health and safety impacts. Features of this chapter include recommendations concerning environmental management systems and the desirability of precaution where there are threats of serious damage to the environment.

VI. Combating Bribery: covers both public and private bribery, and addresses passive and active corruption.

VII. Consumer Interests: recommends that enterprises, when dealing with consumers, act in accordance with fair business, marketing and advertising practices, respect consumer privacy, and take all reasonable steps to ensure the safety and quality of goods or services provided.

VIII. Science and Technology: aims to promote the diffusion by multinational enterprises of the fruits of research and development activities among the countries where they operate, thereby contributing to the innovative capacities of host countries.

IX. Competition: emphasises the importance of an open and competitive business climate.

X. Taxation: calls on enterprises to respect both the letter and spirit of tax laws, and to co-operate with tax authorities.

To whom they apply

The Guidelines' principles apply to multinational enterprises operating in or from the territories of the 38 countries that have adhered to them.[55] These are the OECD countries and 8 others.[56] Other countries that are willing and able to meet the Declaration's requirements are also encouraged to apply

---

[54]   This box is reproduced from The OECD Guidelines for Multinational Enterprises: A Key Corporate Responsibility Instrument, OECD Policy Brief, June 2003.

[55]   I.2 of the Guidelines.

[56]   See note 44 above for the identity of the 8 non-OECD countries.

to adhere. Each additional country that adheres expands the reach of the Guidelines' recommendations to companies that operate in or from the new adhering country.

A precise definition of multinational enterprise is not provided. However, the Guidelines state that multinational enterprises "usually comprise companies or other entities established in more than one country and so linked that they may co-ordinate their operations in various ways."[57] The ownership of the enterprises – whether it is private, state or mixed – is irrelevant.[58] In addition, when the Guidelines apply to a particular multinational enterprise, they generally apply to all the entities it contains – parent companies and/or local entities.[59] Moreover, the Guidelines do not only apply to large enterprises: the adhering governments also encourage small and medium-sized enterprises to observe the Guidelines' recommendations to the fullest possible extent.[60]

Neither are the Guidelines aimed at introducing differences of treatment between multinational and domestic enterprises: they are intended to reflect good practice for all. Multinational enterprises and domestic enterprises are therefore subject to the same expectations in respect of their conduct wherever the Guidelines are relevant to both.[61]

How they are implemented

While the Guidelines are voluntary for companies, they are binding on the governments that have adhered to them. These governments are required to establish a National Contact Point (NCP) – typically a government office[62] – to encourage observance of the Guidelines and ensure that they are known and understood by the national business community and other interested persons.[63] An NCP's responsibilities include: promoting the Guidelines, handling enquiries about them, gathering information on national experiences with the Guidelines, and reporting annually to the OECD Committee on International Investment and Multinational Enterprises (CIME).

The Guidelines' Procedural Guidance provides for a facility that allows interested parties to call a company's alleged non-observance of the Guidelines' recommendations (called a "specific instance") to the attention of an NCP. NCPs are required to offer a forum for discussion and assist the business community, employee organisations, civil society organisations and other parties concerned to deal with the issues raised.[64] The procedures to be used in providing this assistance are set out in Box 26.2.

---

[57]   I.3 of the Guidelines.

[58]   *idem*

[59]   *idem*

[60]   I.5 of the Guidelines.

[61]   I.4 of the Guidelines.

[62]   There are four types of NCP structure presently in use: single government office, multi-departmental government office, tripartite body, and quadripartite body.

[63]   Decision of the OECD Council, June 2000.

[64]   Involving a range of persons and organisations in trying to resolve a problem is consistent the emphasis, in the International Plan of Action to Prevent, Deter and Eliminate Illegal, Unreported and Unregulated Fishing, 9.1 Participation and co-ordination.

Box 26.2. Procedures to be Followed by NCPs in Handling Specific Instances Raised under the OECD Guidelines for Multinational Enterprises

The NCP will:

1. Make an initial assessment of whether the issues raised merit further examination and respond to the party or parties raising them.

2. Where the issues merit further examination, offer good offices to help the parties involved to resolve the issues. For this purpose, the NCP will consult with these parties and where relevant:

   a) Seek advice from relevant authorities, and/or representatives of the business community, employee organisations, other non-governmental organisations, and relevant experts.

   b) Consult the National Contact Points in the other country or countries concerned.

   c) Seek the guidance of the CIME if it has doubts about the interpretation of the Guidelines in particular circumstances.

   d) Offer, and with the agreement of the parties involved, facilitate access to consensual and non-adversarial means, such as conciliation or mediation, to assist in dealing with the issues.

3. If the parties involved do not reach agreement on the issues raised, issue a statement, and make recommendations as appropriate, on the implementation of the Guidelines.

4.   a) In order to facilitate resolution of the issues raised, take appropriate steps to protect sensitive business and other information. While the procedures under paragraph 2 are underway, confidentiality of the proceedings will be maintained. At the conclusion of the procedures, if the parties involved have not agreed on a resolution of the issues raised, they are free to communicate about and discuss these issues. However, information and views provided during the proceedings by another party involved will remain confidential, unless that other party agrees to their disclosure.

   b) After consultation with the parties involved, make publicly available the results of these procedures unless preserving confidentiality would be in the best interests of effective implementation of the Guidelines.

5. If issues arise in non-adhering countries, take steps to develop an understanding of the issues involved, and follow these procedures where relevant and practicable.

In addition to the official OECD Guidelines Procedural Guidance, NGOs and labour organisations have produced manuals to assist those wishing to raise a specific instance to know how to do so and what to expect.[65] Some NCPs have also developed their own more detailed procedures.[66]

---

[65]   See, for example, Friends of the Earth Netherlands, *Using the OECD Guidelines for Multinational Enterprises: A Critical Starter Kit for NGOs*, Amsterdam, Friends of the Earth Netherlands, 2002, available at www.foenl.org/publications/TK_ENG_DEF.PDF; Trade Union Advisory Committee (TUAC) to the OECD, *A User's Guide to the OECD Guidelines for Multinational* Enterprises, available at www.tuac.org/publicat/guidelines-EN.pdf.

The CIME also plays an important role in implementation. It is the OECD body responsible for overseeing the functioning of the Guidelines. It also supervises OECD research projects on the role and use of the Guidelines in particular contexts, and can issue clarifications on their application. It also regularly consults with a range of stakeholders – including business, labour and NGOs – on matters relating to the Guidelines and their implementation, as well as on other issues concerning international investment and multinational enterprises.

How they could be used to tackle IUU fishing

Corporate irresponsibility is clearly an important component of IUU fishing. Corporate entities are implicated in a wide range of IUU-related or IUU-facilitating activities. For example,

- Conducting IUU activities and registration of flags of convenience vessels engaged in IUU fishing.[67]

- Supporting IUU fishing by purchasing illegally caught fish.[68]

- Enabling or facilitating IUU fishing by providing products and services to vessels and persons involved in IUU fishing.

The part of the Guidelines that is most obviously relevant to the problem of IUU fishing is the environmental chapter, which broadly reflects the principles and objectives contained in the Rio Declaration on Environment and Development, in Agenda 21; the (Aarhus) Convention on Access to Information, Public Participation in Decision-making and Access to Justice in Environmental Matters; and the standards in instruments such as the ISO Standard on Environmental Management Systems. However, the Guidelines contain a number of general corporate responsibility principles of potential relevance to the fight against IUU fishing. In some cases, IUU activities may directly contradict a recommendation. For example, the Guidelines recommend that enterprises:

- Comply with local laws and policies: Enterprises are asked to take fully into account established policies in the countries in which they operate, and to consider the views of other stakeholders.[69] IUU activity, especially where it contravenes the law or policies on commercial fishing, may amount to a failure to observe this recommendation.

- Contribute to economic, social and environmental progress with a view to achieving sustainable development.[70] The environmental harm caused by IUU activities is the contrary of environmental progress.

---

[66] For example, the Australian NCP has developed and posted on their website at www.ausncp.gov.au procedures for raising a specific instance. These incorporate, but also build on, the OECD Guidelines Procedural Guidance.

[67] See S.J. Cripps, A. Oliver and J. Cator, "International aspects of the control and eradication of IUU fishing – an NGO's perspective," Fisheries Monitoring, Control and Surveillance, Brussels 24-27 October 2000.

[68] See S.J. Cripps, A. Oliver and J. Cator, "International aspects of the control and eradication of IUU fishing – an NGO's perspective," Fisheries Monitoring, Control and Surveillance, Brussels 24-27 October 2000.

[69] II of the Guidelines.

[70] II.1 of the Guidelines.

- Respect the human rights of those affected by their activities.[71] IUU activities may impact the enjoyment of a range of human rights, especially the economic rights of those whose livelihoods are detrimentally affected by IUU activities.

- Refrain from seeking or accepting exemptions not contemplated in the statutory or regulatory framework related to environmental, health, safety, labour, taxation, financial incentives, or other issues.[72] Some IUU activity may involve bribery and other such conduct that may be inconsistent with this recommendation.

- Within the framework of laws, regulations and administrative practices in the countries in which they operate, and in consideration of relevant international agreements, principles, objectives, and standards, take due account of the need to protect the environment, public health and safety, and generally to conduct their activities in a manner contributing to the wider goal of sustainable development.[73] IUU activity will not normally be consistent with taking due account of the need to protect the environment, or be consistent with sustainable development.

- Continually seek to improve corporate environmental performance.[74] IUU fishing is clearly a step in the opposite direction.

- Encourage, where practicable, business partners, including suppliers and sub-contractors, to apply principles of corporate conduct compatible with the Guidelines.[75] This recommendation, concerning supply chain, is of particular relevance in the IUU fishing area given calls for enterprises to stop doing business with IUU fishing vessels and companies.[76]

- Not, directly or indirectly, offer, promise, give, or demand a bribe or other undue advantage to obtain or retain business or other improper advantage.[77] IUU fishing may require bribery.[78]

- The chapter on disclosure provides, among other things, that enterprises should ensure that timely, regular, reliable and relevant information is disclosed regarding their activities,

---

[71]   II.2 of the Guidelines.

[72]   II.5 of the Guidelines.

[73]   V of the Guidelines.

[74]   V.6 of the Guidelines.

[75]   II.10 of the Guidelines.

[76]   See, for example, M. Gianni "Recommendations to OECD Countries on Measures to Prevent and Eliminate the Problem of Illegal, Unreported and Unregulated Fishing, p. 7. See also Communication from the Commission, Community action plan for the eradication of illegal, unreported and unregulated fishing, 28 May 2002, p.5, action 3 concerning the control of activities associated with IUU fishing. Paragraph 73 of the *International Plan of Action to Prevent, Deter and Eliminate Illegal, Unreported and Unregulated Fishing* also ask States to take measures to ensure that importers, transhippers, buyers, consumers, equipment suppliers, bankers, insurers, and other services suppliers and the public are aware of the detrimental effects of doing business with vessels identified as engaged in IUU fishing.

[77]   VI of the Guidelines.

[78]   D.J. Agnew and C.T. Barnes "The Economic and Social Effects of IUU/FOC Fishing," February 2003, para. 4.6.4. observe that corruption is a significant factor in gaining IUU access to EEZ waters in various parts of the world.

structure, financial situation and performance."[79] It also asks enterprises to disclose basic information showing their name, location, and structure, the name, address and telephone number of the parent enterprise and its main affiliates, its percentage ownership, direct and indirect in these affiliates, including shareholdings between them.[80] The availability of more information about ownership and control of vessels and companies engaged in IUU fishing could help better deal with the problem.[81]

- When dealing with consumers, act in accordance with fair business, marketing and advertising practices and take all reasonable steps to ensure the safety and quality of the goods and services they provide. In particular, they should not make representations or omissions, nor engage in any other practices, that are deceptive, misleading, fraudulent, or unfair.[82] IUU-related activities may entail using deceptive packaging and other practices to deceive consumers and the authorities as to the origin and nature of the fish concerned.[83]

- Contribute to the public finances of host countries by making timely payment of their tax liabilities, complying with the tax laws and regulations of all countries in which they operate, exerting every effort to act in accordance with both the letter and spirit of those laws and regulations, and providing to the relevant authorities the information necessary for the correct determination of taxes to be assessed.[84] Enterprises engaged in IUU fishing are unlikely to be paying their full share of taxes or to be providing to the relevant authorities full information about their activities.

Many IUU-related or IUU-facilitating activities may thus amount to a failure to observe the Guidelines, and could be raised as a specific instance with an NCP. Any interested person could do so – for example, it could be raised by an NGO concerned about IUU fishing, a competitor company, an employee, a concerned coastal community etc. Box 26.2, above, sets out the procedures that an NCP should follow in dealing with a specific instance brought to its attention.[85]

In addition, there are a number of recommendations in the Guidelines that are perhaps less clearly relevant in dealing with IUU fishing, but, if followed, would nevertheless help minimise the likelihood that enterprises would or even could engage in IUU fishing. These recommendations, also drawn from the ten Guidelines chapters, include techniques such as disclosure, communication, training,

---

79   III.1 of the Guidelines.

80   III.3 of the Guidelines.

81   The desirability of such information is made clear by the FAO's Technical Guideline for Responsible Fishing, no. 9 (Implementation of the International Plan of Action to Deter, Prevent, and Eliminate Illegal, Unreported and Unregulated Fishing), which observes that "The beneficial owners of the vessels, who typically have nationalities that differ from those of their vessels, often succeed in preventing fisheries managers and law enforcement officials from ascertaining their identities."

82   VII.4 of the Guidelines.

83   See D.J. Agnew and C.T. Barnes "The Economic and Social Effects of IUU/FOC Fishing" February 2003, observing that enterprises engaged in IUU activities may be motivated to repackage and relabel fish to disguise its origin.

84   X of the Guidelines.

85   See note 68, above, for examples of manuals that civil society groups and labour organisations have prepared to help guide persons wishing to raise a specific instance. Some NCPs have developed their own guidance for persons raising specific instances. See, for example, the guidance referenced in note 69.

management systems, and whistleblower facilities. For example, the Guidelines recommend that enterprises:

- Support and uphold principles of good corporate governance, and develop and apply good corporate governance practices.[86]

- Develop and apply effective self-regulatory practices and management systems that foster a relationship of confidence and mutual trust between enterprises and the societies in which they operate.[87]

- Promote employee awareness of, and compliance with, company policies through appropriate dissemination of these policies, including through training programmes.[88]

- Refrain from discriminatory or disciplinary action against employees who make *bona fide* reports to management or, as appropriate, to the competent public authorities, on practices that contravene the law, the Guidelines or the enterprise's policies.[89] If whistleblowers feel confident that there will be no adverse consequences for reporting IUU activities, they are likely to be more willing to disclose what they know.

- Apply high quality standards for disclosure, accounting and audit, and for non-financial information including environmental and social reporting where they exist.[90]

- Communicate additional information that could include: information on systems for managing risks and complying with laws, and on statements or codes of business conduct.[91]

- Provide the public and employees with adequate and timely information on the potential environment, health and safety impacts of their activities, and engage in adequate and timely communication and consultation with the communities directly affected by the environmental, health and safety policies of the enterprises and by their implementation.[92]

- Maintain contingency plans for preventing, mitigating, and controlling serious environmental and health damage from their operations, and mechanisms for immediate reporting to the competent authorities.[93]

- Provide adequate education and training to employees in environmental health and safety matters, as well as more general environmental management areas.[94]

- Contribute to the development of environmentally meaningful and economically efficient public policy, for example, by means of partnerships or initiatives that will enhance environmental awareness and protection.[95]

---

[86]   II.6 of the Guidelines.

[87]   II.7 of the Guidelines.

[88]   II.8 of the Guidelines.

[89]   II.9 of the Guidelines.

[90]   III.2 of the Guidelines.

[91]   III.5 b) of the Guidelines.

[92]   V.2 of the Guidelines.

[93]   V.5 of the Guidelines.

[94]   V.7 of the Guidelines.

[95]   V.8 of the Guidelines.

- Promote employee awareness of and compliance with company policies against bribery and extortion through appropriate dissemination of these policies and through training programmes and disciplinary procedures.[96]

- Adopt management control systems that discourage bribery and corrupt practices, and adopt financial and tax accounting and auditing practices that prevent the establishment of "off the books" or secret accounts or the creation of documents which do not properly and fairly record the transactions to which they relate.[97]

The standards articulated in the Guidelines are being promoted and reinforced in a variety of ways by adhering governments. For example, in addition to conferences and mailings to business, at least ten countries refer to the Guidelines as a benchmark for companies applying to their investment guarantee, export credit and investment promotion programmes.[98] Other voluntary standards, including those more directly responsive to the problem of IUU fishing, could perhaps be similarly used. In addition, governments could consider referring to the Guidelines in other related contexts, such as in granting licences and permits.

One of the OECD's core strengths is its creation of consensus-based, behavioural norms for governments and private actors.[99] The Guidelines are helping to shape and reinforce norms of good corporate behaviour in many spheres. In addition to promotional activities carried out by adhering countries and the OECD, their profile has been raised through the recognition they have received in high-level political declarations, such as the 2002 OECD ministerial meeting, the G8's 2002 Africa Action Plan, and the G8 finance ministers' statement in May 2003. They were also referenced in a report by a high-level panel of the United Nations Security Council. However, much more can be done by adhering countries to promote observance of the Guidelines by multinational enterprises. Referring to the Guidelines in the context of responses to IUU fishing could simultaneously help further raise the profile and understanding of the Guidelines as an important benchmark for enterprise behaviour, and focus more attention on IUU fishing as a corporate responsibility problem.

Potential limitations to their use in combating IUU fishing

The Guidelines could contribute to the fight against IUU fishing in a number of ways. There are, however, also some limitations.

The Guidelines articulate standards for responsible behaviour, while enterprises engaged in IUU fishing are likely to be the antithesis of responsible. Not only is any IUU activity irresponsible in itself, but the entities concerned also often fail to observe a range of other standards, for example taxation, health and safety, and labour conditions. Enterprises such as these, which deliberately act in an irresponsible way, are unlikely to be moved by the mere existence of the Guidelines. Nevertheless, through the supply chain recommendation, and measures by adhering countries to link the Guidelines to a variety of authorisations etc., there is scope for impacting even these enterprises. In addition, many other companies, perhaps less directly involved in IUU activities, either as an uninformed customer or as a supplier of products or services that unwittingly facilitates IUU activities, may be

---

[96]   VI.4 of the Guidelines.

[97]   V.5 of the Guidelines.

[98]   The OECD Guidelines for Multinational Enterprises: A Key Corporate Responsibility Instrument, OECD Policy Brief, June 2003.

[99]   "Multinational Enterprises and the Quality of Public Governance: A Case Study of Extractive Industries," in *International Investment Perspectives*, No. 1 2002, p. 110.

more susceptible to the pressures that adhering countries could bring to bear on them through the Guidelines process.

For a range of reasons, determining which IUU activities amount to a non-observance of the Guidelines is a matter of interpretation. On some topics – for example Employment and Industrial Relations – the Guidelines are quite specific. However, on many other topics, the Guidelines are very general. IUU fishing is not explicitly mentioned (nor is any other environmental issue) and thus to credibly claim that any particular IUU activity amounts to a failure to observe the Guidelines – and thus invoke the specific instance procedures – is likely to require a detailed analysis of the activity and the Guidelines.

The Guidelines apply to enterprises operating in or from an adhering country. Thus, while they apply to adhering country enterprises wherever they operate, they only apply to non-adhering country enterprises when they are operating within the territory of an adhering country. This means that a non-adhering country enterprise engaged in IUU fishing on the high seas or in the waters of another non-adhering country is beyond the reach of the Guidelines procedures. Since a significant amount of IUU fishing occurs in the waters of developing countries (most of which have not yet adhered to the Guidelines) and is carried out by vessels flying the flags of other countries, which are also not usually adhering countries, this may mean that a lot of IUU activity is beyond the reach of the Guidelines procedures. However, the Guidelines and their procedures could still be relevant if an adhering country enterprise is involved in some way, for example through the supply chain[100] or through its beneficial interest in the IUU activity. The Guidelines might also apply to distribution, that is, to customers, especially insofar as they can be construed as business partners.[101] The principles and standards that the Guidelines contain could thus be used in engaging in dialogue with enterprises and industry associations to encourage them to act responsibly in providing their products and services – such as equipment, banking, insurance – to those likely to be engaging in IUU activities.[102]

Around 64 specific instances have been raised since the Guidelines were reviewed in 2000. It can take several months or even longer for an NCP to handle a specific instance to its resolution. However, NCPs are now focusing on improving the transparency and effectiveness of the Guidelines procedures.[103] Without attracting additional resources, they are unlikely to have the capacity to handle a sudden surge in the number of specific instances. The Guidelines and their procedures can help

---

[100] Some examples might be where an adhering country enterprise has a business partner, including a supplier or sub-contractor, who is involved in IUU and yet does not encourage them to stop/adopt principles of corporate conduct compatible with the Guidelines.

[101] Note, however, that a Roundtable on Corporate Responsibility: Supply Chains and the OECD Guidelines on Multinational Enterprises, held at the OECD in June 2002 did not discuss in any detail concerns about how customers might be using an enterprise's products or services and what the responsibilities of a supplier might be in this context. See OECD Guidelines for Multinational Enterprises: Focus on Responsible Supply Chain Management, Annual Report 2002.

[102] Another potential starting point for a dialogue about customers is contained in the environmental chapter of the Guidelines. It is perhaps most likely to be relevant where the product or service itself may have direct environmental impacts rather than where it only enables an activity like IUU. V.6.c recommends that enterprises should continually seek to improve corporate environmental performance, by encouraging, where appropriate, such activities as promoting higher levels of awareness among customers of the environmental implications of using the products and services of the enterprise.

[103] The OECD Guidelines for Multinational Enterprises: A Key Corporate Responsibility Instrument, OECD Policy Brief, June 2003.

resolve the particular specific instances under consideration, but as a mechanism for bringing about broader change, they suffer from many of the same limitations as other attempts to tackle IUU fishing: lack of resources and lack of political will.

Convention on combating bribery of foreign officials in international business transactions

What it is

The Convention on Combating Bribery of Foreign Officials in International Business Transactions (the Convention) requires states parties to criminalise the bribery of foreign public officials. The offences concerned include intentionally offering, promising or giving a bribe, or complicity in or authorisation of such a bribe.[104] The Convention also requires parties to take measures to prohibit the establishment of off-the-books accounts and other such accounting techniques for the purpose of bribing foreign public officials or of hiding such bribery.[105] A further reinforcing measure is found in a related text, the 1996 Recommendation on the Tax Deductibility of Bribes to Foreign Public Officials (the 1996 Recommendation), which urges member countries that do not disallow the deductibility of bribes to deny this deductibility.[106]

The Convention also contains a number of provisions designed to assist with its implementation. For example, it calls for parties to provide legal assistance to each other to enable investigations and proceedings, and deems bribery of a foreign public official to be an extraditable offence.[107] The Convention, together with the 1996 Recommendation and the 1997 revised Recommendation on Combating Bribery in International Business Transactions, aim to eliminate the supply of bribes to foreign public officials. The Convention entered into force on 15 February 1999.

---

[104]   Article 1(1) and (2) of the Convention provide as follows:

1. Each Party shall take such measures as may be necessary to establish that it is a criminal offence under its law for any person intentionally to offer, promise or give any undue pecuniary or other advantage, whether directly or through intermediaries, to a foreign public official, for that official or for a third party, in order that the official act or refrain from acting in relation to the performance of official duties, in order to obtain or retain business or other improper advantage in the conduct of international business.

2. Each Party shall take any measures necessary to establish that complicity in, including incitement, aiding and abetting, or authorisation of an act of bribery of a foreign public official shall be a criminal offence. Attempt and conspiracy to bribe a foreign public official shall be criminal offences to the same extent as attempt and conspiracy to bribe a public official of that Party.

[105]   Article 8(1) of the Convention provides that: In order to combat bribery of foreign public officials effectively, each Party shall take such measures as may be necessary, within the framework of its laws and regulations regarding the maintenance of books and records, financial statement disclosures, and accounting and auditing standards, to prohibit the establishment of off-the-books accounts, the making of off-the-books or inadequately identified transactions, the recording of non-existent expenditures, the entry of liabilities with incorrect identification of their object, as well as the use of false documents, by companies subject to those laws and regulations, for the purpose of bribing foreign public officials or of hiding such bribery.

[106]   Recommendation of the Council on the Tax Deductibility of Bribes to Foreign Public Officials, adopted by the OECD Council on 11 April 1996.

[107]   Articles 9 and 10 of the Convention.

To whom it applies

The Convention applies to those countries that have ratified it. To date, 35 countries have deposited their instrument of ratification with the Secretary-General of the OECD.[108] Under the Convention, each signatory state is responsible for the activities of its nationals and bribery that occurs on its own territory.[109]

How it is implemented

Compliance with the Convention and implementation of the 1997 revised Recommendation is monitored through country reviews conducted under the supervision of the OECD Working Group on Bribery in International Business Transactions, and is divided into two phases: Phase 1 and Phase 2. Phase 1 evaluated whether the legal texts through which participants implemented the Convention met the standard set by the Convention. Phase 2, which is currently under way, is studying the structures put in place to enforce the laws and rules implementing the Convention and to assess their application in practice. Monitoring is also seen as an opportunity to consult on difficulties in implementation and to learn from the solutions found by other countries.

How it could be used to tackle IUU fishing

Though it does not contain any provisions concerning IUU fishing or fisheries in general, the Convention is nevertheless relevant because of the connection between IUU fishing and corruption. In particular, in order to engage in IUU fishing or related activities, it may be necessary to bribe a foreign official.[110] Where a person directly or indirectly involved in bribing is a national of a country that has ratified the Convention, or bribing occurs on that country's territory, that person may be exposing themselves to the risk of prosecution for bribery, as well as to the possibility of civil or administrative sanctions.[111]

---

[108] These countries are OECD countries plus five others: Argentina, Brazil, Bulgaria, Chile, and Slovenia.

[109] Article 4 of the Convention addresses states parties' jurisdiction to prosecute. It provides that:

1. Each Party shall take such measures as may be necessary to establish its jurisdiction over the bribery of a foreign public official when the offence is committed in whole or in part in its territory.

2. Each Party which has jurisdiction to prosecute its nationals for offences committed abroad shall take such measures as may be necessary to establish its jurisdiction to do so in respect of the bribery of a foreign public official, according to the same principles.

3. When more than one Party has jurisdiction over an alleged offence described in this Convention, the Parties involved shall, at the request of one of them, consult with a view to determining the most appropriate jurisdiction for prosecution.

4. Each Party shall review whether its current basis for jurisdiction is effective in the fight against the bribery of foreign public officials and, if it is not, shall take remedial steps..

[110] D.J. Agnew and C.T. Barnes "The Economic and Social Effects of IUU/FOC Fishing," February 2003, para. 4.6.4. observe that corruption is a significant factor in gaining IUU access to EEZ waters in various parts of the world.

[111] Article 3(4) of the Convention. The Commentary to the Convention indicates that the range of possible civil or administrative sanctions, other than non-criminal fines, includes: exclusion from entitlement to public benefits or aid; temporary or permanent disqualification from participation in public procurement or from the practice of other commercial activities; placing under judicial supervision; and a judicial winding-up order.

Beyond the direct application of the Convention and Recommendations to IUU fishing, there are also a number of aspects about the OECD's approach to dealing with the problem of corruption that could be applied in the fight against IUU fishing. For example, just as the Convention encourages countries to prosecute their nationals for bribery even for conduct occurring outside their own territories, a similar approach could perhaps be adopted with respect to IUU activities. Called the nationality principle, under international law States can generally regulate the conduct of their nationals, even when those nationals are abroad.[112] Given that many of the actual owners of IUU fishing vessels are nationals or residents of OECD countries,[113] the same principle could also be used by States as the basis for creating criminal offences connected with IUU activities engaged in by their nationals, whether at home or overseas. Some examples of possible offences that have been suggested include: owning, operating or knowingly working on a IUU fishing vessel as an officer or fishmaster.[114]

The 1997 revised Recommendation is also an interesting potential model for dealing with IUU fishing. It adopts a multidisciplinary approach to the problem, recommending that member countries take a number of diverse measures. In particular, it recommends that each member country examine seven different areas (with a view to tackling the problem on all fronts) and take concrete and meaningful steps to deter, prevent and combat bribery of foreign public officials. Many of these areas may also be relevant in dealing with the problem of IUU fishing. The seven areas of examination are (paraphrasing):

- Criminal laws and their application so as to criminalise bribery.

- Tax legislation, regulations and practice to eliminate any indirect support of bribery through tax deductions.

- Company and business accounting, external audit and internal control requirements and practices so that they are fully used to prevent and detect bribery of foreign public officials in international business.

- Banking, financial and other relevant provisions, to ensure that adequate records are kept and made available for inspection and investigation.

- Public subsidies, licences, government procurement contracts or other public advantages, so that advantages could be denied as a sanction for bribery in appropriate cases.

- Civil, commercial, and administrative laws and regulations, so that such bribery would be illegal.

- International co-operation in investigations and other legal proceedings.[115]

Since strengthening international co-operation in the detection of IUU fishing is often recommended as an important measure in tackling the problem, lessons learned about how to co-operate in the bribery context may be able to be applied in the context of IUU fishing.

---

[112] See, for example, G. Watson "Offenders Abroad: The Case for Nationality-Based Criminal Jurisdiction" 17 *Yale Journal of International Law* 41 (1992).

[113] M. Gianni "Recommendations to OECD Countries on Measures to Prevent and Eliminate the Problem of Illegal, Unreported and Unregulated Fishing, February 2003.

[114] *idem*

[115] Revised Recommendation of the Council on Combating Bribery in International Business Transactions, Adopted by the Council on 23 May 1997.

The 1997 revised Recommendation provides detailed recommendations and guidance to member countries on the subject of accounting requirements, external audit and internal company controls. It articulates a set of principles and recommends that member countries take the steps necessary to bring their laws, rules and practices into line. Among the principles are (paraphrasing):

- Requiring companies to maintain adequate records of receipts and expenses, including identifying what they relate to, and prohibiting off-the-books transactions and accounts.

- Requiring countries to disclose the full range of material contingent liabilities in their financial statements.

- Adequate sanctioning of accounting omissions, falsifications and fraud.

- Considering whether requirements to submit to external audit are adequate.

- Maintain adequate standards to ensure the independence of external auditors.

- Requiring auditors who discover indications of bribery to report this discovery to management and, as appropriate, corporate monitoring bodies.

- Requiring auditors to report indications of bribery to competent authorities.

- Encouraging the development and adoption of adequate internal company controls, including standards of conduct.[116]

A variant of at least some of these principles, which are aimed at enhanced transparency, could be useful in tackling IUU fishing, for example, requiring certain persons to report to appropriate authorities indications of possible involvement in IUU or related activities. Enhanced information about ownership and control of vessels and companies engaged in IUU fishing could help better deal with the problem.[117] A recent OECD report has made a number of recommendations about measures that governments should consider taking to help combat misuse of the corporate form by acting to ensure the availability of information about ownership and control.[118] Among the suggestions made are that governments should consider taking action to:

- Require up-front disclosure of beneficial ownership and control information to the authorities upon the formation of the corporate vehicle.

- Oblige intermediaries involved in the formation and management of corporate vehicles to maintain such information.

- Develop the appropriate law enforcement infrastructure to enable them to launch investigations into beneficial ownership and control when illicit activity is suspected.

Another way in which the Convention and Recommendations could be of interest in the context of IUU fishing is that the country review mechanism for the Convention and 1997 revised Recommendation, or indeed another model of peer review in use at the OECD, might be useful if applied in an IUU fishing context to ascertain and encourage implementation of measures at the national level to combat the problem. The OECD's concept of peer review was introduced and discussed briefly above, under the section on the Codes of Liberalisation.

---

[116]   *Ibid*, V.

[117]   See note 84 above.

[118]   Behind the Corporate Veil: Using Corporate Entities for Illicit Purposes, OECD, 2001.

Conclusion

The introduction and executive summary highlighted some of the main potential contributions or obstacles of the OECD's investment instruments to the fight against IUU fishing. But what practical steps might the OECD Fisheries Committee wish to take, based on the information and analysis contained in this paper? The Fisheries Committee:

- May find it fruitful to co-operate with the CMIT to better understand the work each group is doing and thus harmonise their activities to ensure that any potential for mutual support be realised.

- May wish to explore, in more detail, efforts being taken by other OECD bodies and individual member countries to refer to the OECD Guidelines for Multinational Enterprises as a benchmark for companies applying to investment guarantee, export credit and investment promotion programmes. There are potentially many other IUU-related areas where similar linkages could be made.

- Could use the principles and standards contained in the Guidelines in engaging in dialogue about corporate responsibility for IUU activities.

- May wish to promote the existence of the specific instance procedures under the Guidelines to civil society organisations and other interested persons and groups.

- Could offer to assist any NCP to deal with an IUU-related specific instance raised under the OECD Guidelines for Multinational Enterprises (in the event that one is raised in the future).

- May wish to indicate their support for outreach efforts to expand the reach of the Guidelines by encouraging more non-member countries to recommend respect for the principles and standards contained in the Guidelines to the companies operating in and from their territories.

- Could reference the OECD Declaration and Decisions on International Investment and Multinational Enterprises in encouraging adhering countries to address the impact of investment incentives and disincentives on the drivers of IUU fishing.

- May, using the OECD Convention on Combating Bribery of Foreign Public Officials in International Business Transactions as a model, wish to consider encouraging countries to hold their own nationals responsible for their IUU-related conduct whether engaged in at home or abroad.[119]

- Could encourage prosecution of instances of IUU-related corruption in accordance with the Convention.

- May wish to consider the possibility of developing a peer review mechanism – using already existing OECD peer review mechanisms as a model – to help countries with their efforts to fight IUU fishing.

---

[119] This would be consistent with provisions in the International Plan of Action to Prevent, Deter and Eliminate Illegal, Unreported and Unregulated Fishing, 18, 19 (State Control over Nationals) and 21 (Sanctions).

REFERENCES

Agnew, D.J. and C.T. Barnes (2003), *The Economic and Social Effects of IUU/FOC Fishing.*

Carr, C.J. and H.N. Scheiber (2003), "Dealing with a Resource Crisis: Regulatory Regimes for Managing the World's Marine Fisheries" UCIAS Edited Volume 1, *Dynamics of Regulatory Change: How Globalization Affects National Regulatory Policies.*

Cripps, S.J., A. Oliver and J. Cator, (2000), "International aspects of the control and eradication of IUU fishing – an NGO's perspective," Fisheries Monitoring, Control and Surveillance, Brussels 24-27.

Commentaries on the Convention on Combating Bribery of Officials in International Business Transactions, adopted by the Negotiating Conference on 21 November 1997.

FAO (2001), (Food and Agriculture Organization), *International Plan of Action to Prevent, Deter and Eliminate Illegal, Unreported and Unregulated Fishing*, Rome 2001.

FAO, (2002), *Implementation of the International Plan of Action to Deter, Prevent, and Eliminate Illegal, Unreported and Unregulated Fishing*, Technical Guideline for Responsible Fishing, no. 9.

Friends of the Earth Netherlands, (2002), *Using the OECD Guidelines for Multinational Enterprises: A Critical Starter Kit for NGOs*, Amsterdam, Friends of the Earth Netherlands, available at www.foenl.org/publications/TK_ENG_DEF.PDF.

Gianni, M. (2003), "Recommendations to OECD Countries on Measures to Prevent and Eliminate the Problem of Illegal, Unreported and Unregulated Fishing" Endangered Seas Programme.

MAI and the Environment, Note by the Chairman, Negotiating Group on the Multilateral Agreement on Investment (MAI), 9 October 1996 (DAFFE/MAI(96)30).

OECD (1999), Convention on Combating Bribery of Foreign Public Officials in International Business Transactions and related Documents, (DAFFE/IME/BR(97)20.

OECD (2000), *No Longer Business as Usual: Fighting Bribery and Corruption*, OECD, Paris.

OECD (2001), *Behind the Corporate Veil: Using Corporate Entities for Illicit Purposes*, OECD Paris

OECD (2001), *Corporate Responsibility: Private Initiatives and Public Goals*, OECD, Paris.

OECD (2001), *Guidelines for Multinational Enterprises: Global Instruments for Corporate Responsibility, Annual Report 2001*, OECD, Paris

OECD (2002), *Guidelines for Multinational Enterprises: Focus on Responsible Supply Chain Management, Annual Report 2002*, OECD, Paris.

OECD (2002), "Successful Capital Movements Liberalisation: A Question of Governance – Recent OECD Experience" in *International Investment Perspectives*, No. 1, OECD, Paris.

OECD (2002), "Multinational Enterprises and the Quality of Public Governance: A Case Study of Extractive Industries" in *International Investment Perspectives*, No. 1, OECD, Paris.

OECD (2003), Code of Liberalisation of Capital Movements, OECD Paris.

OECD (2003), *Codes of Liberalisation of Capital Movements and Current Invisible Operations: Users' Guide*, OECD Paris.

OECD (2003), *Guidelines for Multinational Enterprises: A Key Corporate Responsibility Instrument, OECD Policy Brief, June 2003*, OECD, Paris.

OECD (2003), Peer Review: A Tool for Co-operation and Change – An Analysis of an OECD Working Method, OECD, (SG/LEG(2002)1).

OECD (2003), "Steps Taken and Planned Future Actions by Participating Countries to Ratify and Implement the Convention on Combating Bribery of Foreign Public Officials in International Business Transactions".

OECD (2003), The Environmental, Economic and Social Issues and Effects of IUU/FOC Fishing Activities in the High Seas, (AGR/FI(2003)5).

OECD Declaration and Decisions on International Investment and Multinational Enterprises: Basic Texts, OECD 2000 (DAFFE/IME(2000)20.

TUAC (Trade Union Advisory Committee to the OECD), *A User's Guide to the OECD Guidelines for Multinational* Enterprises, available at www.tuac.org/publicat/guidelines-EN.pdf.

von Moltke, K. "The Environment and Non-Discrimination in Investment Regimes: International and Domestic Institutions" International Sustainable and Ethical Investment Rules Project, November 2002.

Watson, G. (1992), "Offenders Abroad: The Case for Nationality-Based Criminal Jurisdiction" *Yale Journal of International Law*.

CHAPTER 27

MEASURES TAKEN BY CHINESE TAIPEI IN COMBATING FOC/IUU FISHING[1]

David Chang, Fisheries Development Council International, Chinese Taipei

Introduction

On June 23, 2001 the FAO Council endorsed an International Plan of Action to Prevent, Deter and Eliminate Illegal, Unreported and Unregulated Fishing (IPOA-IUU). The content of IPOA-IUU in principle emphasises the implementation of the relevant international agreements and observance of the Code of Conduct for Responsible Fisheries by flag States, coastal states and port states. It also recommends that through participation in various regional fisheries management bodies and exchange of information, States co-operate with each other in combating IUU fishing activities. As the success of IPOA-IUU depends greatly on the extent of co-operation by individual States, it is therefore important to provide a framework for States that are willing to exert their efforts to combat and eliminate IUU fishing. Recently the issue of FOC/IUU fishing of large-scale tuna longline vessels has focused to an extent on Chinese Taipei. This paper tries to provide the background information of FOC/IUU fishing of large-scale tuna longline vessels, and summarise measures taken by the Chinese Taipei Government in combating and eliminating FOC/IUU fishing by large-scale tuna longline vessels.

Background of FOC/IUU fishing of large-scale tuna longline vessels

The large-scale tuna longline fishery in Chinese Taipei has a history going back nearly half a century. According to the existing fisheries law in Chinese Taipei, all the fishing vessels are owned by its citizens, who are required to apply for fishing licenses and observe fisheries laws and regulations as promulgated by the government. In 1989, a policy on limited entry was implemented. In other words, building of vessels is only permitted after the scrapping of an old vessel or decommission of a vessel due to an incident, on a one-ton-to-one-ton basis. The total tonnage of vessels is therefore controlled at a certain level and will not increase. In order to reduce the fleet size, between 1991 and 1995 an overall vessel reduction programme was launched, during which a total of 2,337 fishing vessels of various sizes was scrapped, among which 136 were longliners over 100 GRT. In 1995 a further measure was adopted to forbid new vessel building when a licensed vessel has been exported, to avoid a further increase of the global size of the tuna longline fleet.

---

[1]   This paper was submitted as a background document to the Workshop.

However, the price of tuna in Japan increased significantly during the period from the end of 1980 and the beginning of 1990. To increase the supply of tuna to meet the demand in the Japanese market, the export of secondhand fishing vessels by Japan triggered the Chinese Taipei idea of buying these secondhand vessels, and operating on the high seas. As the exportation of Japanese secondhand longline vessels could not meet the growing demand, in 1995 the operators in Chinese Taipei started building new tuna longline vessels in the local shipyards.

Most of these vessels – including secondhand Japanese vessels and Chinese Taipei-built new vessels – were registered in countries such as Belize, Cambodia, Honduras and Equatorial Guinea, as flag of convenience (FOC) vessels, where the management and control of vessels were lenient or even non-existent. The operators were not required to provide catch reports to the flag States. Even worse, the flag States of FOC vessels did not join regional fisheries management organisations and hence did not comply with the conservation and management measures adopted by these organisations, causing problems in the management of fisheries resources and unfair competition in fish trade. Such irresponsible FOC/IUU fishing activities have become a focal point of international community concern, especially the regional fisheries management organisations. To achieve the goal of sustainable utilization of tuna resources in the Atlantic, ICCAT adopted a resolution calling Japan and Chinese Taipei to work together to combat and eliminate the FOC/IUU fishing by large-scale tuna longline vessels.

Efforts in combating FOC/IUU fishing by large-scale tuna longline vessels

In view of the serious impact of FOC/IUU fishing by large-scale tuna longline vessels on fisheries resources, which undermines the effectiveness of management measures adopted by regional fisheries management organisations, for the past few years the Chinese Taipei government has taken the following measures to effectively combat and eliminate FOC/IUU fishing:

Joint effort of Chinese Taipei and Japan on the elimination of FOC/IUU fishing by large-scale tuna longline vessels

In February 1999, in response to the ICCAT resolution, Chinese Taipei and Japan signed an Action Plan under which Japan was to scrap those secondhand longline vessels it exported, and Chinese Taipei was to encourage those longline vessels built in its shipyards to acquire registration so that they would be properly managed and controlled. To effectively implement the content of the Action Plan, the Chinese Taipei government has taken the following efforts and measures:

Establishment of a mechanism for Chinese Taipei-built FOC/IUU vessels under Chinese Taipei registration

In 2001 and 2003, the Chinese Taipei Fisheries Agency amended the fisheries law and regulation prohibiting the importation of any type of fishing vessels, making allowance for the Chinese Taipei–built FOC/IUU vessels to acquire Chinese Taipei registration. Later, the Chinese Taipei Fisheries Agency promulgated the procedure for the importation of non-Chinese Taipei registered fishing vessels which were built in Chinese Taipei and operated by Chinese Taipei, to allow those Chinese Taipei-built FOC/IUU vessels to apply for registration.

During the transition period before these Chinese Taipei-built vessels complete the registration process, they are required to submit catch reports and install VMS on board, in order to provide a linkage with our fisheries authority in preparation for genuine control over these vessels. To avoid further growth of the Chinese Taipei tuna longline fleet, such re-registered Chinese Taipei-built tuna longline vessels still must comply with the policy of limited entry. In other words, it is required to scrap an old vessel when any re-registered vessel comes back to Chinese Taipei. In this way, the total number of Chinese Taipei large-scale tuna longline vessels will not increase.

Providing financial support to Japan's scrapping programme and assisting Japan in consulting with FOC/IUU vessel owners

To assist Japan in achieving the goals of its scrapping programme, at Japan's request, the Chinese Taipei Fisheries Agency agreed with the commitment made by the owners of legitimate tuna longline vessels under Chinese Taipei registration to provide financial support to Japan's scrapping programme. In addition, again at the request of Japan, Chinese Taipei also arranged at least nine rounds of consultations between Japanese delegates and Japan-built longline vessel owners, encouraging them to join the scrapping programme.

At the moment, 48 Chinese Taipei large-scale tuna longline vessels have obtained registration in Chinese Taipei, while Japan has purchased 42 secondhand longline vessels it exported for scrapping. In addition, a new joint action plan between Chinese Taipei and Japan, agreed in April 2003, concluded a special arrangement between the two countries, in co-operation with Vanuatu and Seychelles, to legitimize 69 FOC/IUU tuna longline vessels. And so, after years of joint efforts by Chinese Taipei and Japan, almost all the FOC/IUU large-scale tuna longline vessels have been scrapped, registered or legitimized.

Measures taken domestically in combating FOC/IUU fishing by large-scale tuna longline vessels

1. Prohibiting the export of fishing vessels to countries that are subject to trade sanctions imposed by regional fisheries management organisations, due to the operation of IUU fisheries by means of FOC vessels.

2. Prohibiting fishing vessels on the IUU list or registered under countries subject to trade sanctions to enter into the port of Chinese Taipei.

3. In addition, the Chinese Taipei Fisheries Agency has also made efforts to combat FOC/IUU fishing, such as educating the general public, vessel owners and shipyards against becoming involved in FOC/IUU fishing activities, and fishermen have been advised not to work on a FOC/IUU vessel. Information also has been provided to local banks, convincing them not to provide credits for the construction of new FOC/IUU vessels.

Co-operation with regional fisheries management organisations

Despite the efforts exerted by both Chinese Taipei and Japan in solving the problem of FOC/IUU fishing, without effective global constraint to discourage such activities, FOC/IUU fishing by large-scale tuna longline vessels will continue. Therefore, unless the international community takes appropriate action to refuse imports by market countries on tuna catch from IUU fishing, refuse port entry by port states to IUU vessels and refuse registration of IUU vessels by all States to prevent them from flag-hopping, the joint efforts of Chinese Taipei and Japan will not be effective. Therefore, curbing IUU fishing by large-scale tuna longline vessels requires concerted efforts in compliance with the management measures adopted by regional fisheries management organisations. In this respect, Chinese Taipei has not only co-operated with regional fisheries management organisations, but has also made its greatest effort in implementing those management measures as adopted by the respective regional fisheries management organisations.

For example, Chinese Taipei has provided information on FOC/IUU fishing vessels to the relevant regional fisheries management organisations, and complied with management measures such as the mechanism of a "white list", IUU list, catch certificate and trade documentation, adopted by the respective regional fisheries management organisations. Such information exchange has enabled the parties concerned to work together to combat FOC/IUU fishing. Compliance with management measures also discourages the owners of FOC/IUU vessel to continue their IUU fishing activities.

To sum up, with the efforts made by Chinese Taipei, Japan and respective regional fisheries management organisations, almost all the FOC/IUU large-scale tuna longline vessels have been scrapped, registered or legitimized. In my opinion, the major reason for this was that regional fisheries management organisations allowed Chinese Taipei to participate in their work, creating an opportunity for Chinese Taipei to understand the importance of combating and eliminating FOC/IUU fishing. Such increased participation in the work of a regional fisheries management organisation also helped the Chinese Taipei Fisheries Agency to obtain support from Congress.

At the moment, Chinese Taipei is a member of the Commission of WCPFC and IATTC. At present Chinese Taipei is a member of the Extended Commission of CCSBT, and a co-operating non-contracting member of ICCAT. However, for political reasons, the Chinese Taipei fisheries authority could not join and participate in the activities of IOTC.

Conclusion

Chinese Taipei has invested enormous efforts to effectively combat and eliminate FOC/IUU fishing. From its experience in dealing with FOC/IUU fishing by large-scale tuna longline vessels, Chinese Taipei learned that the issue of FOC/IUU fishing is extremely complicated. Not only flag States are involved, but port States (including the state whose transport vessels delivered the catch from IUU fishing) and the market States concerned are all involved, too. Teamwork among states can solve this problem. It is therefore very important that the international community, especially the international and regional fisheries management organisations, continue to support Chinese Taipei's participation in their work, creating more scope for Chinese Taipei participation, especially in the field of fisheries issues, under the auspices of the FAO.

CHAPTER 28

HALTING IUU FISHING: ENFORCING INTERNATIONAL FISHERIES AGREEMENTS[1]

Kelly Rigg, Rémi Parmentier and Duncan Currie,
The Varda Group

Executive summary

The world's fisheries are in crisis. Experts report that 75% are significantly depleted, over-exploited or fully exploited. Behind these statistics are the stories of countless families whose livelihoods have been destroyed as the once-bountiful resources of the oceans have dwindled. Governments generally recognise that there is little time left to act decisively to reverse the trends of the last decades. The question is whether the political will exists (and by extension, whether sufficient resources will be made available) to take the necessary measures to do so.

The most important factor undermining the effectiveness of international co-operation and management of fisheries on straddling and highly migratory stocks and fisheries on the high seas is the prevalence of illegal, unregulated and unreported (IUU) fishing.

Oceana has conducted a detailed study (of which this paper is a summary version) into the legal and regulatory frameworks governing fishing on the high seas which aim to ensure the sustainable management of fisheries resources, but which ultimately perpetuate IUU fishing. It can be concluded from this study that, on paper, there is a complex network of binding and non-binding agreements ('hard' and 'soft' law) which forms a solid basis in international law for promoting the development of sustainable fisheries, and for preventing or eliminating IUU fishing.

In practice, however, there are weaknesses and loopholes, the most important ones being:

- Flags of Convenience (FOC), or open registries, allow unscrupulous operators to avoid any regulation of their activities. They fish anywhere and anytime they want to, in contravention of the regulations put in place by Regional Fisheries Management Organisations (RFMOs) to manage and conserve fish stocks.

---

[1]   This paper was submitted as a background document to the Workshop.

- As one country or region more aggressively acts to deter IUU fishing, activities are displaced to another which is less willing or able to do so. As one flag tightens its registry, vessels simply re-flag to another less restrictive State. And as more States tighten their registers, new FOC countries emerge.

- Transhipping at sea means that vessels need never enter ports with their illegally caught fish. The mingling of illegally and legally caught fish onboard reefers essentially serves to whitewash the contraband fish.

- Monitoring, control and surveillance of the high seas and within the Exclusive Economic Zones (EEZs) of many countries (particularly poorer developing countries) are insufficient to ensure that illegal fishers will be apprehended. Even when they do get caught, bonds and fines are set too low to serve as any kind of deterrent. Such fines are simply considered a cost of doing business; vessels invariably return to the fishing grounds, and carry on as before.

The solutions to these problems are not all easy to implement, but they are clearly identifiable.

The single most effective step to combat IUU fishing would be to close the loophole in international law that allows States to issue flags of convenience to vessels with which they have no genuine link and then fail to exercise control over those vessels. A combination of existing instruments, the negotiation of new instruments, and litigation at the International Tribunal for the Law of the Sea could be used to accomplish this.

1. Unless and until the FOC system is effectively eliminated, it is important that States do everything in their power to prevent, deter and eliminate IUU fishing through the following means:

2. Port State controls: port States must prevent IUU fishing and support vessels from using their harbours for transhipment, resupply and other activities and/or must where possible take action to arrest or detain IUU vessels in the event such vessels enter their ports.

3. Market measures: States must adopt and enforce legislation to make it illegal to import or trade in IUU-caught fish. Moreover, States should make it illegal or otherwise discourage companies (*e.g.* insurers, resuppliers, fishing gear manufacturers) from doing business with companies engaged in IUU fishing.

4. At-sea transhipment: Flag States must make it illegal for their transport vessels to tranship fish caught by vessels engaged in IUU fishing.

5. Companies and nationals: States must make it illegal for their nationals and for companies within their jurisdiction to engage in IUU fishing, including the use of fines, penalties and, as necessary, prison sentences of sufficient severity to deter IUU fishing activities.

6. Comprehensive management regime for the high seas: IUU fishing not only involves illegally fishing within an EEZ or in contravention of any regional fisheries management organisation (RFMO) agreements in place on the high seas. It also includes fishing on the high seas in regions where there is no fisheries management regime in place at all. Fishing (mainly bottom trawling) on seamounts and other deep-sea areas on the high seas, which is largely free of any international management agreement to date, has recently become an issue of international

concern. The UN General Assembly is now calling attention to the problem, and its urgency has been widely recognised by fisheries experts.

This paper is derived from the study, focusing in particular on the issues under discussion at the OECD workshop on IUU fishing and providing a wide variety of policy recommendations to help provide fisheries with a sustainable future.

The existing legal and political framework

An impressive array of conventions, agreements, organisations, laws and other international instruments provides for a system in which sustainable fisheries management should be possible, yet weaknesses inherent in each must still be overcome:

The Law of the Sea Convention: UNCLOS aimed to establish a legal order for the seas and oceans which would facilitate international communication, and promote the peaceful uses of the seas and oceans, the equitable and efficient utilisation of their resources, the conservation of their living resources, and the study, protection and preservation of the marine environment.[2] It also initiated important dispute resolution provisions and in particular established the International Tribunal for the Law of the Sea.[3]

However, it focuses primarily on fishing within the 200 mile EEZ, which was a significant innovation at the time it was negotiated. But now much fishing – particularly of migratory stocks such as tuna and swordfish, and straddling stocks such as cod and turbot as well as deep sea fish such as orange roughy – takes place in international waters. It placed great reliance on the concept of the maximum sustainable yield in managing fisheries, whereas it has become clear that other paradigms are required, and in particular the precautionary principle and a more ecosystem-oriented approach have evolved. Possibly its greatest shortcoming is its heavy reliance on flag States for enforcement of environmental and maritime protection provisions, when it has become evident that some flag States have neither the capacity nor the intention of exercising that control.

FAO Compliance Agreement: It was the first international legally-binding instrument to directly deal with reflagging and other FOC issues, focusing on flag State compliance issues and in particular on strengthening flag State responsibility. It requires parties to control the activities of their flag vessels on the high seas, and ensure that its vessels do not undermine international fishery conservation and management measures. Additionally, flag States must give information to the FAO about high seas fishing vessels.[4]

The Agreement has failed to gain widespread acceptance, which explains why it only came into force in 2003, ten years after its conclusion. It is largely restricted to actions taken by flag States rather than port States, and does not address catches. Its efficacy is limited due to the small number of ratifications, and in particular the failure to ratify of FOC States and other States whose vessels may be involved in IUU fishing.

---

[2] Law of the Sea Convention Preamble.

[3] Law of the Sea Convention Annex VI establishes the Statute of ITLOS.

[4] FAO Compliance Agreement Article VI.

FAO Code of Conduct for Responsible Fisheries:[5] The Code of Conduct, concluded in 1995, is voluntary or 'soft' law. Pursuant to the Code, four International Plans of Action (IPOA) have been developed on seabirds, sharks, managing fishing capacity, and IUU fishing.[6] **The IPOA- IUU**[7] adopted in 2001 aims to prevent, deter and eliminate IUU fishing,[8] and addresses the problem of FOCs particularly in relation to RFMOs. It goes further and is more detailed than the Compliance Agreement and calls on States to take measures to ensure that nationals subject to their jurisdiction do not support or engage in IUU fishing. However, it is still soft law and not legally binding.

The 1995 Fish Stocks Agreement: For the first time, it was established that a precautionary approach is expressly required in fisheries management.[9] States must take conservation measures such as assessing[10] and managing[11] species in the same ecosystem and species associated with or dependent on the target species to protect biodiversity,[12] addressing overfishing and excess fishing capacity,[13] and monitoring and controlling fisheries.[14] It allows for boarding and inspecting vessels on the high seas under certain circumstances, and provides for measures which may be taken by a port State[15] including inspections and prohibition of landings and transhipments.

Finally, it requires States which are not parties to sub-regional or regional fisheries management organisations to nonetheless co-operate in the conservation and management of the relevant fish stocks. Moreover, States parties to the Fish Stocks Agreement, which are not members of the relevant RFMO, may not authorise their flagged vessels to engage in fishing operations for straddling or highly migratory fish stocks.[16]

UN General Assembly (UNGA): UNGA Resolutions which call for a halt to IUU fishing, including FOC practices, are not binding, but they do provide some measure of the recognition of the seriousness of the problem by the international community.

---

[5]   FAO Code of Conduct for Responsible Fisheries. http://www.fao.org/fi/agreem/codecond/ficonde.asp.

[6]   See http://www.fao.org/fi/ipa/ipae.asp. International Plan of Action for Reducing Incidental Catch of Seabirds in Longline Fisheries - 1999, International Plan of Action for the Conservation and Management of Sharks - 1999 and International Plan of Action for the Management of Fishing Capacity - 1999. All three of these texts can be found at:

   http://www.fao.org/docrep/006/x3170e/X3170E00.HTM.

[7]   Food and Agriculture Organization "International Plan Of Action To Prevent, Deter And Eliminate Illegal, Unreported And Unregulated Fishing", (IPOA-IUU) adopted by consensus at the Twenty-fourth Session of COFI on 2 March 2001 and endorsed by the Hundred and Twentieth Session of the FAO Council on 23 June 2001, at:

   http://www.fao.org/DOCREP/003/y1224e/y1224e00.HTM.

[8]   IPOA-IUU III, para. 8.

[9]   Fish Stocks Agreement Articles 5(c) and 6.

[10] Fish Stocks Agreement Articles 5(d).

[11] Fish Stocks Agreement Articles 5(e).

[12] Fish Stocks Agreement Article 5(g).

[13] Fish Stocks Agreement Article 5(g\h).

[14] Fish Stocks Agreement Article 5(l).

[15] Fish Stocks Agreement Article 23.

[16] Fish Stocks Agreement Article 17.

Most recently at the 58th Session of the UN General Assembly, two resolutions on the oceans were passed (available at:

http://www.un.org/Depts/los/general_assembly/general_assembly_resolutions.htm).

The resolution on the Law of the Sea has clear language on flag and port State control including a call for the IMO to further examine and clarify the meaning of establishing a 'genuine link.'

In the resolution on sustainable fisheries, the UN General Assembly has called on States to take action on IUU fishing, in particular to implement the IPOA-IUU. It also calls on the Secretary General, in consultation with the FAO, RFMOs and States to consider the risks to the biodiversity of seamounts and other deep ocean areas.

Legal issues and financial incentives regarding IUU fishing and flags of convenience

The FAO Technical Guidelines on Responsible Fisheries sums up the situation quite succinctly: "IUU fishers must evade detection in order to succeed. As noted above, the operators of IUU vessels often conduct fishing operations in areas where MCS is lacking, particularly in remote high seas regions or in waters under the jurisdiction of coastal States, particularly developing States that do not have the ability to stop such fishing. The owners of these vessels also seek to avoid detection through deceptive business practices. For example, they create extended and complex corporate arrangements to hamper investigators, they repeatedly change the names and call signs of their vessels and they regularly re-flag the vessels in States that continue to maintain open registries."[17]

The UNCLOS requirement that there be a 'genuine link' between the flag State and the vessel or operator is ignored or circumvented under the FOC system. A fishing interest wishing to engage in IUU fishing will usually incorporate a shell company in the flag State, often with bearer shares. Shares in the shell company will then be held by other shell interests, with the real beneficial owner being hidden. Thus even if the State of the national had the will to exercise jurisdiction over the national, the interest of the owner may be well hidden. A look at www.flagsofconvenience.com shows a one-stop shop for flag registration and incorporation of shell companies in offshore jurisdictions.

Beneficial ownership is often in Chinese Taipei, Japan, Korea and European countries. According to a Greenpeace report,[18] Lloyds data for 1999 showed that the greatest number of beneficial ownerships of FOC vessels was held by Chinese Taipei companies, followed by the EU (of which the vast majority was held by Spain/Canary Islands), Singapore, South Korea, Japan and China (leaving aside beneficial interest showing to reside in FOC countries).

Thus control over vessels through the flag is essentially negated by lack of control of FOC flag States and by lack of control over the owners.

IUU fishing is not restricted to traditional FOC countries. Vessels caught in IUU fishing activities for Patagonian toothfish have been sailing under the flags of Russia and Uruguay, as well as Panama.

---

[17]   http://www.fao.org/docrep/005/y3536e/y3536e06.htm#bm06.2.5 Section 3.2.5.

[18]   Greenpeace International, "Pirate Fishing Plundering the Oceans," February 2001, at http://www.greenpeace.org/~oceans/reports/pirateen.pdf, page 20.

Yet despite international concern about illegal fishing activities, and associated effects such as the by-catch of albatross[19] (all 21 species of which are now on the IUCN endangered list), positive action to bring FOC practices to an end has not been forthcoming. Calls to close ports to vessels engaged in IUU fishing and their support vessels, to close markets to fish caught from IUU fishing activities, and to take enforcement action on the international level against such activities have not been sufficiently heeded.

In the meantime, the FOC fishing fleet continues to grow. An International Transport Workers' Federation (ITF) report stated that in the 20 years from 1980 to 2000, the number of open registers grew from 11 to 29.[20] An FAO report from 2002 examines the data from 1997 to 2001, which shows that the number of vessels registered on open registries increased by 208 vessels, to just over 1500 vessels in total (though it is not clear whether the proportion of FOC as a percentage of the global fleet has increased). [21] The shift in specific countries was in some cases dramatic: For example the number of fishing vessels on Belize's register more than tripled during this period, while Panama's decreased by 54% (or 70% by percentage).

In 1999, Greenpeace listed the worst offenders of the FOC countries, accounting for 80% of the flags of convenience, as being Belize (with 404 vessels), Honduras (with 395 vessels), Panama (with 214 vessels), and St. Vincent & the Grenadines (with 108 vessels). Smaller flags were Equatorial Guinea (56 vessels), Cyprus (45 vessels), Vanuatu (34 vessels), Sierra Leone (27 vessels), Mauritius (22 vessels) and the Netherlands Antilles (18 vessels). [22]

FOC vessels undermine fishing conservation and management regimes by taking fish outside quotas, not reporting catches (making assessment difficult), taking by-catch such as non-target birds and species including albatrosses and dolphins, and poaching fish in EEZs which are difficult to police due to isolation or lack of capacity by developing coastal states.

Under the FOC system, there is nothing to prevent ships from changing registries as often as they like, for example in response to countries' efforts to curtail IUU fishing or to better implement the decisions of RFMOs. And this is exactly what IUU vessels regularly do.

States wishing to put a stop to this could impose strict conditions on deregistration of vessels flying their flags. Under Article 91 of the Law of the Sea Convention, every State is required to fix the conditions for granting its nationality to ships, for registering ships in its territory, and for granting them the right to fly its flag. Ships have the nationality of the State whose flag they are entitled to fly. Putting stringent conditions on deregistering ships (as opposed to registering, which is where much of the discussion has been focused) could amount to an implementation of the IPOA-IUU which provides that flag States should deter vessels from re-flagging for the purposes of non-compliance with conservation and management measures or provisions adopted at a national, regional or global level[23]

---

[19]   Greenpeace has estimated that in 2002 alone, up to 93,000 Southern Ocean seabirds – including endangered species of albatross – have been caught and drowned as by-catch by pirate fishers. http://www.greenpeace.org/international_en//press/release?item_id=89498&campaign_id=4022

[20]   As reported in Swan, 2002.

[21]   Swan, Judith, FAO Fisheries Circular No. 980 FIPP/C980, "Fishing Vessels Operating Under Open Registers and the Exercise of Flag State Responsibilities – Information and Options. Rome, 2002. The figures used in the paper were obtained from Lloyd's Maritime Information Services.

[22]   Greenpeace, Dodging the Rules: flags of Convenience fishing, at http://archive.greenpeace.org/oceans/piratefishing/dodgingrules.html

[23]   IPOA-IUU para 38.

and that States should take all practicable steps to prevent "flag hopping".[24] While those measures are directed at flag acquisition, there is nothing to prevent them being applied to deregistration as well.

Means of avoiding detection

Vessels flying flags of convenience, such as the *Salvora*, often carry concealed or no markings to mask their identities at sea.

In areas where VMS systems are in place, hardware, software and data are frequently tampered with.

Transfer of catch on the high seas

Another means by which IUU fishing remains undetected – arguably the biggest loophole in fisheries management agreements – is by vessels rarely or never entering the ports of countries which maintain adequate port State control measures. The largest vessels are able to remain at sea for months at a time (or even years if they are re-supplied at sea), taking more than half of the annual global catch of fish which is simply offloaded to reefer (transport) ships.[25] Transhipment of the catch in this way allows, in essence, a 'whitewashing' of illegal fish by the time it arrives on the market.

Avoiding serious punishment

Penalties for owners, operators, captains and crew of IUU are at present largely financial. This means that the decision to engage in IUU activities is reduced to a cost/benefit analysis, where the calculus involves the probabilities of getting caught, the entry cost, the potential rewards and the penalties if the vessel is caught. In the case of the owner, the probability of any penalty other than the loss of the fishing boat is negligible. In the unlikely event that a fishing boat is arrested, the owner can demand release of the vessel and if the bond set by the arresting state is significant, engage counsel to take a case to ITLOS to have the bond reduced. Most such cases have succeeded, the most recent being the *Volga* in late 2002 where the bond was reduced from AUD 3 332 500 to AUD 1 920 000.[26]

A large bond would help, and in this respect large financial penalties would enable arresting States to justify a higher bond, but ultimately jail time not only for captains, but for beneficial owners, is necessary to act as a real deterrent. At present, the Law of the Sea Convention prohibits imprisonment for violations of fisheries laws and regulations in the EEZ, in the absence of agreements to the contrary by the States concerned.[27] However, this does not preclude States from imposing prison terms for violation of national laws by beneficial owners and those who aid and abet IUU fishing, and imprisonment for captains can be agreed in an MOU or other document between States. UNCLOS Article 73 does not necessarily require agreement by the flag State: agreement by the flag of the national who is to be imprisoned should suffice. Increasing fines is another and a very simple means to increase deterrence (see related recommendations at the end of this chapter).

---

24   IPOA-IUU para 39.

25   Bours, Gianni, Mather, "Pirate Fishing Plundering the Oceans," Greenpeace, February 2001.

26   See discussion of the *Volga* case in section 3.

27   Law of the Sea convention Article 73(3).

Incentives and disincentives for IUU fishing

Incentives

The scale of the problem, and by extension the amount of money which is being made by IUU fishing operations, is poorly understood – given that these people obviously do not report on their activities. Estimates of the scale of IUU fishing can be compiled on the basis of reports by RFMOs (for example 39% of total fishing in the CCAMLR region, 18% for ICCAT) and then extrapolated to estimate global figures.[28] An alternative means of assessing the scale is to compare trade figures (which include IUU fish) and catch data (which does not). This approach suggests that the problem is even worse than is being reported by the RFMOs.[29]

Financial benefits from IUU fishing through FOC practices accrue to at least three different parties: flag States, port States, and fishing companies/vessels

Flag States

The FOC countries, which on the whole are smaller developing countries, earn revenue by charging fishing boats fees to fly their flag. In return, FOC countries turn a blind eye to IUU fishing activities, leaving fishing boats largely free to ignore international laws.

According to a 2002 FAO report[30], the total revenue from registering fishing vessels in 21 countries operating open registries amounted to just over USD 3 million, although this is likely to be an underestimate. While this figure may seem relatively small, it should be noted that fishing vessels represented only 7% of all the vessels registered in these States, and only 4.9% of the income. Given that these States incur few costs from implementing international agreements, the FOC system is clearly a lucrative one from the standpoint of open registry States.

Port States

Las Palmas de Gran Canaria is one of the major ports of convenience.[31] It serves as the main distribution centre for fish caught off Africa, provides services to IUU fleets, and hosts a number of companies which operate pirate vessels.[32] Other such ports include Port Louis, in Mauritius and (historically) Cape Town.

---

[28]   Upton, Simon and Vitalis, Vangelis, "Stopping the High Seas Robbers: Coming to Grips with Illegal, Unreported and Unregulated Fishing on the High Seas," OECD, 2003 at: http://www.oecd.org/dataoecd/15/16/16801381.pdf

[29]   Upton, Simon and Vitalis, Vangelis, *op cit.*

[30]   Swan, Judith, FAO Fisheries Circular No. 980 FIPP/C980, "Fishing Vessels Operating Under Open Registers and the Exercise of Flag State Responsibilities – Information and Options", Rome, 2002. The figures used in the paper were obtained from Lloyd's Maritime Information Services.

[31]   See *The European Union Action Plan to Eradicate IUU Fishing: A Greenpeace Critique*, at: http://web.greenpeace.org/multimedia/download/1/40628/0/pirate_fishing_critique.rtf.

[32]   Greenpeace, "Witnessing the Plunder: A Report on the *MV Greenpeace* Expedition Investigating Pirate Fishing in West Africa," November 2001.

Fishing companies/vessels

Operators have a variety of incentives to engage in IUU fishing in general, and to operate under flags of convenience in particular:

- Avoiding regulatory or legal obligations: The IMO is increasingly stringent in its requirements for the safe operation of vessels. These requirements, including the acquisition of specialised safety gear, accident insurance, and crew training, can be very costly. FOC registration helps keep those costs to a minimum. In addition, by sailing under an FOC flag, operators do not have to pay for licences, VMS, observers, or the administration of Catch Documentation Systems.[33]

- RFMO decisions to restrict access to fishing areas seasonally or year-round means that the most prized fish species are unavailable for certain periods.[34] This is not a problem for FOC vessels.

- FOC registration is quick, easy and cheap: A couple of clicks at www.flagsofconvenience.com and a few hundred or thousand dollars will buy a registration.

- Short-term profit: Bluefin tuna, for example, currently brings fishermen between USD 2 and USD 17 per pound, depending on a variety of factors (quality of the fish, fat content, value of the yen since Japan is the primary market, etc.). These fish weigh upwards of 500 pounds, so a single fish can bring in USD 1 000 to USD 8 500 or more.[35] In the not too distant past, however, a high quality tuna would bring in as much as USD 50-60 per pound, or USD 25 000-30 000 per fish.[36]

- Another highly sought after species, Patagonian toothfish, sells for up to USD 1 000 per fish. In 1997, illegally caught Patagonian toothfish was valued at over USD 500 million.[37]

- To give an example of the scale of the catch taken by individual vessels, the largest super trawlers can process 50-80 tons of fish per day, and have nets capable of catching 400 tons of fish.[38]

---

[33]   Upton, Simon and Vitalis, Vangelis, "Stopping the High Seas Robbers: Coming to Grips with Illegal, Unreported and Unregulated Fishing on the High Seas," OECD, 2003 at: http://www.oecd.org/dataoecd/15/16/16801381.pdf

[34]   Swan, Judith, FAO Fisheries Circular No. 980 FIPP/C980, "Fishing Vessels Operating Under Open Registers and the Exercise of Flag State Responsibilities – Information and Options. Rome, 2002. However, according to this report, it is not always the case that vessels re-flag to countries which are not bound to RFMOs. Spanish fishing vessels, for example, flag out primarily to Honduras, Panama and Morocco, which are members of ICCAT. The main issue appears to be whether a country actively implements those agreements or not.

[35]   http://www.capecodonline.com/cctimes/biz/tunaprices14.htm. The current average price is USD 6-8 per pound. Prices are under pressure due to the increase of tuna-penning: fish are caught, penned, and fed until they are fat enough to bring a good price on the Japanese market.

[36]   http://www.eagletribune.com/news/stories/19980927/FP_001.htm

[37]   Greenpeace International, "Mauritius, Indian Ocean Haven for Pirate Fishing Vessels", March 2000

Disincentives for IUU fishing

At present, there are unfortunately few disincentives for IUU fishing. As RFMOs and their member States tighten agreements, including through the application of trade sanctions, pirate fishing vessels simply change registries, or operate under no flag at all.

Clearly the main disincentive to fish legally is the knowledge that the vessel is unlikely to be caught, and even if it is, that it is unlikely to incur a fine large enough to hurt. For many older fishing vessels, even the threat of impoundment provides little disincentive because their value is minimal.[39] A recent FAO study demonstrates this problem quite clearly:[40]

Table 28.1. Estimating* the Probability of Being Penalised for a Violation at Sea in an OECD Country

Sampled vessels	
Average number of fishing days/yr	257
Perceived average boardings/vessel/year (from interviews)	4
Probability of being boarded/day (from interview/MCS records)	1.56%
All vessels	
Total fishing vessel targets/day (av. of samples from high level radar)*...(a)	195
Boardings per patrol day (1999)..(b)	0.98
Probability of being boarded/day (all vessels) (b/a).........................(A)	0.5%
Probability of detection of violation per boarding (from MCS records)	15%
Probability of detention (arrest) at sea...(B)	3%
Probability of penalty if detained (ratio prosecutions/penalties).............(C)	66%
Probability of paying a penalty in a given fishing day.............(A*B*C)	0.01%

*Actual example from an OECD country.

*Source:*http://www.fao.org/DOCREP/005/Y3780E/y3780e00.htm#Contents

The role of subsidies

The depletion of global fisheries is largely due to over-capacity or overcapitalisation – too many (high-tech) boats catching too few fish. Overcapitalisation is exacerbated by direct and indirect government subsidies to the fishing industry. No distinction has been made up until now between legal and pirate fishermen when it comes to providing subsidies. It is therefore safe to assume that governments are subsidising IUU fishing.

Various studies have attempted to calculate the global level of fisheries subsidies. This is not an easy task, partly because there is no agreed definition of what constitutes a fisheries subsidy (for

---

[38]   Porter, Gareth, "Fisheries Subsidies, Overfishing and Trade" at:
       http://www.sdnbd.org/sdi/issues/environment/article/1.pdf

[39]   ITF, Greenpeace, "More Troubled Waters fishing, pollution and FOCs" August 2002 at:
       http://www.itf.org.uk/publications/pdf/more_troubled_waters.pdf

[40]   Kelleher, Kieran, "The Costs of Monitoring, Control and Surveillance of Fisheries in Developing Countries", FAO Fisheries Circular 976, Rome, 2002
       http://www.fao.org/DOCREP/005/Y3780E/y3780e00.htm#Contents

example should fuel subsidies for all sectors – which are enormous – or port improvements be counted?). Most researchers cite the results of a 1998 study by M. Milazzo which estimates the level at USD14-20 billion per year, or approximately 17-25% of global fishing industry revenues.[41] The worst offenders are reportedly the EU, Japan, and China.[42] Another study breaks it down as follows (reportedly in line with Milazzo's results, as the combined figures suggest a global level of USD 15 billion per year):[43]

- Asia-Pacific Economic Co-operation (APEC) with 21 countries along the Pacific Rim, accounting for 85% of the world's fish catch on a tonnage basis: USD 13 billion (study published in 2000);

- OECD, with 24 fishing countries out of its 30 members: 6.3 billion, with Japan accounting for 2.9, the EU for 1.4, the US for 0.877, Spain 0.345 and Korea 0.342 billion (1997 data).

Even subsidies which purport to promote responsible fishing by encouraging vessel retirement have contributed to overcapacity. Subsidies granted to fishers to retire their old boats are often reinvested in more modern boats with even greater capacity. Even if the total number of boats decreases, there will still be an increase in capacity. This is because the level of capacity of the fleet is not measured by the number of boats, but by fleet tonnage, engine power, and the advanced nature of the fishing gear. Large super trawlers (greater than 1,000 gross tons) with powerful engines can travel greater distances, in worse weather, reaching areas which would otherwise be inaccessible. They are assisted by planes, satellite images and sonar systems which identify concentrations of fish even in depleted fisheries.[44] Moreover, boats which have been retired from one registry or fishery may simply be re-flagged and/or displaced to another.[45, 46]

As Porter describes it: "The main cause of overcapitalisation may be the 'open access' nature of most of the world's marine fisheries. An open access system of management for any resource is one in which no individual producer has the right to exclude any other producer from harvesting or otherwise using any part of the resource. Fishers continue to enter the fishing industry because there are no effective limits on access to the resource. And they maximise their fishing effort because, without any effective property right to the resource, they calculate that fish left in the water will be caught by someone else. Eventually this expansion of aggregate fishing reduces the fish stock, and catch per unit of effort declines, along with economic returns to producers. Producers will continue to increase fishing effort, however, as long as they have hopes of achieving some level of profit. Finally, stocks are reduced to the point that total fishing costs are equal to the value of the harvest and profitability in

41   Milazzo, M. World Bank Technical Paper No. 406 "Subsidies in World Fisheries: A Reexamination," Washington, D.C. 1998.

42   Arnason, Ragnar, "Fisheries subsidies, overcapitalisation and economic losses".

43   Steenblik, Ronald P. and Wallis, Paul F. "Subsidies to Marine Capture Fisheries: The International Information Gap", http://biodiversityeconomics.org/incentives/toics-340-00.htm

44   Porter, Gareth, "Fisheries Subsidies, Overfishing and Trade" http://www.sdnbd.org/sdi/issues/environment/article/1.pdf

45   Swan, Judith, FAO Fisheries Circular No. 980 FIPP/C980, "Fishing Vessels Operating Under Open Registers and the Exercise of Flag State Responsibilities – Information and Options". Rome, 2002.

46   Arnason, Ragnar, "Fisheries subsidies, overcapitalisation and economic losses"

the fishery is zero. Then fishing capacity cease [sic] to increase. But by that time, the fishery is already in a state of serious depletion."[47] IUU fishing is by definition 'open access.'

Moreover, once this process plays out, and the fishing industry heads for crisis, additional subsidies are often granted to ensure survival and thus discourage fishers from withdrawing from the industry. Indeed, while the short-term financial benefits to fishers may be substantial, they are inevitably negated by the loss of profit due to unsustainably high fishing levels.[48]

The EU in particular is saddled with an enormous problem of over-capacity, largely as a result of subsidies for fleet modernisation in the 1970s and 80s. One solution has been for the EU to ensure access for its fleets to distant water fisheries, for example by buying access to the fishing grounds of African countries for example. This in itself represents a subsidy – by 1996, the EU was paying USD 193 million a year to 15 African countries. [49]

With regard to the Mediterranean, a 2001 English Nature report[50] notes that aid under the Financial Instrument for Fisheries Guidance (FIFG) continues to be available to encourage the adoption of more selective fishing gear, but there is no explicit linkage with compliance with technical compliance rules. The FIFG has thus worked to increase fishing capacity, where it should be used to encourage technical measures.

Subsidy reforms

Major fishing countries have been wrestling with options for dealing with the subsidies problem for the last ten years, impeded in part by the problem of defining the term 'subsidy'. An FAO Expert Consultation on the subject was held in December 2000, and concluded that no single definition could be agreed.[51] They did, however, identify four different types of subsidies which could be used as a standard for classification purposes, which have been summarised as follows:

- "Set 1 Subsidies: Government financial transfers that reduce the costs and/or increase the revenues of producers in the short term.

- Set 2 Subsidies: Government interventions – regardless of whether or not they involve financial transfers – that reduce the costs and/or increase the revenues of producers in the short term.

- Set 3 Subsidies: Subsidies in set 3 are set 2 subsidies plus the short-term benefits to producers that result from the absence or lack of interventions by government to correct distortions (imperfections) in production and markets, which can potentially affect fishery resources and trade.

---

47   Porter, Gareth, "Fisheries Subsidies, Overfishing and Trade" http://www.sdnbd.org/sdi/issues/environment/article/1.pdf, page 12-13.

48   Arnason, Ragnar, "Fisheries subsidies, overcapitalisation and economic losses".

49   Porter, Gareth, "Fisheries Subsidies, Overfishing and Trade" http://www.sdnbd.org/sdi/issues/environment/article/1.pdf

50   Clare Coffey for English Nature, Mediterranean Issues: Towards Effective Fisheries Management, 12, at http://www.jncc.gov.uk/marine/fisheries/pdf/Mediterranean2.pdf, 12.

51   FAO, "The State of World Fisheries and Aquaculture 2002, Part 2 Selected issues facing fishers and aquaculturists" at http://www.fao.org/docrep/005/y7300e/y7300e00.htm

- Set 4 Subsidies: Subsidies in set 4 are government interventions, or the absence of correcting interventions, that affect the costs and/or revenues of producing and marketing fish and fish products in the short, medium, or long term."[52]

On the basis of these guidelines, fishing countries are now working to classify subsidies and assess their impacts.

Note: many developing countries are opposed to the elimination of subsidies that they consider necessary for the development of their fishing capacity and industry in general. Developing countries are negatively affected by highly subsidised distant water fishing fleets. At the same time, developing countries are facing a situation where fish stocks are declining, ever stricter management rules and standards are being imposed (which increase the costs of management) and industrialised countries have a quasi-monopoly on access to resources. It is therefore no surprise that they perceive moves to eliminate subsidies as yet another obstacle in getting what they consider to be an equitable a share of the resources.[53]

WTO

The GATS Agreement on Subsidies and Countervailing Measures of 1994, generally known as the Agreement on Subsidies, arose out of the Uruguay Round and provided for the first time a clear definition of a subsidy.[54] Article VI deals with the use of subsidies and the actions that countries may take to counter the adverse effects of subsidies from a third party (countervailing measures). Under the Agreement, a country can use the WTO's dispute-settlement procedure to seek the withdrawal of a subsidy, or it can launch its own review and ultimately charge extra duty on subsidised imports that are found to be distorting its domestic market ("countervailing duty").

The Agreement makes a distinction between Prohibited Subsidies (subsidies that require the recipients to meet certain export targets or to use domestic goods instead of imported ones, which are designed to distort international trade) and Actionable Subsidies (subsidies that have an adverse effect on the interests of the plaintiff – environmental harm is not currently listed as a potentially adverse effect.)[55] Prohibited subsidies can be challenged in the WTO dispute settlement procedure under an accelerated time-table, and if it is determined that the subsidy is indeed prohibited it must be withdrawn immediately. In the case of actionable subsidies, if it is determined that the subsidy has an adverse effect, the country must withdraw it, or modify it so that the adverse effect disappears.[56]

There is a provision in the Agreement which states that countries should not cause 'serious prejudice' to the interests of other members.[57] One of the conditions representing 'serious prejudice' would be for the subsidisation of a product to exceed 5 per cent of the value of the product exported by

---

[52] FAO, "The State of World Fisheries and Aquaculture 2002, Part 2 Selected issues facing fishers and aquaculturists" pages 93-95 at http://www.fao.org/docrep/005/y7300e/y7300e00.htm

[53] Bours, Hélène, personal communication.

[54] Agreement on Subsidies and Countervailing Measures at:
http://www.wto.org/english/docs_e/legal_e/24-scm_01_e.htm

[55] For a comprehensive consideration of the Doha Agenda, see:
http://www.wto.org/english/tratop_e/dda_e/dda_e.htm

[56] "Understanding the WTO: The Agreements" at:
http://www.wto.org/english/thewto_e/whatis_e/tif_e/agrm8_e.htm

[57] Article 5c.

that country – a condition which applies to many fishing subsidies.[58] However, solving the problem by bringing isolated cases before the WTO based on the Subsidies Agreement would be time consuming, costly, and inefficient.

Within the WTO, for several years a group of countries (known in Geneva as the "Friends of the Fish" made up of Australia, Chile, Iceland, New Zealand, Peru, the Philippines and the USA), have been promoting the reduction and/or elimination of fisheries subsidies on the basis that these are trade-distorting, environmentally harmful, and inconsistent with the free-trade *mantra* of the WTO.

Japan, Korea and the European Union – three delegations from countries with highly subsidised fishing fleets and which have consistently denied the existence of a link between over-capacity and high levels of subsidies – have been on the opposing side of this discussion.[59]

At its Fourth Ministerial Conference held in Doha in November 2001, the WTO agreed to put fisheries subsidies on the agenda of the *Doha Round* of trade liberalisation, scheduled (at least before the failed Cancun stock-taking ministerial conference of September 2003) to be concluded on 1 January 2005. Reference is made twice to the elimination of fisheries in the Doha Declaration:

In Paragraph 28 (emphasis added):

> In the light of experience and of the increasing application of these instruments by members, **we agree to negotiations aimed at clarifying and improving disciplines under the Agreements on Implementation of Article VI of the GATT 1994 and on Subsidies and Countervailing Measures**, while preserving the basic concepts, principles and effectiveness of these Agreements and their instruments and objectives, and taking into account the needs of developing and least-developed participants. **In the initial phase of the negotiations, participants will indicate the provisions, including disciplines on trade distorting practices, that they seek to clarify and improve in the subsequent phase**. In the context of these negotiations, participants shall also aim to clarify and improve WTO disciplines on fisheries subsidies, taking into account the importance of this sector to developing countries. We note that **fisheries subsidies are also referred to in paragraph 31**.[60]

And in Paragraph 31, "Trade and Environment", in order to emphasise that the environmentally harmful aspect of fisheries subsidies also forms part of their review (emphasis added):

> With a view to **enhancing the mutual supportiveness of trade and environment**, we agree to negotiations, without prejudging their outcome, on:
>
> (i) the relationship between existing WTO rules and specific trade obligations set out in multilateral environmental agreements (MEAs). The negotiations shall be limited in scope to the applicability of such existing WTO rules as among parties to the MEA in question. The

---

[58]   Article 6, and see discussion in Porter, "Fisheries Subsidies, Overfishing and Trade" http://www.sdnbd.org/sdi/issues/environment/article/1.pdf

[59]   Although after (and in part as a result of) the Doha WTO Conference, the EU undertook its Common Fisheries Policy reform, which provides for the progressive elimination of some EU subsidies in the fisheries sector.

[60]   Abstracted from Paragraph 28 of the Doha Main Ministerial Declaration, Paragraph 28, WTO Rules, emphasis added.

negotiations shall not prejudice the WTO rights of any Member that is not a party to the MEA in question;

(ii) procedures for regular information exchange between MEA Secretariats and the relevant WTO committees, and the criteria for the granting of observer status;

(iii) the reduction or, as appropriate, elimination of tariff and non-tariff barriers to environmental goods and services.

We note that fisheries subsidies form part of the negotiations provided for in paragraph 28.[61]

Testing the ability of the WTO to recognise the importance of environmental harm in the framework of the negotiation launched with Paragraph 28 of the Doha Declaration can be of paramount importance for countries (in Southern and West Africa for example) whose fisheries are being deprived from their fisheries resources by EU and other subsidised fleets. However, the comment above about the perceptions of developing countries with regard to subsidies should be borne in mind.

EU

Although it is commonly accepted that fishing capacity must be brought into balance with available resources (one of the major objectives of the EU Common Fisheries Policy), governments are reluctant to effectively reduce fishing fleets.

In June 2003, the European Commission wrote: "While fishing capacity (defined in terms of vessels' tonnage and engine power) has been somewhat reduced through Multi-Annual Guidance Programmes (MAGPs), recent reduction targets under MAGP IV have been too modest. Moreover, increasing fleet efficiency and dwindling stocks have meant that, in some segments, the fleet is still too large for the stocks it is targeting."[62] Fleet reductions have also been achieved through the transfer of vessels to other flags, including flags of convenience. Under the current EU fisheries subsidy policy (Financial Instrument for Fisheries Guidance - FIFG), the premium to transfer a vessel to another country in the framework of a joint enterprise can be up to 80% of the premium to scrap the vessel. But the owner has been able to keep his vessel and continue to fish.

In December 2001, EU fisheries ministers agreed to amend the FIFG to prohibit the use of EU subsidies to transfer an EU-flagged vessel to certain countries such as those operating open registries.

Council Regulation (EC) No 179/2002 of 28 January 2002 amending Regulation (EC) No 2792 laying down the detailed rules and arrangements regarding Community structural assistance in the fisheries sector[63] provides that Article 7(3)(b) be amended with the following addition:

"(iv) If the third country to which the vessel is to be transferred is not a Contracting or Cooperating party to relevant regional fisheries organisations, that country has not been identified by such organisations as one which permits fishing in a manner which jeopardises the effectiveness of international conservation measures. The Commission shall publish a list

---

[61]   Doha Main Declaration, Paragraph 31, *Trade and Environment*, emphasis added.

[62]   http://europa.eu.int/comm/fisheries/scoreboard/fleet_en.htm

[63]   http://europa.eu.int/eur-lex/en/archive/2002/l_03120020201en.html

of the countries concerned on a regular basis in the series C of the Official Journal of the European Communities."

Bilateral fishing agreements

Developing countries are often heavily dependent on the revenues stemming from distant water fishing fleets. Revenues are obtained through bilateral fishing agreements, which provide for the licensing of foreign vessels to fish in a country's EEZ. Major distant water fishing countries include Japan, Chinese Taipei, Korea, US, and Spain. Despite the fact that the revenues generated by licensing fees are extremely low in relation to the value of the landed catch by foreign vessels (for example in the Pacific, it is roughly 5%[64]), countries are under constant pressure to reduce them.

Revenues may be augmented by tied aid. Japan is widely cited as providing aid to coastal developing countries in exchange for access by its distant water fleet to important fishing grounds.

Most developing countries with which the EU, Japan, and others have bilateral fisheries agreements do not have the means to control activities in their EEZ. That results in widespread IUU fishing and the destruction of fish stocks, the marine environment and coastal communities' livelihoods.

The EU is in the process of adapting its policy on fisheries agreements (now branded "partnership agreements") to make it look more coherent with its own environmental and development policies as well as international commitments (for example those made at the WSSD). It remains to be seen whether this is a real change or simply a means of hiding the "business as usual" effort to over-exploit other countries' waters to keep their own fleets active and to supply the large EU market.

The EU is also claiming to help fight IUU fishing in developing countries' waters by allocating some of the financial contribution paid for access to what they call "targeted actions". The amounts vary significantly between agreements. In the case of Guinea Conakry, for example, EU money supposedly dedicated to control and surveillance is very obviously not used for that purpose. The EU Commission has admitted that it has no way to demand or even ensure that the money is used for the agreed purpose. The bottom line is that EU public money is used to subsidise the access of EU fleets to developing countries' waters, with no way to ensure that the waters where the fleets operate are properly controlled.[65]

Policy recommendations

Deterring FOC practices

There are essentially four ways of deterring FOC practices under existing laws:

1. Deter reflagging;

2. Increase controls over vessels in ports;

---

[64] Teaiwa, Tarte, Maclellan, Penjueli, "Turning the Tide: Towards a Pacific Solution to Conditional Aid," Greenpeace Australia/Pacific, June 2002
http://www.greenpeace.org.au/features/pdf/Turning_the_Tide_FINAL_large.pdf

[65] Bours, Hélène, personal communication.

3. Apply market and other sanctions to encourage flag States to *i)* join relevant fisheries agreements and *ii)* force their flagged vessels to comply or remove (or specifically NOT remove them as discussed on page 374) from their registries; or

4. Increase control over nationals.

In addition, there are a variety of measures which could be taken to strengthen existing laws, such as the elaboration of 'the genuine link' and new agreements to strengthen port State controls. Litigation under ITLOS is another avenue which could be more creatively approached.

Deter re-flagging

In practical terms, designing measures which cannot be circumvented under existing law to deter nationals from re-flagging will be difficult. However financial incentives and taxation measures can be used to deter the re-flagging of vessels. It is more straightforward legally to impose controls over nationals' (including corporations') fishing activities, but a number of measures could be taken in line with both approaches.

Recommendations on re-flagging

1) States should require (such as in taxation legislation) nationals to disclose beneficial interests in foreign flagged vessels.

2) States should negotiate agreements for information sharing between flag States as to beneficial ownership of vessels.

3) Information sharing should be promoted between flag States and RFMOs to increase transparency in ship-owning arrangements.

4) Port States should co-operate to acquire, exchange and make available to enforcement authorities detailed information which would reveal the true beneficial ownership of fishing companies and vessels. For instance, details of vendors of fish catches, purchases of bunkers and stores, agents of vessels, bank accounts, etc. should be logged and kept in a central register.

5) States should impose stringent conditions on vessel deregistration.

6) The 1986 UN Convention on the Registration of Ships could be cloned, and applied to fishing vessels.[66] [In doing so, however, the provision requiring a specific number of fishing States to ratify should be eliminated: a set number of ratifications (*e.g.* 40) should be sufficient to bring it into force.]

7) The IMO initiative requiring a 'continuous synopsis' (showing a complete history of owners and flags) should be extended to cover fishing vessels.[67]

---

[66] Currently a ship is defined as any self-propelled sea-going vessel used in international seaborne trade for the transport of goods, passengers, or both – with the exception of vessels of less than 500 gross registered tons: Article 2.

[67] Proposed by Swan, in FAO Fisheries Circular No. 980 FIPP/C980, "Fishing Vessels Operating Under Open Registers and the Exercise of Flag State Responsibilities – Information and Options." Rome, 2002.

Increasing controls over vessels in port

The legal basis of port State jurisdiction is complex. The starting point is that a port State has sovereignty over its own territory and that a vessel subjects itself to that sovereignty by entering its port. An argument can be made that by voluntarily seeking admission to the port of a State, a vessel accepts the jurisdiction of that State. The question then arises as to how far that jurisdiction goes. It seems clear that a state can deny facilities, with the possible exception of vessels in distress, subject to non-discrimination requirements. Much legal discussion surrounds the issue of any arrest and detention of a vessel. This distinction must be borne in mind: denial of port access, and, even more so, of offloading or other facilities, are much more straightforward from a legal perspective than the detention or arrest of a ship. Forfeiture of catch is somewhat more akin to the detention of a vessel, but may be less problematic legally.

An FAO Expert Consultation to Review State Measures to Combat IUU Fishing, held in November 2002, concluded[68] that a Memorandum of Understanding on Port State measures would constitute one of the numerous useful tools to prevent, deter and eliminate IUU fishing.[69] Suggested elements[70] included provisions for inspections, prior notice of port access and exchange of information. A draft MOU was included.[71] It also suggested possible sanctions for IUU vessels, such as denial of permission to land fish or fishery products, forfeiture of fish or fishery products, and refusal to permit a vessel to leave port pending consultation with the flag State of the vessel.

An inspection and detention regime using the Paris MOU as a model clearly has benefits. It certainly would be of more value than the proposed EU conference to negotiate an agreement on the rights and responsibilities of port States concerning access by fishing vessels to port facilities, although it is possible that the proposed conference could be used as a vehicle to negotiate the MOU. The point of an MOU would be to go further than existing law and allow detention of suspected IUU vessels. It would also improve co-operation measures and put the legality of inspection and denial of port facilities beyond doubt. An MOU could, as the FAO has suggested, improve the current permissive approach and make port State controls mandatory, and in addition could help harmonise the various port State controls. Improving the linkages with regional fisheries management organisations would allow States to benefit from the knowledge and experience of their secretariats as well as to provide a two-way flow of information.

Recommendations on port State controls

To start with, port States should conduct rigorous inspections of all open registry ships which aim to use port facilities. If such inspections reveal evidence of IUU fishing, (or if a vessel is blacklisted by an RFMO) a number of specific measures could be taken (1-4 below). In addition, further measures should be adopted:

---

[68]   FAO Report of the Expert Consultation to Review Port State Measures to Combat Illegal, Unreported and Unregulated Fishing - Rome, 4-6 November 2002 at:
http://www.fao.org/DOCREP/005/Y8104E/Y8104E00.HTM.

[69]   http://www.fao.org/DOCREP/005/Y8104E/y8104e06.htm#bm06

[70]   http://www.fao.org/DOCREP/005/Y8104E/y8104e07.htm#bm07

[71]   http://www.fao.org/DOCREP/005/Y8104E/y8104e0b.htm#bm11.5

1) IUU vessels should be prevented from bunkering and discharging their catches.[72]

2) Such sanctions should be extended to support vessels including cargo vessels and tankers.

3) All such vessels should be inspected, and port States should co-operate with other states to verify the status of any fish on board.

4) States should implement provisions in national legislation for penalties on vessels fishing in the port State's EEZ, including inspection and forfeiture of any catch and deterrent penalties, and with respect to vessels fishing in the high seas, and implement any measures agreed in any MOU on port State control.

5) Concerned states should negotiate an MOU on port State control.

6) States should adopt new legally binding instruments at the national or regional level to implement the IPOA-IUU recommendations on port State control. Individual states or regional groupings such as the EU should implement a system of prior notification before entry into port, inspections, and denial of port facilities including bunkering and catch unloading to *i)* vessels which inspections find to have engaged in IUU activities and *ii)* vessels on an IUU blacklist. Such a blacklist could be adapted from the CCAMLR or ICCAT lists. (A step further based on the precautionary principle would be to deny access to facilities to all fishing boats NOT listed as being legal and responsible operators.)

7) State legislation should make it an offence simply to be in port with IUU fish on board. This would not include a reference to where the fish was caught, and would thus avoid a number of jurisdictional problems.

8) States should prohibit the landing of IUU fish. This will probably require a catch documentation scheme to be in place. For instance, in the EU, Control Regulation 2847/93[73] should be amended accordingly. This Regulation currently allows vessels from third countries to offload fish that were caught on the high seas as long as the species were caught outside the regulatory areas of the relevant RFMOs of which the EU is a member, so does not necessarily prevent IUU fishing.

9) States should provide for the forfeiture of catches of IUU vessels. This can be achieved a) for nationals under a state's jurisdiction and vessels flying its flag and b) otherwise, in an MOU with relevant states.

---

[72] See with respect to Antarctica: ASOC, "The Application of Port State Jurisdiction, " attaching paper "Port State Jurisdiction: An Appropriate International Law Mechanism To Regulate Vessels Engaged In Antarctic Tourism" (8 October 2002), at: http://www.asoc.org/Documents/XXIICCAMLR/ASOC.Port%20State.doc. The paper proposes a memorandum of understanding modelled on the Paris MOU to implement an effective port state control regime to regulate vessels engaged in Antarctic tourism.

[73] Council Regulation (EEC) No 2847/93 of 12 October 1993 establishing a control system applicable to the common fisheries policy, at: http://europa.eu.int/smartapi/cgi/sga_doc?smartapi!celexapi!prod!CELEXnumdoc&lg=EN&numdoc=31993R2847&model=guichett

Market sanctions

Market-based sanctions have proven effective. ICCAT import controls on FOC states such as Honduras and Belize doubtless had an influence on the reduction of fishing boats on their registries and on efforts to reduce IUU fishing activities. It should be noted, however, that in order for trade sanctions not to violate WTO rules, they must be non-discriminatory, transparent, and linked to a policy of 'conserving an exhaustible natural resource'.[74]

Recommendations on market sanctions

1. Other RFMOs should adopt the ICCAT system so that member states prohibit the import of fish products from non-complying parties.

2. States should impose higher tariffs for fish and fish products from identified states where vessels have frequently engaged in IUU fishing.[75] The tools to do this are already in place in the EU and should be used more.[76]

3. States and/or RFMOs should take measures to deter companies from doing business with IUU operations, as recommended in the IPOA-IUU.[77] Companies identified in the IPOA include: importers, transhippers, buyers, consumers, equipment suppliers, bankers, insurers, other services suppliers and the public.

Control over nationals

Increasing control over nationals requires increased transparency in registries and corporate shareholding so that states are in fact able to monitor and control the activities of their nationals who own, crew and supply IUU fishing vessels, regardless of the flag under which they sail. This paper has already recommended that taxation policy be used to force nationals to disclose their beneficial interests in foreign flagged vessels. In addition, there are a number of specific things which can be done in the context of implementing the IPOA-IUU:

Recommendations on control of nationals

1) IPOA-IUU language on control of nationals should be implemented to ensure that nationals subject to a State's jurisdiction do not support or engage in IUU fishing. Measures include introducing prison sanctions for IUU fishing, including aiding and abetting, to prevent, deter and eliminate IUU fishing and depriving offenders of the benefits from IUU fishing. Sanctions could be extended to companies that do business with IUU operations, as provided for in IPOA-IUU paragraph 73. In other words, states should adopt measures to make it illegal to own or otherwise participate in any aspect of IUU fishing.

---

[74]   Upton, Simon and Vitalis, Vangelis, "Stopping the High Seas Robbers: Coming to Grips with Illegal, Unreported and Unregulated Fishing on the High Seas," OECD, 2003 at:
http://www.oecd.org/dataoecd/15/16/16801381.pdf

[75]   See EU Parliament draft report on the role of flags of convenience in the fisheries sector (2000/2302)(INI), 23 September 2001, at:
http://www.europarl.eu.int/meetdocs/committees/pech/20011008/439060EN.pdf

[76]   Already in the EU Regulation 2820/98 article 22 allows for temporary withdrawal of tariff preferences in case of manifest infringement of the objectives or RFMOs.

[77]   IPOA-IUU Paragraph 73 http://www.fao.org/DOCREP/003/y1224e/y1224e00.HTM

2) A system for penalizing those nationals benefiting from IUU fishing should be implemented as suggested in the IPOA-IUU, to deprive them of benefits of such fishing and act as a deterrent.

3) EU Regulations[78] already require member states to ensure that appropriate measures are taken, including administrative action or criminal proceedings according to their national law, against natural or legal persons responsible. But the regulations only apply to vessels in EU waters and EU vessels in the high seas. They should be modified to apply to EU citizens wherever the vessel and whatever the flag.

Elaborate the definition of "genuine link"

ITLOS appears to be favourable to upholding the requirement to establish a genuine link between the flag State and the vessel. After two cases (the *Camouco* and *Monte Confurco* cases) wherein ITLOS reduced the amount of the bond levied by the French government, ITLOS reached a turning point with the *Grand Prince* case by declining jurisdiction and holding that "in the view of the tribunal, the assertion that the vessel is 'still considered as registered in Belize' contains an element of fiction, and does not provide sufficient basis for holding that Belize was the flag State of the vessel for the purposes of making an application under article 292 of the convention."[79] In other words, ITLOS did not accept that the vessel properly was entitled to the protection of Belize despite the fact that it was flying the Belize flag at the time it was arrested. ITLOS therefore let the bond of EUR 1.74 million set by the French government stand. Of course the entire flag of convenience system contains an element of fiction, and while the ITLOS decision turned on the facts of that case where the status of the registration was in doubt, the *Grand Prince* decision showed a welcome readiness to move back to requiring a genuine link.

Further elaboration of the concept of 'genuine link' would help to ensure that the flag State does its duty to force vessels to comply with the rules.

The IPOA-IUU could serve as a starting point. It goes some way towards cutting off the supply of vessels to be flagged under FOCs by preventing re-flagging: "19. States should discourage their nationals from flagging fishing vessels under the jurisdiction of a state that does not meet its flag State responsibilities."

It then lays out responsibilities for flag States:

35. A flag State should ensure, before it registers a fishing vessel, that it can exercise its responsibility to ensure that the vessel does not engage in IUU fishing.

36. Flag States should avoid flagging vessels with a history of non-compliance except where:

36.1 the ownership of the vessel has subsequently changed and the new owner has provided sufficient evidence demonstrating that the previous owner or operator has no further legal, beneficial or financial interest in, or control of, the vessel; or

---

[78] Regulation 2847/93.

[79] Grand Prince (Belize v France), Judgment of 20 April 2001, at:
http://www.itlos.org/case_documents/2001/document_en_88.doc, paragraph 85.

36.2 having taken into account all relevant facts, the flag State determines that flagging the vessel would not result in IUU fishing.

37. All states involved in a chartering arrangement, including flag States and other states that accept such an arrangement, should, within the limits of their respective jurisdictions, take measures to ensure that chartered vessels do not engage in IUU fishing.

38. Flag States should deter vessels from re-flagging for the purposes of non-compliance with conservation and management measures or provisions adopted at a national, regional or global level. To the extent practicable, the actions and standards flag States adopt should be uniform, to avoid creating incentives for vessel owners to re-flag their vessels to other states.

39. States should take all practicable steps, including denial to a vessel of an authorisation to fish and the entitlement to fly that State's flag, to prevent "flag hopping"; that is to say, the practice of repeated and rapid changes of a vessel's flag for the purposes of circumventing conservation and management measures or provisions adopted at a national, regional or global level or of facilitating non-compliance with such measures or provisions.

40. Although the functions of registration of a vessel and issuing of an authorisation to fish are separate, flag States should consider conducting these functions in a manner which ensures each gives appropriate consideration to the other. Flag States should ensure appropriate links between the operation of their vessel registers and the record those states keep of their fishing vessels. Where such functions are not undertaken by one agency, states should ensure sufficient co-operation and information sharing between the agencies responsible for those functions.

41. A Flag State should consider making its decision to register a fishing vessel conditional upon its being prepared to provide to the vessel an authorisation to fish in waters under its jurisdiction, or on the high seas, or conditional upon an authorisation to fish being issued by a coastal State to the vessel when it is under the control of that flag State.

All of these provisions assume the will and capacity of FOC States to undertake these actions. Where, as is likely to be the case, the will or capacity is lacking, there must be the ability to pierce the corporate veil and apply sanctions to the true beneficial owner.

Recommendations on elaborating genuine link

1) Legally binding measures to implement paragraphs 19 and 35-41 of the IPOA-IUU should be adopted.

2) One or more States should take a case to ITLOS to elaborate the requirements for a genuine link as well as flag State (and even national State) responsibilities.

Monitoring, control and surveillance

Implementing the best available systems for Monitoring, Control, and Surveillance (MCS) is key to enforcing existing agreements to prevent IUU fishing. The IPOA-IUU (paragraph 24 for example) contains numerous references to the myriad of tools available to fisheries managers, "including (but not limited to) vessel monitoring systems (VMS), observer programmes, catch documentation schemes, inspections of vessels in port and at sea, denial of port access and/or privileges to suspected IUU vessels, maintenance of "black" and "white" lists, and the creation of presumptions against the legitimacy of catches by Non-Party fishing vessels in areas regulated by RFMOs."[80] The exchange of information between management and enforcement officials, within and between regions, is also critical.

Unfortunately, MCS is not carried out globally. According to a 2001 Greenpeace report, "...fisheries control and surveillance are virtually non-existent on the high seas of the Atlantic Ocean. Most of the national exclusive economic zones (EEZs) off the west coast of Africa, where both legal and illegal foreign distant-water fishing fleets operate, are not sufficiently controlled either."[81] Existing agreements recognise the need for states to exercise their responsibilities to inspect, and ultimately to prosecute, those who violate the rules. While many states have invested in building up their capacity to do so, others, particularly poorer developing countries, do not have the resources to do so.

The UN Fish Stocks Agreement contains provisions for assistance, including financial, to developing country Parties.[82] This is being implemented for example through the creation of an Assistance Fund in collaboration with the FAO, bilateral partnerships between developed and developing countries, and assistance from the World Bank.[83] One positive example of a bilateral partnership is the support from the government of Luxembourg for the Sub-Regional Fisheries Commission (based in Senegal) and the Surveillance Operations Co-ordination Unit (Gambia) which are co-operating to develop an MCS programme.[84] Germany's GTZ has also provided support to Mauritania in developing its MCS programme.[85]

Recommendations on MCS

1) All States should introduce and/or expand their use of VMS systems as a cost-effective means of monitoring and surveillance, and to participate in the International MCS Network.

2) The current system used by some RFMOs to establish 'white' and 'black' lists of fishing vessels should be expanded. The precautionary principle, which has already been agreed in the UN Fish Stocks Agreement and other instruments, suggests that the burden of proof should be shifted to vessel owners. A new type of list, which identifies vessels which are known NOT to be engaged in IUU fishing, should be drawn up and used by fisheries management authorities. This would inherently require the use of VMS to demonstrate innocence.

3) Monitoring systems should be improved, for example by ensuring that devices cannot be disabled, or the data tampered with. NGOs attending CCAMLR meetings repeatedly call for

---

[80]   http://www.fao.org/docrep/005/y3536e/y3536e06.htm#bm06.2.5 Section 3.2.5

[81]   Bours, Gianni, Mather, "Pirate Fishing Plundering the Oceans," Greenpeace, February 2001, page 9.

[82]   Fish Stocks Agreement, Part VII, Article 26:
http://www.un.org/Depts/los/convention_agreements/texts/fish_stocks_agreement/CONF164_37.htm

[83]   "Second Informal Consultations of the States Parties to the Agreement for the Implementation of the Provisions of the United Nations Convention on the Law of the Sea of 10 December 1982 Relating to the Conservation and Management of Straddling Fish Stocks and Highly Migratory Fish Stocks (New York 23-25 July 2003) - Report" at:
http://www.un.org/Depts/los/convention_agreements/FishStocksMeetings/UNFSTA_ICSP2003_Rep.pdf

[84]   Greenpeace, "Pirate Fishing: Plundering West Africa," September 2001.

[85]   Greenpeace, "Witnessing the Plunder: A Report on the *MV Greenpeace* Expedition Investigating Pirate Fishing in West Africa," November 2001.

the Commission to require centralised VMS systems which transmit data in real time back to the Secretariat, arguing that flag State vessel monitoring is insufficient.[86] They cite the fact that NAFO already uses a centralised system. They propose that CCAMLR look at adopting the comprehensive measures to prevent tampering which Spain has put in place. Such measures should be adopted by CCAMLR, and extended to other RFMOs.

4) Developing countries should be assisted to increase their capacity to carry out MCS by providing assistance and funding through whatever means possible. Such support should not be contingent upon the developed country getting (increased) access to the recipient country's fishing grounds. The results of such assistance should be monitored to ensure that assistance achieves its intended result.

Recommendations on catch and trade documentation systems

1) Catch documentation schemes should be implemented more widely to help resolve the problems of transhipments. Catch documentation schemes must not be reliant on the filling in of forms by fishing captains, as is the case with the CCAMLR model, but must include verification and inspection protocols by national fishing officers in ports in co-operation with RFMOs.

2) Likewise, trade documentation schemes should be implemented more widely, which would provide for documentation to accompany fish in trade starting from the point it is caught, all the way through to the time it reaches the consumer. There should be a widespread system implemented to include important markets (such as Japan and Chinese Taipei) and ports (especially ports of convenience such as Las Palmas and Mauritius) to put into place effective labelling and tracing of fish products.

3) Consumers should be dissuaded from purchasing non-certified fish and fish products. In addition to ongoing campaigns *e.g.* not to buy Chilean sea bass, or the wallet guides to sustainably caught fish which many groups publish, consumers could be educated only to buy certified fish and fish products. This would of course be contingent on effective tracing and labelling regimes being in place.

4) States should make the import or export of non-certified fish products a criminal act under their domestic legislation, based on the CITES model.

Recognise a formal role for NGO vessels

Coastal states could engage in discussions with NGOs and RFMOs to co-operate in information and evidence gathering and could for instance nominate authorised inspectors to go on board private vessels such as those operated by NGOs and ensure that evidence gathered by NGOs can be used against apprehended IUU vessels. Close co-operation with NGOs will enable fisheries enforcement vessels to react to reports by NGOs and arrest IUU vessels. In some cases NGO vessels could be authorised to be on government service and thus even engage in inspections, boarding and arrest, under supervision of the inspectors, of vessels found fishing illegally. Such vessels would need to be marked as being on government service[87] and would enjoy immunity as government vessels.[88] It

---

[86]   ECO, 3 November 2003.

[87]   Law of the Sea Convention Article 111.

[88]   Law of the Sea Convention Article 96.

should be noted that the requirements of hot pursuit when chasing, boarding and arresting vessels are exacting and should be followed.[89] For instance there must be a visual or auditory message to stop, and the hot pursuit must begin when the fishing boat (or one of its boats) is within the EEZ, and may not be interrupted.[90]

Such an approach is in keeping with the recommendations of COLTO, the legal toothfish operators' coalition: "Effective surveillance and enforcement can only come, we believe, by legal operators, conservation groups and government agencies working in partnership to combat IUU fishing."[91]

Recommendation on NGO involvement in MCS

1) In many developing countries, there is a will to undertake effective MCS activities, but the capacity is simply not there, or not sufficient. Such countries should attempt to negotiate MOUs with NGOs, where appropriate, to assist in patrolling the EEZ.

Criminalise IUU fishing

Revise laws on arrest of fishing vessels to work better as a deterrent

International law prohibits imprisonment for captains and crew of vessels fishing illegally and fines are often seen as a cost of doing business. It favours the release of fishing vessels, and states may be forced to release arrested vessels which then reflag and carry on with IUU fishing. The negotiation of regional or even international agreements, such as MOUs on port State control, would go some way towards introducing new controls. But international agreements would need to specifically involve FOC States as parties to ensure truly effective deterrents such as confiscation of fishing vessels and imprisonment, as well as to provide for imprisonment of beneficial owners. Until such agreements are in place, it would assist considerably if states were to implement penalties which considerably exceed the value of the vessel and potential profits: in the millions or even tens of millions of dollars. This would allow ITLOS to sanction large bonds and would act as a significant deterrent. Legislation should ensure that catches or the value of catches are confiscated.

Recommendation on arrest of fishing vessels

1) States, particularly Spain where many IUU beneficial owners hold their nationality, should enact laws requiring prison sentences for beneficial owners and operators of IUU fishing vessels and for those who aid and abct them.

2) Coastal States should provide for penalties under their domestic legislation which will exceed the value of fishing vessels and their catch. Such penalties will be in the several millions of dollars and should in addition ensure that catches or the value of catches are confiscated.

3) Coastal States should negotiate agreements with other states, both within RFMOs and with other states such as EU member states, to allow for prison sentences for captains, owners and operators of IUU fishing vessels, and to allow permanent confiscation of IUU fishing boats.

---

[89]   Law of the Sea Convention Article 111.

[90]   Law of the Sea Convention Article 111.

[91]   http://www.colto.org/About_Us.htm

When a vessel is arrested and the arresting State wishes to detain the vessel, the Law of the Sea Convention requires that the arresting State set a reasonable bond for the release of the vessel.[92] The flag State may then apply to ITLOS for prompt release of the vessel, and in effect for a reduction in bond, claiming the bond set by the arresting State is not reasonable.[93]

ITLOS has frequently been asked to decide applications for prompt release of vessels under article 292 of the Law of the Sea Convention, for example the previously mentioned cases involving the *Camouco*[94] *Monte Confurco,*[95] and *Grand Prince.*[96]

The IPOA-IUU stresses deterrence, however.[97] The need for deterrence has yet to be fully implemented by ITLOS in its assessment of the reasonableness of the bond. A bond should not be held to be unreasonable if it is at a level necessary for a coastal State to ensure the effective enforcement of fisheries laws. Judge Anderson noted in the *Monte Confurco* case that "where there is persistent non-observance of the law, deterrent fines serve a legitimate purpose."

Recently the *Volga* case *(Russian Federation v Australia)*[98] involved a longline fishing vessel flying the Russian flag which was boarded in February 2002 by the Australian navy outside the EEZ of the Australian Territory of Heard Island and the McDonald Islands with over 131 tonnes of Patagonian toothfish (*Dissostichus eleginoides*). Australia sought a bond of AUD 3 332 500 which included AUD 1 920 000 as security to cover the assessed value of the vessel, fuel, lubricants and fishing equipment, AUD 412 500 to secure payment of potential fines and a security of AUD 1 000 000 related to the carriage of a fully operational VMS and observance of CCAMLR conservation measures.

The Tribunal held the first to be reasonable, and decided that the second would serve no practical purpose, since the crew had been granted bail so they could return to their native Spain. In doing so, the Tribunal held that a "good behaviour bond" to prevent future violations of the laws of a coastal State cannot be considered as a bond or security within the meaning of article 73(2) of the Convention,

---

[92]   Law of the Sea Convention Article 73(2).

[93]   Law of the Sea Convention Article 292.

[94]   The bond in *Camouco (Panama v France),* 7 February 2000, of 20 million FF was reduced to FF 8 million (about EUR 1.2 million) at: http://www.itlos.org/case_documents/2001/document_en_129.doc,

[95]   The bond of FF 56 400 000 in *Monte Confurco (Seychelles v France)* 18 December 2000 was reduced to FF 18 million (about EUR 2.7 million), at: http://www.itlos.org/case_documents/2001/document_en_115.doc,

[96]   Grand Prince (Belize v France), Judgment of 20 April 2001, at: http://www.itlos.org/case_documents/2001/document_en_88.doc, paragraph 85.

[97]   IPOA-IUU Para. 21 provides that "States should ensure that sanctions for IUU fishing by vessels and, to the greatest extent possible, nationals under its jurisdiction are of sufficient severity to effectively prevent, deter and eliminate IUU fishing and to deprive offenders of the benefits accruing from such fishing." Paragraph 22 states that "All possible steps should be taken, consistent with international law, to prevent, deter and eliminate the activities of non-cooperating States to a relevant regional fisheries management organisation which engage in IUU fishing."

[98]   See judgment at http://www.itlos.org/case_documents/2002/document_en_215.doc.

read in conjunction with article 292 of the Convention.[99] The Russian Federation argued that the proceeds of the sale of the catch should suffice as security given by the owner for the release of the vessel and its crew.

If accepted, this argument would have been analogous to the fruits of an alleged crime being considered as security.[100] ITLOS, however, held that the proceeds have no relevance to the bond to be set for the release of the vessel and the members of the crew. In doing so ITLOS moved forward from its previous position in *Monte Confurco*.

ITLOS also expressly noted that it "understands the international concerns about illegal, unregulated and unreported fishing and appreciates the objectives behind the measures taken by States, including the States Parties to CCAMLR, to deal with the problem."[101] In his dissenting opinion, Judge Anderson stated that "In my opinion, the duty of the coastal State to ensure the conservation of the living resources of the EEZ contained in article 61 of the Convention, as well as the obligations of Contracting Parties to CCAMLR to protect the Antarctic ecosystem, are relevant factors when determining in a case under article 292 whether or not the amount of the bail money demanded for the release of a vessel such as the *Volga* is 'reasonable'."[102]

Judge Anderson found that Article 73 contains no explicit restriction upon the imposition of non-financial conditions for release of arrested vessels. Indeed, the reasonableness of a good behaviour bond, bearing in mind the risk of re-offending, does seem fully consistent with the object and purpose of Article 73 and of the Convention as a whole. If the gravity of the alleged offences is a factor to be taken into account in assessing reasonableness, as it was in the *Monte Confurco* judgment and recognised in the *Volga* judgment[103] then *a fortiori* the imposition of a good behaviour bond should not be considered as unreasonable. Indeed, Article 73(1) itself empowers coastal States to take such measures as are "necessity to ensure compliance" with its laws and regulations.[104]

Similarly, Judge Nelson in his separate opinion[105] in the *Camouco* case said that "in my opinion, this Tribunal …should also take account of what, in the introduction to the Statement in Response of the French Republic, was referred to as "the context of illegal, uncontrolled and undeclared fishing in the Antarctic Ocean and more especially in the exclusive economic zone of the Crozet Islands where the facts of the case occurred". This material constitutes part of the "factual matrix" of the present case – the factual background surrounding the case. In my view this factor ought to have played some part, not by any means a dominant part, but a part nevertheless in the determination of a reasonable bond."

Judge Nelson was right to be concerned about deterrence. After its bond was reduced by ITLOS, the *Camouco* was reflagged under the Uruguay flag and renamed the *Arvisa* 1 and continued to fish

---

99   *The "Volga" Case* (Russian Federation v Australia), Judgment of 23 December 2002, at: http://www.itlos.org/cgi-bin/cases/case_detail.pl?id=11&lang=en#judgement , para. 80.

100   Judge Shearer accepted this in his dissent at para. 15, at: http://www.itlos.org/case_documents/2002/document_en_220.doc.

101   *Volga* Judgment, para. 68.

102   Anderson dissenting opinion in *Volga,* para. 2, *at:* *http://www.itlos.org/case_documents/2002/document_en_219.doc*

103   Volga Judgment, para. 63.

104   See Anderson dissenting opinion, note 102, paragraph 16.

105   *Camouco (Panama v France)* , Prompt Release, Judgement of 7 February 2000, Vice President Nelson separate opinion, at http://www.itlos.org/case_documents/2001/document_en_129.doc

for Patagonian toothfish. Arvisa 1 was one of two vessels found fishing inside the CCAMLR Area by an Australian research vessel in January 2002 and was caught yet again, this time by the French Navy, in July 2002, this time having apparently been reflagged to the Netherlands Antilles. Clearly, its owners have not been deterred by the previous arrests.

There is already sufficient authority in the Law of the Sea Convention for ITLOS to treat the need for deterrence, prevention and innovative bonding arrangements as relevant matters for assessing whether bonds are reasonable under Article 73. Nonetheless, additional compliance mechanisms are required, such as including increased powers for port States, better regulation of markets, enforcement of the genuine link requirement of flag States and mechanisms to ensure the application of fisheries laws to flags of convenience.

Recommendations on bonding procedures

1) States should implement measures which set the maximum permissible fines for infringement of fisheries laws high enough to serve as a credible deterrent. This will allow States which have arrested IUU fishing boats to set a correspondingly high bond.

2) States should work together to discuss arrest and bonding procedures and devise effective and legal bonding arrangements to act as a deterrent and prevent vessels from re-offending. This will help ensure that such decisions are upheld by ITLOS.

Subsidy reforms

The Johannesburg Plan of Implementation called on States to eliminate subsidies that contribute to IUU fishing and over-capacity, even in advance of the WTO completing its efforts in this area.

With regard to the WTO, one option would be to use the WTO dispute settlement procedure by a country wishing to protect its own fisheries from the activities of foreign subsidised fleets.

Recommendations on subsidy reforms

1) Subsidies which promote IUU or otherwise unsustainable fishing activity should be identified. On the basis of the language in the Johannesburg Plan of Implementation and the Doha Declaration, such subsidies should be eliminated or redirected (*e.g.* to scrap vessels, or help developing countries to develop control capacity or local, sustainable fishing capacity).

2) States wishing to protect their fisheries from the activities of foreign subsidised fleets should consider the possibility of launching a WTO dispute.

Strengthen and harmonise national legislation

Paragraph 30(d) of the Johannesburg Plan of Action calls on States to put into effect the IPOA-IUU by 2004. This deadline was agreed by consensus.

Recommendations on national legislation

1) National legislation should be strengthened and harmonised on the basis of the measures included in the IPOA-IUU.

While it is beyond the scope of this paper to provide a detailed analysis of national legislation, it is important to touch on the legislation governing the European Union given that it is one of the major markets for IUU fish, has a major distant-water fishing fleet, and hosts a major port of convenience (Las Palmas).

The European Common Fisheries Policy (CFP) provides the framework for common EU positions in four areas: conservation, structures, markets, and relations with the outside world.[106] A revised CFP has been in effect since January, 2003. A number of changes in the CFP have bearing on the subject of this paper:

- It aims to take a long-term approach to fisheries management (as opposed to previously, when measures were adopted annually), and attempts to conserve the ecosystem as a whole, rather than individual fish stocks.

- It addresses fishing capacity, and in particular prohibits subsidies for renewing or modernizing fishing vessels.

- It aims to harmonise and strengthen measures at the national level on controls and sanctions.

Within the framework of the reform of the CFP, a number of action plans have been adopted, developed or proposed, including a plan to eradicate IUU fishing.[107] (Additional action plans include: Community Action Plan for the Conservation and Sustainable Exploitation of Fisheries Resources in the Mediterranean Sea Under the Common Fisheries Policy,[108] A Council Regulation Laying Down Measures Concerning Incidental Catches of Cetaceans in Fisheries and Amending Regulation (EC) No 88/98,[109] Strategy for Sustainable Development of European Aquaculture,[110] Integration of Environmental Protection Requirements into the CFP,[111] Measures to Counter the Social, Economic and Regional Consequences of Fleet Restructuring,[112] Plan to Reduce Discards of Fish,[113] and a Communication Towards Uniform and Effective Implementation of the CFP (*i.e.* plan to introduce a uniform CMS system).[114] However, the CFP has not yet been adapted to implement the IPOA-IUU.

---

[106] Introduction to the CFP at http://europa.eu.int/comm/fisheries/doc_et_publ/ctp_en.htm

[107] Plan available at:
http://europa.eu.int/comm/fisheries/doc_et_publ/factsheets/legal_texts/docscom/en/com_02_180_en.pdf

[108] http://europa.eu.int/comm/fisheries/doc_et_publ/factsheets/legal_texts/docscom/en/com_02_535_en.pdf

[109] http://europa.eu.int/eur-lex/en/com/pdf/2003/com2003_0451en01.pdf

[110] http://www.europa.eu.int/eur-lex/en/com/cnc/2002/com2002_0511en01.pdf

[111]   http://europa.eu.int/comm/fisheries/doc_et_publ/factsheets/legal_texts/docscom/en/com_02_186_en.pdf

[112] http://europa.eu.int/comm/fisheries/doc_et_publ/factsheets/legal_texts/docscom/en/com_02_600_en.pdf

[113] http://europa.eu.int/comm/fisheries/doc_et_publ/factsheets/legal_texts/docscom/en/com_02_656_en.pdf

[114] http://europa.eu.int/comm/fisheries/doc_et_publ/factsheets/legal_texts/docscom/en/com_03_130_en.pdf

Final conclusions

On paper, there is a complex network of binding and non-binding agreements ('hard' and 'soft' law) which forms a solid basis in international law for promoting the development of sustainable fisheries, and for preventing or eliminating IUU fishing.

In practice, however, there are weaknesses and loopholes, the most important ones being:

- Flags of Convenience (FOC), or open registries, allow unscrupulous operators to avoid any regulation of their activities. They fish anywhere and anytime they want to, in contravention of the regulations put in place by Regional Fisheries Management Organisations (RFMOs) to manage and conserve fish stocks.

- As one country or region more aggressively acts to deter IUU fishing, activities are displaced to another which is less willing or able to do so. As one flag tightens its registry, vessels simply reflag to another less restrictive state. And as more states tighten their registers, new FOC countries emerge.

- Transhipping at sea means that vessels need never enter ports with their illegally caught fish. The mingling of illegally and legally caught fish on board reefers essentially serves to whitewash the contraband fish.

- Monitoring, control and surveillance of the high seas and within the Exclusive Economic Zones (EEZs) of many countries (particularly poorer developing countries) are insufficient to ensure that illegal fishers will be apprehended. Even when they do get caught, bonds and fines are set too low to serve as any kind of deterrent. Such fines are simply considered a cost of doing business; vessels invariably return to the fishing grounds, and carry on as before.

The solutions to these problems are not all easy to implement, but they are clearly identifiable.

The single most effective step to combat IUU fishing would be to close the loophole in international law that allows states to issue flags of convenience to vessels with which they have no genuine link and then fail to exercise control over those vessels. A combination of existing instruments, the negotiation of new instruments, and the litigation at the International Tribunal for the Law of the Sea could be used to accomplish this.

Unless and until the FOC system is effectively eliminated, it is important that States do everything in their power to prevent, deter and eliminate IUU fishing through the following means:

- Port State controls: port States must prevent IUU fishing and support vessels from using their harbours for transhipment, resupply and other activities and/or must where possible take action to arrest or detain IUU vessels in the event such vessels enter their ports.

- Market measures: states must adopt and enforce legislation to make it illegal to import or trade in IUU-caught fish. Moreover, states should make it illegal or otherwise discourage companies (*e.g.* insurers, re-suppliers, fishing gear manufacturers) from doing business with companies engaged in IUU fishing.

- At-sea transhipment: flag States must make it illegal for their transport vessels to tranship fish caught by vessels engaged in IUU fishing.

- Companies and nationals: states must make it illegal for their nationals and for companies within their jurisdiction to engage in IUU fishing, including the use of fines, penalties and, as necessary, prison sentences of sufficient severity to deter IUU fishing activities.

- Comprehensive management regime for the high seas: IUU fishing not only involves illegally fishing within an EEZ or in contravention of any regional fisheries management organisation (RFMO) agreements in place on the high seas. It also includes fishing on the high seas in regions where there is no fisheries management regime in place at all. The problem of fishing (mainly bottom trawling) on seamounts and other deep-sea areas on the high seas, which is largely free of any international management agreement to date, has recently become an issue of international concern. The UN General Assembly is now calling attention to the problem, and its urgency has been widely recognised by fisheries experts.

ANNEX

BIOGRAPHIES OF AUTHORS, SPEAKERS AND SESSION CHAIRS

David **AGNEW** (Marine Resource Assessment Group, London) is a research Lecturer at Imperial College in the UK, and Head of the Falklands Fisheries Research Group. He has extensive experience in the conservation and management of marine living resources, specialising in international fisheries agreements in both tropical and high latitude fisheries. Previously, Mr. Agnew worked for CCAMLR as a senior manager dealing with the conservation and rational use of Antarctic marine living resources.

Alejandro **ANGANUZZI** (IOTC, Seychelles) is the Deputy Secretary of IOTC, where his main responsibility lies with the co-ordination of the technical activities of the Secretariat, particularly in providing support to the Working Parties and the Scientific Committee. In addition to technical activities, he has participated in drawing up budgets and work plans for the Secretariat since 1998. He developed the first two generations of the IOTC main database and co-designed the third generation. In setting up the database, he developed a computerised data-tracking system to link documentation of the data transactions, the basic data and relevant correspondence with liaison officers.

David A. **BALTON** (US Department of State, Washington D.C.) is Deputy Assistant Secretary for Oceans and Fisheries in the Bureau of Oceans and International Environmental and Scientific Affairs, US Department of State. In that capacity, Mr. Balton is responsible for co-ordinating US foreign policy on a broad range of issues concerning the oceans as well as the Arctic and Antarctic regions.

Colin **BARNES** (Marine Resources Assessment Group, London) is a natural resource economist and an associate with the MRAG in London. Mr. Barnes has a Master's degree in development economics from the University of East Anglia (UEA) and a PhD in economics from the University of Manchester. He has worked extensively in various aspects of fisheries policy, fisheries economics and coastal management, including on projects dealing with the impact of distant water fishing fleet; subsidies on the fisheries of low-income countries and appraisal of fisheries research programmes.

Serge **BESLIER** (DG FISH, EU Commission, Brussels) is the Head of Section in the DG FISH, dealing with general external relations. Mr. Beslier has represented the EU Commission at numerous meetings and conferences and has been the EU fisheries representative in various FAO, OECD and UN bodies. Mr. Beslier has extensive experience in bilateral fisheries relations and has been a key figure in devising the external fisheries policy in the newly revamped Common Fisheries Policy.

David **CHANG** (Chinese Taipei Overseas Fisheries Development Council, Chinese Taipei) is the Director of Fisheries at the Overseas Fisheries Development Council. This Council is an international non-governmental organisation, with the main objective of promoting sustainable utilisation of fisheries resources through international co-operation and information exchange. The Council's recent work has focused on assisting the fisheries authorities in Chinese Taipei to formulate fisheries policies that will ensure sustainable utilisation of fisheries resources; urging fishermen, operators and fisheries associations to apply fishing practices which are consistent with the guidelines as set forth in the FAO Code of Conducts for Responsible Fisheries as well as the conservation and management measures adopted by the regional fisheries management organisations.

Duncan **CURRIE**, Barrister, has an Honours degree in Law from the University of Canterbury in New Zealand and a Masters from the University of Toronto in Canada. He has specialised in environmental and international law for the past 18 years. From 1985 to 1996 he was in-house legal adviser to Greenpeace International, where he advised extensively on law of the sea issues. Since then he has advised various NGOs on international and environmental law.

Ignacio **ESCOBAR** (Ministry of Agriculture and Fisheries, Madrid) is Head of Unit for Regional Multilateral Fisheries Organisations at the Spanish Ministry for Agriculture and Fisheries. He participates in all of the RFMOs to which the European Union is a member (ICCAT, IOTC, IATTC, NAFO, NEAFC, CGPM, CCAMLR, SEAFO, SWIOFC, WCPFC, FAO, IWC, etc.). He is also responsible for bilateral fisheries agreements with EU and Norway, Greenland and the Faeroe Islands. Mr. Escobar has a Degree in Law and a Masters in Financial Markets. As a career diplomat, he has served in the Spanish Embassies in Kuwait and Panama, and was Consul General in Agadir (Morocco).

Martin **EXEL** (COLTO, Australia) has been working since 1997 as the General Manager, Environment and Policy, with Newfishing Australia Pty. Ltd. one of Australia's leading fishing companies, with operations as diverse as toothfish fishing in the Antarctic regions, to prawn trawling in the northern regions of Australia. In response to concerns about IUU fishing for toothfish, he has most recently worked with other legal industry members to implement the Coalition of Legal Toothfish Operators (COLTO), a group comprising 26 member companies, with operations in 10 countries around the southern hemisphere. Mr. Exel is the interim Chairman of COLTO.

Matthew **GIANNI** (Discussant, International Oceans Network, Amsterdam) works as an independent Consultant and is the Director of International Oceans Network. Established in 2004, the International Oceans Network, an NGO based in the Netherlands, promotes fisheries conservation and the sustainable use of ocean resources through working with a network of concerned individuals and organisations from around the world. As a consultant, Matt has provided political, legal and technical advice relating to fisheries conservation and management, marine biodiversity and oceans governance. Major clients have included the World Bank, IUCN/World Conservation Union, Swedish Commission on the Marine Environment, WWF International, Natural Resources Defence Council (US), Oceana, Greenpeace International and Greenpeace UK.

Kathryn **GORDON** (OECD, Paris) is a senior economist in the International Investment Division of the OECD. She is responsible for the work on the OECD Guidelines for Multinational Enterprises, a multilaterally endorsed code of conduct with a government-backed follow-up mechanism. Ms. Gordon was one of the main OECD Secretariat participants in the negotiations that led to the successful review of the Guidelines that was adopted by governments in June 2000. She is also in charge of research on corporate responsibility that supports implementation of the Guidelines.

Katsuma **HANAFUSA** (MAFF, Tokyo) currently holds the position of Director for International Negotiations, International Affairs Division, Resources Management Department, Fisheries Agency, Ministry of Agriculture, Forestry and Fisheries, Government of Japan. He is also a Japanese Commissioner to the Inter-American Tropical Tuna Commission (IATTC) as well as the Head Delegate to the Indian Ocean Tuna Commission (IOTC) and the alternate head delegate to the International Commission for the Conservation of Atlantic Tunas (ICCAT).

Yuichiro **HARADA** (OPRT, Japan) worked for the Japanese Tuna Fishing Industry for 25 years (1977 to 2002). He was responsible for dealing with international affairs in securing overseas fishing grounds for Japanese tuna fishing fleets, and in recent years, also for environmental affairs related to fisheries. Mr Harada was elected as Managing Director of OPRT in May 2002 to promote responsible tuna fisheries.

Aaron **HATCHER** (CEMARE, Portsmouth) is a Senior Research Fellow at the Centre for the Economics and Management of Aquatic Resources in the Department of Economics, University of Portsmouth, UK. His research interests focus on the microeconomics of fisheries management, in particular the implications of non-compliance for fisheries management. He has also written on various aspects of UK and EU fisheries policy. Mr. Hatcher has led a number of EU-funded research projects and has undertaken work for the FAO, the DG FISH and the UK Government, as well as for the OECD.

Kjartan **HOYDAL** (NEAFC, London). Before becoming Secretary of NEAFC, Mr. Hoydal was the head of the Faroese Fisheries Administration and represented the Faroese Government as a Head of Delegation, mediator and observer in a variety of negotiations. He has worked with the Nordic Council of Ministers as Secretary for Nordic Atlantic Co-operation, and has served on the board of various organisations including the Danish Institute of Marine Research and the Faroese Aquaculture Research Station. He chaired the North Atlantic Marine Mammals Commission, NAMMCO and also served as President of the Northwest Atlantic Fisheries Organization, NAFO. He trained in Biology and Marine Ecology at the University of Copenhagen and subsequently specialised in the areas of fish biology, fish stock assessment and fisheries management.

Michele **KURUC** (Discussant, NOAA, Washington D.C.) is the Assistant General Counsel for Enforcement and Litigation where she serves as NOAA's chief prosecutor. Prior to coming to NOAA, Ms. Kuruc was a trial attorney at the U.S. Department of Justice where she worked in the Wildlife and Marine Resources section. Michele is also the chair of the International Network for Co-operation and Co-ordination of Fisheries Related Monitoring, Control and Surveillance Activities, a global organisation of law enforcement professionals dedicated to combating illegal fishing and degradation of the oceans.

Terje **LOBACH** (Ministry of Fisheries, Norway) currently works as legal advisor in the Norwegian Ministry of Fisheries. His experience includes bilateral and multilateral negotiations and consultancy work for a number of countries and organisations. He has done legal work in relation to most aspects of the management of marine resources, including drafting and implementation of fisheries legislation. In this regard Mr. Lobach has been special adviser on matters related to the 1982 UN Convention on the Law of the Sea, the 1995 UN Fish Stocks Agreement, the FAO Code of Conduct for Responsible Fisheries and the FAO Compliance Agreement.

Brice **MARTIN-CASTEX** (International Maritime Organisation, London) is a French Administrator of Maritime Affairs and Commander of the French Navy. Since 1993, Mr. Martin-Castex has been seconded to the International Maritime Organization where he is Senior Technical Officer within the Maritime Safety Division, Secretary of the Sub-Committee on Flag State Implementation (FSI) dealing with implementation-related issues including IUU fishing, self-assessment of flag State compliance, IMO Voluntary Member State Audit Scheme, piracy and armed robbery against ships, stowaways, illegal migrants, maritime casualties, port State control and technical co-operation.

Frank **MEERE** (Discussant) is representative of the Australian government's commitment to the OECD Ministerial Task Force on IUU Fishing where Mr. Meere has worked since arriving at the OECD in February 2004. Prior to his appointment to the IUU Task Force, Mr. Meere was Managing Director of the Australian Fisheries Management Authority (AFMA) for five years.

Denzil **MILLER** (CCALMR, Hobart) has been Executive Secretary of CCAMLR since February 2002. He has published more than 60 peer-reviewed papers on Antarctic science, fisheries management, marine biology, and marine policy in general. In 2002, Mr. Miller was appointed an Honorary Research Professor in the Institute of Antarctic and Southern Ocean Studies at the University of Tasmania.

Rémi **PARMENTIER** (Varda Group, Australia) is a specialist in political strategy and advocacy, and analysis and development of organisational and campaign strategies. In his current post as a Varda Group Director, and previously as representative of Greenpeace International (of which he was a founding member in the 1970s), Rémi has promoted environmental approaches within many intergovernmental organisations, governments, corporate entities, NGOs and other stakeholders.

Jean Francois **PULVENIS DE SELIGNY MAUREL** (Chair, FAO, Rome) joined the FAO Fisheries Department in August 2002 as Director of the Fishery Policy and Planning Division. Since he started working on fisheries issues, Mr. Pulvenis has participated in numerous negotiations and conferences, at the bilateral, regional and global levels, not least at the *United Nations Conference on Straddling Fish Stocks and Highly Migratory Fish Stocks*.

Victor **RESTREPO** (ICCAT, Madrid) has a Ph.D. in Fisheries Science from the University of Miami where he worked as professor until 1999 when he joined the International Commission for the Conservation of Atlantic Tunas (ICCAT) to work on assessments of the major Atlantic tuna species. Since 2002, Mr. Restrepo has been Assistant Executive Secretary of ICCAT. In this position he helps co-ordinate the work of the Standing Committee on Research and Statistics; he also helps facilitate the implementation of all decisions made by ICCAT, including those related to IUU fishing.

Lori **RIDGEWAY** (Chair, Department of Fisheries and Oceans, Ottawa) is Director General of the Economic and Policy Analysis Branch in the Canadian Department of Fisheries and Oceans. She has been head of Delegation to numerous international meetings including the FAO, WTO, and the APEC Ocean Ministerial meeting in 2002. Ms. Ridgeway is the Chair of the OECD Committee for Fisheries, a post she has held since 2000.

Kelly **RIGG** (Varda Group, Australia) was a founding Director of the Varda Group, after more than 20 years with Greenpeace and other organisations co-ordinating campaigns on oceans and other environmental issues. Ms. Rigg led the Greenpeace campaign to save Antarctica, and the organisation's efforts in 2002 in the run-up to the Johannesburg Earth Summit. She played a major role in the establishment of the Greenpeace Amazon campaign. Through her current work with the Varda Group, she serves as a policy advisor to Oceana in Europe.

Hiroya **SANO** (OPRT, Japan) has been President of the Organisation for the Promotion of Responsible Tuna Fisheries since December 2000. He was also President of the Japan Fisheries Association until May 2003. He previously held the position of Director General of the Fisheries Agency in Japan.

Olav **SCHRAM STOKKE** (Fridtjof Nansen Institute, Norway) is a senior research fellow from the Fridtjof Nansen Institute in Norway. Olav has worked for years on many aspects of IUU fishing, and is a well known expert in this field. Olav has recently been working on a book entitled *Governing High Seas Fisheries: The Interplay of Global and Regional Regimes.*

Walt **SIMPSON** has a 30-year maritime background including 21 years at sea on research, cargo and fishing vessels and holds Master Mariner qualifications from the United States and the United Kingdom. From 1988 until 2002, Walt was employed by Greenpeace International, serving as Captain and Mate onboard their ships and as Marine Operations Manager and Marine Services Co-ordinator at Greenpeace International in Amsterdam. In the early 1990s he co-ordinated the development of techniques and equipment which allowed Greenpeace to locate, track and document vessels engaged in IUU fishing, transportation of toxic waste, and illegally harvested forest products. He holds a Bachelor of Science degree from Delta State University and has completed studies in nautical science in the US and UK. He is currently based in the Netherlands.

Rashid **SUMAILA** (University of British Columbia, Canada) is assistant professor at the University of British Columbia, a post he has held since January 2002. He has written extensively on fisheries issues, in particular on over-fishing and over-capacity. Mr. Sumaila has worked as a consultant for a number of agencies including the Canadian Research Council, Environment Canada and the Research Council of Norway.

Davor **VIDAS** (Fridtjof Nansen Institute, Norway) is Programme Director and Senior Research Fellow at FNI where he has worked since 1992. Mr. Vidas is also Director of the International project "Ballast Water Issues for Croatia: Possible Usefulness of Norwegian Experience and Expertise" and is currently Project leader for "Illegal, Unregulated and Unreported (IUU) Fishing: Implementing Mechanisms for Jurisdiction, Control and Enforcement", as well as Director of the international project "Implementation of Legal Measures for Protection of Regional Marine Environment: Croatia and Norway" in co-operation with the University of Rijeka, Faculty of Law, Croatia.

Brandt **WAGNER** (ILO, Geneva) is Maritime Specialist with the International Labour Office. Mr. Wagner is the principal author of the ILO report *Safety and health in the fishing industry.* He has most recently been responsible for the preparation of *Conditions of work in the fishing sector: A comprehensive standard* (a Convention supplemented by a Recommendation) on work in the fishing sector, and *Conditions of work in the fishing sector: The constituents' views*, the two discussion documents before the 92nd Session of the International Labour Conference (Geneva, June 2004).

Jon **WHITLOW** (ITF, London) is working with the International Transport Workers' Federation where he is the Secretary of the Seafarers Section. Mr. Whitlow has participated in many international negotiations regarding the welfare of seafarers and is very concerned about the social status of fishers on IUU fishing vessels.

Jane **WILLING** (Chair, Ministry of Fisheries, Wellington) is the Director of the International and Bio-security branch of the New Zealand Ministry of Fisheries. Ms. Willing has participated at numerous international meetings and has held the post of Chair of the FAO Sub-committee on Fisheries Trade. Ms. Willing has also served on the Bureau of the Fisheries Committee of the OECD.

Anna **WILLOCK** (TRAFFIC, Sydney) is the Senior Fisheries Advisor in TRAFFIC International, responsible for overseeing the TRAFFIC Network's fisheries programme. Key work areas include research on the impact of international trade on the sustainability of fisheries, review legislation and management regimes, and monitoring international agreements. Ms. Willock represents the TRAFFIC Network at relevant international meetings, including the FAO Committee on Fisheries and its associated expert consultations and various regional fisheries organisations.

Sachi **WIMMER** (Discussant) is the Manager of Illegal, Unreported and Unregulated Fishing Branch of the Australian Government Department of Agriculture, Fisheries and Forestry. She currently has responsibility for developing and managing the Australian Government's response to Illegal, Unreported and Unregulated (IUU) fishing. This involves the implementation of a comprehensive strategy that was agreed by the Australian Government in 2002. Prior to this appointment, Ms. Wimmer worked as the Fisheries Management Adviser on an Asian Development Bank project to strengthen Papua New Guinea's National Fisheries Authority.

Ursula **WYNHOVEN** (UN, New York) is the Special Assistant to the Executive Head of the United Nations Global Compact Office, where she works on the UN Secretary-General's voluntary corporate citizenship initiative for human rights, labour and the environment. Prior to this appointment, she worked with Kathryn Gordon on the OECD Guidelines for Multinational Enterprises as a Visiting Fellow.

Nobuyuki **YAGI** (Chair, MAFF, Tokyo) is the Deputy Director, Processing Industries and Marketing Division, Fisheries Agency, Government of Japan where he is responsible for trade-related issues at FAO, OECD, WTO, and for bilateral FTA negotiations. Mr. Yagi has been a member of the Bureau of the OECD Fisheries Committee since 2003.